AF595324

Shedding Light to the Dark Sides of the Universe: Cosmology from Strong Interactions

Shedding Light to the Dark Sides of the Universe: Cosmology from Strong Interactions

Editors

Roman Pasechnik
Michal Šumbera

MDPI • Basel • Beijing • Wuhan • Barcelona • Belgrade • Manchester • Tokyo • Cluj • Tianjin

Editors
Roman Pasechnik
Lund University
Sweden

Michal Šumbera
Nuclear Physics Institute, The
Czech Academy of Sciences
Czech Republic

Editorial Office
MDPI
St. Alban-Anlage 66
4052 Basel, Switzerland

This is a reprint of articles from the Special Issue published online in the open access journal *Universe* (ISSN 2218-1997) (available at: https://www.mdpi.com/journal/universe/special_issues/QCD_cosmology).

For citation purposes, cite each article independently as indicated on the article page online and as indicated below:

LastName, A.A.; LastName, B.B.; LastName, C.C. Article Title. *Journal Name* **Year**, *Volume Number*, Page Range.

ISBN 978-3-0365-5869-1 (Hbk)
ISBN 978-3-0365-5870-7 (PDF)

Cover image courtesy of Unsplash License

© 2022 by the authors. Articles in this book are Open Access and distributed under the Creative Commons Attribution (CC BY) license, which allows users to download, copy and build upon published articles, as long as the author and publisher are properly credited, which ensures maximum dissemination and a wider impact of our publications.
The book as a whole is distributed by MDPI under the terms and conditions of the Creative Commons license CC BY-NC-ND.

Contents

About the Editors

Roman Pasechnik

Roman Pasechnik, born 1983, is an associate professor at the Department of Astronomy and Theoretical Physics at Lund University in Lund. His research is focused on theoretical studies of strongly interacting field theories and their implications in collider physics and in the early universe as well as on New Physics model building and its applications in collider phenomenology and astroparticle physics. He obtained his PhD at JINR, Dubna, for studies of non-perturbative aspects of QCD.

Michal Šumbera

Michal Šumbera, born 1951, is a senior physicist at the NPI CAS and lecturer at the Czech Technical University in Prague. His scientific work focuses mainly on ultrarelativistic heavy ion collisions at the LHC and RHIC (experiments ALICE and STAR) and on high-energy phenomenology. In the past, he was involved in CERN SPS experiments NA45 and WA98 and also in TAPS experiment at GSI. He obtained his PhD at JINR, Dubna, for studies of light nuclei interactions from Synchrophasotron with photoemulsion nuclei

 MDPI

Editorial

Shedding Light to the Dark Sides of the Universe: Cosmology from Strong Interactions

Roman Pasechnik [1,*] and Michal Šumbera [2]

[1] Department of Astronomy and Theoretical Physics, Lund University, SE-223 62 Lund, Sweden
[2] Nuclear Physics Institute CAS, 25068 Řež, Czech Republic
* Correspondence: roman.pasechnik@thep.lu.se

Citation: Pasechnik, R.; Šumbera, M. Shedding Light to the Dark Sides of the Universe: Cosmology from Strong Interactions. *Universe* **2022**, *8*, 545. https://doi.org/10.3390/universe8100545

Received: 9 October 2022
Accepted: 14 October 2022
Published: 19 October 2022

Publisher's Note: MDPI stays neutral with regard to jurisdictional claims in published maps and institutional affiliations.

Copyright: © 2022 by the authors. Licensee MDPI, Basel, Switzerland. This article is an open access article distributed under the terms and conditions of the Creative Commons Attribution (CC BY) license (https://creativecommons.org/licenses/by/4.0/).

The basic aim of this Special Issue was to reflect upon the modern status of research on strong interactions and their implications in Cosmology. In addtion to a few important original research articles, our Special Issue comprises a collection of reviews on existing implications of the physics of strongly coupled and non-perturbative phenomena such as those in Quantum Chromodynamics (QCD) to the universe evolution in different epochs. Such implications concern, in particular, cosmological aspects of quark–gluon plasma and phase transition dynamics, the physics of neutron stars, QCD equations of state (EoS), as well as the origins of Dark Matter (DM) and Dark Energy (DE) from both dark sectors and QCD, inflationary cosmology, etc.

A broad review [1] by the Guest Editors and their collaborators provides a comprehensive outlook of the key research results in the field of confined and de-confined QCD dynamics and their implications in physics of the early Universe. In addition, it briefly covers the basic methodology for studies of quantum field theories in the strongly coupled regime on the non-stationary background of the expanding Universe. A few rather important connections between currently pursued research in particle physics and possible dynamics of the early Universe have been identified and elaborated upon, potentially addressing such fundamental questions as particle production mechanisms in the early Universe, the origins of cosmic acceleration, and the non-perturbative real-time dynamics of the QCD ground state, among others. Given the broadly inter-disciplinary coverage of a variety of different and challenging topics, this review represents an important, but not exhaustive, reference for frontier research at an intersection of particle physics and cosmology.

The review [2] by Vitaly Beylin, Maxim Khlopov, Vladimir Kuksa, and Nikolay Volchanskiy discusses some of the basic cosmological effects of strongly coupled New Physics focusing on the possible nontrivial role of strong interactions. One particular example is the presence of new stable colored particles, such as exotic quarks, which could give rise to a composite DM candidate. In addition, an overview of new stable DM composite candidates was given in the context of QCD-like interactions in various scenarios of New Physics, broadly referred to as Techni (or Hyper)-Color. A particular emphasis was given on possible interactions of new stable particles with those in the Standard Model, with interesting implications for cosmic-ray and high-energy neutrino astrophysics and for the phenomenology of stable fractionally charged particles.

Another review [3] by Ralf Hofmann overviews possible cosmological consequences of a scenario when the SU(2) gauge principle governs the Cosmic Microwave Background (CMB) instead of U(1) one attempting to address the existing tensions between local and global cosmological measurements in the framework of ΛCDM. These consequences concern possible changes in the radiation and dark sector components of the Universe, as well as in the structure formation assisted by the de-percolation of condensed ultralight axion configurations with the Peccei–Quinn scale close to the Planck mass. It has been demonstrated that a compatibility of the axionic field profiles with typical DM galactic halos emerges if cosmological Yang–Mills fields confining at much higher energy scales than that of $SU(2)_{CMB}$ are responsible for the generation of the axion mass.

An extensive review [4] by G. Fiorella Burgio and Isaac Vidaña considers possible correlations between astrophysical observables of neutron stars (NS) and the properties of atomic nuclei. This review focuses in particular on the major role of the EoS of nuclear matter in defining the main characteristics of NS, including the existing experimental constraints. Such aspects of the NS phenomenology as the NS tidal deformability and its correlations with the stellar radius, the stiffness of the symmetry energy, and the neutron-skin thickness have been discussed.

The article [5] by Petr Jizba, Lesław Rachwał, Stefano G. Giaccari, and Jaroslav Kňap addresses the issue of a dynamical breakdown of scale invariance in quantum Weyl gravity (QWG), together with the related cosmological implications. In their study, which is based on the approach of functional renormalization group, the authors investigate the infrared (IR) physics of the quantum Weyl gravity and find it to be surprisingly rich and interesting, in particular in connection with the structure of its IR fixed point and the corresponding cosmological consequences. In a cosmology that is implied by the broken phase of QWG, they are able to map the broken-phase's effective action on a two-field hybrid inflationary model that, in its low-energy phase, approaches the Starobinsky $f(R)$ model with a gravi-cosmological constant that has a negative sign in comparison to the usual matter-induced cosmological constant. The implications of this finding for cosmic inflation are also discussed.

In the article [6] by Roland Kirschner and George Savvidy, it has been considered a new possibility that inside hadrons inhabitated by standard $SU(3)_c$ partons—quarks and gluons—there are additional partons–tensorgluons, which can carry a part of the proton momentum. Since the gluons do not couple directly to the photon in deep-inelastic scattering measurements, their density inside the hadrons is one of the least constrained functions. Here comes the opportunity for the tensorgluons: although their existence does not predict a new hadronic state, it leads to a modification of the parton distribution functions of a proton. Moreover, because tensorgluons have a larger spin than ordinary gluons, they can influence the spin structure of the nucleon.

The study [7] by Andrea Addazi, Stephon Alexander and Antonino Marcianò addresses an important fundamental issue of origin of the late-time cosmological acceleration due to the possible existence of an addition dark QCD-like matter sector. Using the arguments of strong dynamics, such as the formation of dark gluon and dark quark condensates breaking the chiral symmetry in the dark sector, the authors draw a conclusion that the interaction energy between the dark condensates may cause late-time cosmic acceleration, reproducing the observable effect of the cosmological constant.

Last but not least, the article [8] by Dmitri N. Voskresensky is devoted to evolution of quasiperiodic structures in a non-ideal hydrodynamic description of phase transitions. It starts with a general introduction to first- and second-order phase transitions. The latter could have taken place either in the early universe, in the course of heavy-ion collisions and supernova explosions, and also in proto-neutron stars, in cold compact stars, and in the condensed matter at terrestrial conditions. The author presents some novel solutions of non-ideal hydrodynamics describing the evolution of quasiperiodic structures that are formed in the course of the phase transitions. The most important result of this work concerns the finding that viscosity and thermal conductivity are the driving forces of the first-order liquid–gas and quark–hadron phase transitions to the state characterized by the zeroth wave number and by the instability occurring for temperatures below the isothermal spinodal region.

Funding: This Guest Editors' activity received no external funding.

Data Availability Statement: Not applicable.

Conflicts of Interest: The author declares no conflict of interest.

References

1. Addazi, A.; Lundberg, T.; Marcianò, A.; Pasechnik, R.; Šumbera, M. Cosmology from Strong Interactions. *Universe* **2022**, *8*, 451. [CrossRef]
2. Beylin, V.; Khlopov, M.; Kuksa, V.; Volchanskiy, N. New physics of strong interaction and Dark Universe. *Universe* **2020**, *6*, 196. [CrossRef]
3. Hofmann, R. An SU(2) Gauge Principle for the Cosmic Microwave Background: Perspectives on the Dark Sector of the Cosmological Model. *Universe* **2020**, *24*, 135. [CrossRef]
4. Burgio, G.F.; Vidana, I. The Equation of State of Nuclear Matter: from Finite Nuclei to Neutron Stars. *Universe* **2020**, *6*, 119. [CrossRef]
5. Jizba, P.; Rachwał, L.; Giaccari, S.G.; Kňap, J. Dark side of Weyl gravity. *Universe* **2020**, *6*, 123. [CrossRef]
6. Kirschner, R.; Savvidy, G. Parton Distribution Functions and Tensorgluons. *Universe* **2020**, *6*, 88. [CrossRef]
7. Addazi, A.; Alexander, S.; Marcianò, A. Invisible QCD as Dark Energy. *Universe* **2020**, *6*, 75. [CrossRef]
8. Voskresensky, D.N. Evolution of Quasiperiodic Structures in a Non-Ideal Hydrodynamic Description of Phase Transitions. *arXiv* **2020**, arXiv:2001.10841.

MDPI

Review

Cosmology from Strong Interactions

Andrea Addazi [1,2], Torbjörn Lundberg [3], Antonino Marcianò [2,4], Roman Pasechnik [3,*] and Michal Šumbera [5]

1 Center for Theoretical Physics, College of Physics Science and Technology, Sichuan University, Chengdu 610065, China
2 Laboratori Nazionali di Frascati INFN, Via Enrico Fermi 54, 00044 Frascati, RM, Italy
3 Department of Astronomy and Theoretical Physics, Lund University, SE-223 62 Lund, Sweden
4 Center for Field Theory and Particle Physics, Department of Physics, Fudan University, Shanghai 200433, China
5 Nuclear Physics Institute CAS, 25068 Řež, Czech Republic
* Correspondence: roman.pasechnik@thep.lu.se

Abstract: The wealth of theoretical and phenomenological information about Quantum Chromodynamics at short and long distances collected so far in major collider measurements has profound implications in cosmology. We provide a brief discussion on the major implications of the strongly coupled dynamics of quarks and gluons as well as on effects due to their collective motion on the physics of the early universe and in astrophysics.

Keywords: QCD in the early universe; phase transitions; hydrodynamical evolution; equation of state of super-dense matter; classical Yang-Mills fields; Dark Energy; Dark Matter; gluon condensate; effective Yang-Mills action; cosmic inflation

PACS: 98.80.Qc; 98.80.Jk; 98.80.Cq; 98.80.Es

Citation: Addazi, A.; Lundberg, T.; Marcianò, A.; Pasechnik, R.; Šumbera, M. Cosmology from Strong Interactions. *Universe* **2022**, *8*, 451. https://doi.org/10.3390/universe8090451

Academic Editor: Antonino Del Popolo

Received: 27 April 2022
Accepted: 20 August 2022
Published: 29 August 2022

Publisher's Note: MDPI stays neutral with regard to jurisdictional claims in published maps and institutional affiliations.

Copyright: © 2022 by the authors. Licensee MDPI, Basel, Switzerland. This article is an open access article distributed under the terms and conditions of the Creative Commons Attribution (CC BY) license (https://creativecommons.org/licenses/by/4.0/).

1. Introduction and Historical Perspective

The strongly coupled dynamics of quarks and gluons has many important implications in particle physics, astrophysics, and cosmology [1–16]. The fundamental theory of strong interactions, known as Quantum Chromodynamics (QCD), provides a successful description of a variety of observables in high-energy hadronic collisions [17], hadronic masses [18], and, to a lesser extent, also of the properties of phases of the QCD matter [19,20]. While QCD is successful in the interpretation of short-distance phenomena (i.e., in the weakly coupled regime), a long-standing theoretical problem is a dynamical description of the color confinement phenomenon. The latter appears in the infrared (strongly coupled) regime of QCD and still remains the major unsolved problem of the Standard Model (SM) of particle physics [21,22].

Due to confinement, color-charged particles do not exist as free states at large spacetime separations. They are instead bound together into colorless collective excitations that evolve into a gas of hadrons. No exact dynamical transition in spacetime between the fundamental (parton) and the composite (hadron) states of QCD is known to date despite the wealth of phenomenological information available from particle and heavy-ion collision experiments. Therefore, one usually resorts to a heuristic description using the concept of quark-hadron duality [23,24] together with effective field theoretical (EFT) approaches [25]; this is used also in the framework of thermal field theory (for recent reviews of the latter, see also Refs. [26–28]). On the theory side, effective (typically, static or equilibrium) approaches, such as lattice QCD (LQCD) [19,21], are commonly being exploited while very little has been done on first-principle real-time evolution of QCD states [8].

The term "quark matter" was first used in 1970 by Itoh [29] in the context of neutron stars. Even before then, in 1965, Ivanenko and Kurdgelaidze [30] considered a star made of quarks. Since the mechanism for quark confinement was unknown at that time, they had to assume that the quark masses are much larger than the masses of ordinary baryons. A few years later, however, following the realization that QCD exhibits asymptotic freedom [31,32], several authors have suggested that the transition from a hadronic phase to a one dominated by quarks and gluons may be relevant to describe the state of matter in the early universe or inside the neutron stars with a possibility to re-create such a condition also in the laboratory by colliding heavy ions [3,5,33–38].

The terms "hadronic plasma" [35] and "quark-gluon plasma" (or QGP) [36] were coined by Shuryak to describe a hypothetical state of matter existing at temperatures of order 100 MeV. The corresponding phenomena were expected to occur at a characteristic energy density close to 1 $\mathrm{GeV/fm^3}$. This makes a good analogy with a classical gaseous plasma in which electrically neutral gas at high enough temperatures turns into a statistical system of mobile charged particles [39]. While for such plasma the particle interactions obey the $\mathrm{U(1)}_{em}$ gauge symmetry of Quantum Electrodynamics (QED), in the former QCD case, the interactions between plasma constituents are driven by their $\mathrm{SU(3)_c}$ color charges. For an exhaustive collection of key references tracing the development of theoretical ideas on the QGP up to 1990, see e.g., Ref. [2]. For a summary of later developments, see more recent reviews [8,10,40].

Let us note that contrary to initial oversimplified expectations [2], strongly interacting multi-particle systems feature numerous emergent phenomena that are difficult to predict from the underlying QCD theory, just like in condensed matter and atomic systems where the interactions are controlled by QED. In addition to the hot QGP phase, several additional phases of QCD matter were predicted to exist [15,41]. In particular, the long-range attraction between the quarks in the color anti-triplet ($\bar{\mathbf{3}}$) channel was predicted to lead to the color superconductivity (CSC) phase with condensation of $^1S_0$ Cooper pairs [42,43]. This result was anticipated, though using a different reasoning, already in 1969 by Ivanenko and Kurdgelaidze [44], who predicted that the superconducting quark phase may be relevant for the super-dense star interiors. At high baryon density, an interesting symmetry breaking pattern $\mathrm{SU(3)_c \times SU(3)_L \times SU(3)_R \times U(1)_B \to SU(3)_{c+L+R} \times Z(2)}$ leading to the formation of quark Cooper pairs was found in QCD with three massless quark flavors (i.e., under an assumption that $m_u = m_d = m_s = 0$) [41,45]. This breaking of color and flavor symmetries down to the diagonal subgroup $\mathrm{SU(3)_{c+L+R}}$ implies a simultaneous rotation of color and flavor called the color-flavor locking (CFL). It is expected that CSC and CFL phases might play important role in the equation of state (EoS) of neutron stars [46].

Another interesting phase of QCD matter called quarkyonic matter situated in the QCD phase diagram between the chirally restored and the confined phases was proposed in Ref. [47]. The quarkyonic matter is expected to exist at densities parametrically large compared to $\sim\Lambda_{\mathrm{QCD}}^4$ when the number of colors N_c is large. Since gluons are in the adjoint representation of $\mathrm{SU(3)}_c$, their effects are scaled as $\sim N_c^2$, and so, they dominate all quark-induced $\sim N_c$ effects. This provides the binding of gluons into quark-free states, so-called glueballs, and so, the quarkyonic matter has only N_c degrees of freedom (DoFs). This form of matter is expected to play some role in the structure of neutron stars [48]. The existence of another peculiar form of hadronic matter—the pion condensate—was suggested by Migdal already in the 1970s [49].

The rich phase structure of QCD at nonzero temperature and baryon chemical potential was recently reaffirmed by the proposed existence of phases with spatial modulations; see [50] and references therein. Their moat-shaped energy spectrum with a minimum of the energy over a sphere at nonzero momentum leads to a characteristic peak. In heavy-ion collisions at low energy, these new QCD phases are expected to leave their imprints in particle spectra and their correlations. Their cosmological implications are so far unexplored.

Our current knowledge of the QCD phase diagram is illustrated in Figure 1. Comparing this diagram to the phase diagram of water, see e.g., Ref. [8], one notices that (at least, theoretically) the complexity of the former approaches the latter.

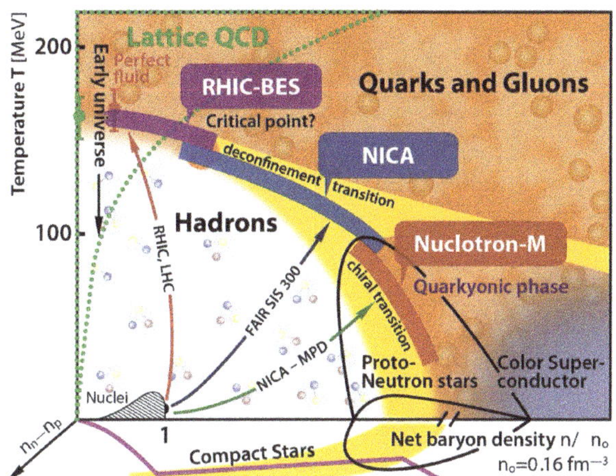

Figure 1. The schematic phase diagram for QCD matter in terms of the temperature T and net baryon density n normalized to the cold nuclei baryon density n_0. From https://nica.jinr.ru/physics.php accessed on 27 April 2022 (see also Ref. [51]).

Experimental study of the QCD phase diagram at high temperatures, see Figure 2, dates back to the CERN SPS fixed-target program with the lead ion beams in 1995–2000 and covers the domain of the baryon chemical potential $\mu_B = 200 - 400$ MeV [8]. With the advent of a first heavy-ion collider in 2000, the investigation of the $\mu_B \simeq 0$ region soon led to a discovery of the strongly interacting quark-gluon plasma (sQGP) at RHIC in 2005 [52–55]. The existence of this new phase of hot and strongly interacting QCD matter was five years later confirmed at order-of-magnitude higher energies of the LHC at CERN [8,40].

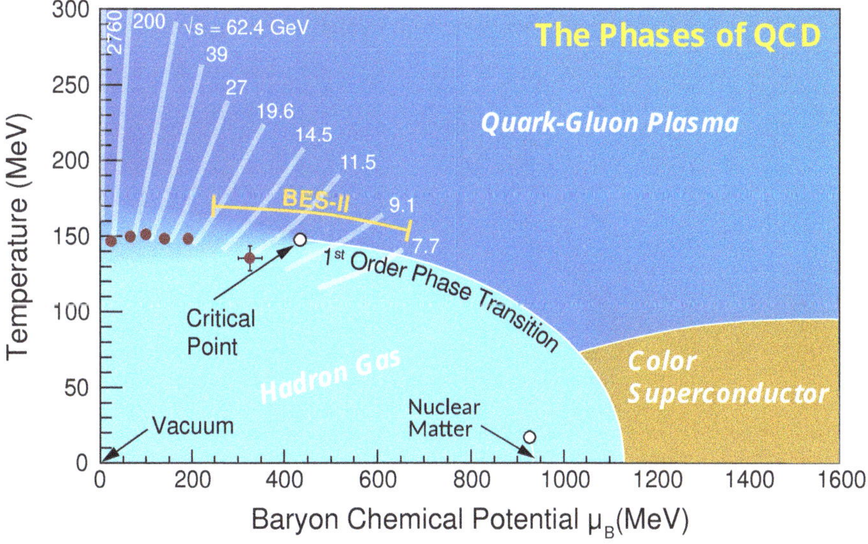

Figure 2. A schematic QCD phase diagram in the thermodynamic parameter space spanned by the temperature T and the baryonic chemical potential μ_B. The corresponding (center-of-mass) collision energy ranges for different accelerator facilities, especially the RHIC Beam Energy Scan (BES II) program, are indicated in the figure. Adapted from Ref. [56] (see also Ref. [57]).

Starting from 2010, it became possible to explore systematically the phase structure of hot and dense matter at nonzero baryon density and, in particular, to search systematically for the critical endpoint (CEP) of the QCD phase diagram. The CEP, a postulated second-order phase transition point, is an expected endpoint of a line of the first-order phase transitions (FOPTs) that separates the low-temperature, low-density hadronic phase from a low-temperature, large-baryon number density QGP phase. Similarly to the water-steam transition where at the critical point, one finds bubbles of steam and drops of water intermixed at all length-scales from macroscopic, visible sizes down to atomic scales (with drops and bubbles near micron scale causing the strong light scattering called "critical opalescence" [58]), several interesting phenomena are also expected to occur near the CEP of the QCD phase diagram [57,59–62]. The search for the CEP is conducted by the STAR collaboration at RHIC within its Beam Energy Scan (BES) program at the energies indicated in Figure 1.

Current experimental and theoretical studies of the QCD phase diagram thus cover a wide region in temperature and baryon chemical potential $(T, \mu_{\rm B})$, particularly, at small $\mu_{\rm B} \simeq 0$ [19,20,63,64] and large $\mu_{\rm B} \simeq 100 - 600$ MeV [20,41,57,60], see Figure 2. The red and black full circles denote the critical endpoints of the chiral and nuclear liquid-gas phase transitions, respectively. The (dashed) freeze-out curve indicates where the hadro-chemical equilibrium is attained at the final stage of the collision. The nuclear matter ground-state at $T = 0$ and $\mu_{\rm B}$ = 0.93 GeV and the approximate position of the QCD critical point at $\mu_{\rm B} \sim 0.4$ GeV are also indicated. The dashed line is the chiral pseudo-critical line associated with the crossover transition at low temperatures.

The hot and dense QCD matter is considered to be a dominant ingredient of the early universe evolution in its first few microseconds. Physics of heavy-ion collisions (HIC), therefore, provides necessary means for theoretical understanding of the cosmological processes at those time scales. In HIC theory, an important progress has been made when relativistic viscous fluid dynamics was formulated starting from the first principles in an EFT framework, which was based entirely on the knowledge of symmetries and long-lived degrees of freedom, see e.g., Ref. [25] and Appendix B of this review. However, for proper understanding of the cosmological evolution, at least in a vicinity of the QCD phase transition epoch, the precise dynamical information on color-field media at finite temperatures is mandatory. Ongoing precision tests of QCD under extreme conditions, in particular those at the Large Hadron Collider (LHC) at CERN and the Relativistic Heavy Ion Collider (RHIC) at Brookhaven National Laboratory (BNL), are currently pushing the energy, temperature and density frontiers, opening up new largely unexplored possibilities for understanding also the cosmological QCD phase transition. There is strong hope that the growing amount of data and phenomenological concepts will eventually boost theoretical developments on infrared and finite-temperature dynamics of QCD. The latter is particularly relevant for understanding the real-time evolution of its ground state and the associated phase transitions as well as hadronization processes relevant for the dynamics of the early universe.

The QCD dimensional transmutation mechanism, breaking the conformal symmetry of the classical QCD action, has deep implications for the early universe evolution. Indeed, from higher to lower primordial plasma temperatures, QCD crosses a phase transition to a chiral symmetry-breaking ground state related to the color confinement phase. Thus, an attractive possibility is that the QCD vacuum energy may provide a source of universe acceleration and Dark Energy (DE) [65,66]. For the current status of this problem, see also Ref. [67] and references therein.

At high temperatures above the confinement scale $\Lambda_{\rm QCD}$, i.e., during the first microseconds after the Big Bang, the thermal bath in the early universe was dominated by the primordial QGP [11,12,68,69]. When the temperature of the universe decreases down to $\Lambda_{\rm QCD}$, the QGP dissolves out through collective hadronization phenomena. It is worth remarking that the QGP formation can be highly favoured under the very high-density conditions, where the matter chemical potential starts to be comparable to the QCD critical scale. Indeed, this can happen in high-density objects such as in the core of neutron stars as an effect of the gravitational potential [9,70,71]. Nevertheless, the issue of whether the QGP exists inside the neutron stars is still highly controversial and is under intense debate in the literature. In fact, the critical phenomena of QCD dynamics and hadronization and their connections to the QCD ground-state evolution in real time have paramount consequences on the whole cosmological scene, which may also shed light on the late-time cosmic acceleration [65], the formation of Dark Matter (DM) [72], primordial black holes (PBHs) [73] and could be imprinted in the spectra of primordial density perturbations and gravitational waves (GWs) [74].

The main aim of our review is to provide a new critical sight on our current picture of quantum Yang-Mills (YM) field theories in the strong-coupling regime in a dynamical (i.e., non-stationary) spacetime background and in cosmology, in connection to the empirical knowledge that comes from particle physics measurements and cosmological data. In the following, unless otherwise noted, we will mainly exploit the standard natural units $\hbar = c = k_B = 1$, where k_B is Boltzmann constant, c is speed of light in the vacuum and $\hbar = h/(2\pi)$, with h is the Planck constant.

The review is organized as follows. In Section 2, we discuss the nowadays standard scenario of the phase transitions in the early universe, making a connection to the production of primordial black holes and to super-dense weakly interacting saturated QCD matter. We also discuss possible applications of the axion dynamics to the early universe and close with the possible role of non-perturbative QCD ground-state cosmological evolution. For completeness, we also mention the possible role of the phase transitions in grand unified theories of particle interactions. In Section 3, we first introduce the basic notions of the hydrodynamical description of an expanding universe. There, we discuss simple models with constant speed of sound and then move on to a more complicated equation of states for the early universe. We also present current progress in the description of the dissipative effects in relativistic hydrodynamics. The section is finalized by an overview of the problematics regarding the Cosmological Constant and the Vacuum Catastrophe. Section 4 is devoted to a brief discussion of the real-time dynamics of the ground state in an effective action approach to quantum YM theories. We first discuss the YM ground state as time crystal; then, we develop an effective action approach providing the equation of state of the quantum ground state of the universe. The section is closed with a discussion of cosmological attractors—the solutions of the YM-Einstein equations using the Renormalization Group (RG) methods. Section 5 provides an overview of basic concepts of cosmic inflation models driven by YM dynamics in the early universe. Finally, a short summary is given in Section 6.

2. The Phase Transitions in the Early Universe

2.1. The Phase Transitions in the Standard Model

In the SM of elementary particle interactions, the dynamics of fireball expansion is based on the asymptotic freedom property of underlying non-Abelian gauge theories [31,32]. QCD is a quantum non-Abelian field theory, an important part of the SM, that describes the fundamental interactions between colored quarks and gluons. The generalization of classical electrodynamics to non-Abelian gauge theories was first studied and exemplified in SU(2) by Yang and Mills in 1954 [75]. The classical Lagrangian density of an SU(N) gauge theory reads,

$$\mathcal{L}_{\rm cl} = -\frac{1}{4} F^a_{\mu\nu} F^{a\,\mu\nu}\,, \tag{1}$$

in terms of the field strength tensor defined in terms of YM fields A^a_ν as $F^a_{\mu\nu} = \partial_\mu A^a_\nu - \partial_\nu A^a_\mu +$ $g_{\rm YM} f^{abc} A^b_\mu A^c_\nu$, where f^{abc} are the structure constants of the SU(N) group. Throughout, $a, b, \ldots$ denote internal indices of SU(N) in the adjoint representation. Here, the parameter $g_{\rm YM}$ is known as the YM coupling constant. Gauge theories based on SU(N) are known as YM theories, and they became the target of a wider interest prompted by the discovery that massless particles may acquire a mass and a longitudinal polarization through spontaneous symmetry breaking (or Higgs) mechanism of a massless YM theory [76–78]. The latter is a vital part of the SM framework realizing the classical Higgs mechanism of EW symmetry breaking that has found an excellent confirmation through the discovery of the Higgs boson [79,80].

In the framework of the quantum field theoretical approach, the YM field fluctuations are quantized around a given ground state through the first quantization procedure à la QED. However, due to self-interactions of the YM quanta, manifested via the term $\propto A^2$ in the field strength tensor, the quantum YM ground state acquires, in general, a very non-trivial structure. This structure is well understood in the weakly coupled (perturbative) regime of the theory, implying $g_{\rm YM} \ll 1$, which is the case of the EW theory or in the UV regime of QCD with the so-called asymptotic freedom of color charges. The strongly coupled (non-perturbative) regime, in which $g_{\rm YM} \gtrsim 1$ that is realized in particular in the infrared limit of QCD, corresponds to the color confinement phenomenon and has remained the subject of active research over the last few decades. In recent years, significant progress has been made in understanding of the quantum YM ground state in SU(N) gauge theories at finite temperatures, see e.g., Refs. [81,82].

As a result of a series of cosmological phase transitions that occurred in early universe during the first few microseconds after the Big Bang, new vacuum subsystems associated with breaking of the fundamental symmetries were formed. In the early universe, the SM predicts that cooling proceeds as a series of two phase transitions associated with the various spontaneous symmetry breakings of the corresponding gauge symmetries [12,68,83–86]. One at the temperature $T_c^{\rm EW} = 160$ GeV [87] is responsible for spontaneous breaking of the EW symmetry providing masses to the elementary particles; see the left panel on Figure 3. Due to the large value of its critical temperature T_c, it is not amenable to experimental study under the laboratory conditions. The second and the only one accessible in the laboratory, QGP-to-hadronic matter phase transition happening at $T_c^{\rm QCD} \approx 160$ MeV [88], is related to the spontaneous breaking of the chiral symmetry and manifesting itself in the massless quark limit of the QCD Lagrangian. Since both phase transitions are considered to be analytic crossovers, the bulk motion of the corresponding plasmas did not depart from thermal equilibrium. Therefore, such transitions, if realized in nature, are not expected to generate cosmological relics [86,89,90] or to be helpful for a baryogenesis mechanism.

The QCD phase transition has occurred at characteristic temperatures of above 200 MeV that correspond to a cosmological time-scale of above 10^{-5} s and the Hubble length-scale of approximately 10 km. The nature of the QCD phase transition is still a matter of intense debates in the literature [74,91–103], with results derived so far heavily relying either on lattice field theory methods applied to QCD [92,93], i.e., lattice QCD, or on holographic analyses of QCD at early cosmology [103]. For a thorough review on various aspects of the QCD transition epoch, see e.g., Refs. [12,104–108].

It is undeniable that an abrupt QCD phase transition occurring reversibly in the early universe would lead to a promising cosmological scenario, according to which a large part of the quark excess would be condensed into invisible quark nuggets—a possible explanation for DM only relying on QCD. As Witten suggested in Ref. [5], this would happen only if quark matter retains an energy per baryon which is less than 938 MeV: then, neutron stars might generate a quark matter component for cosmic rays, and detectable gravitational radiation could be produced during the QCD phase transition. Conversely, several recent studies drew a different conclusion, pointing toward the realization of second-order or crossover phase transition scenarios [101,102].

In a hot and dense QCD matter, the u, d and to some extent, depending on the temperature, also the s quarks become nearly massless, and the QCD Lagrangian acquires an approximate chiral symmetry ${\rm SU}(N_F)_{\rm L} \times {\rm SU}(N_F)_{\rm R}$, with the number of massless quark flavors $N_F = 2$ (u,d) or 3 (u,d,s). At low $T < T_c^{\rm QCD}$, the QCD vacuum becomes unstable, and this symmetry is spontaneously broken by $q-\bar{q}$ pairing. The corresponding order parameter $\langle \bar{q}q \rangle = -(245\ {\rm MeV})^3$, known as the chiral quark condensate, gives rise to masses of light hadrons as well as to constituent masses of u, d quarks and to some extent also to the s quark; see the left panel of Figure 3. Recent lattice calculations with $m_u = m_d = 0$ and strange quark having its physical mass reveal that the chiral phase transition occurs at $T_\chi^{\rm QCD} = 132^{+3}_{-6}$ MeV [109].

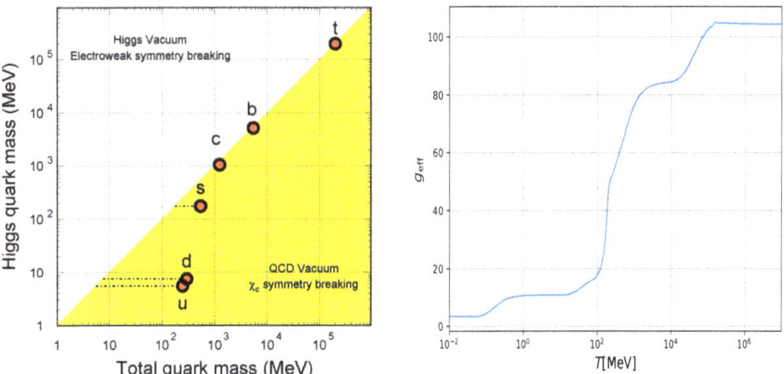

Figure 3. (**Left**) Quark masses in the QCD vacuum and the Higgs vacuum. A large fraction of the light quark (u, d, s) mass is due the chiral (χ_c) symmetry breaking in the QCD vacuum, with numerical values from Ref. [110] (see also Ref. [111]). (**Right**) The effective number of relativistic DoFs $g_{\rm eff}$ in the cosmological plasma in the SM as a function of temperature T, taking into account interactions between particles, obtained with both perturbative and lattice methods. From Ref. [112].

Let us note that in addition to the above scenario when the thermal history of the universe proceeds by a sequence of phase transitions from a more symmetric to a less symmetric state of matter, there is also a possibility of the reverse evolution when part of the zero-temperature unbroken gauge group of the SM or other gauge theory might have been broken in the early universe by thermal effects. As first noted by Weinberg [113] in the context of an $O(n) \times O(n)$ gauge theory, with decreasing T, one may encounter a transition to a state of higher symmetry $O(n) \times O(n-1) \to O(n) \times O(n)$. Within the minimal extensions of the SM containing an additional color triplet scalar field, the scenario in which the early universe underwent an epoch when SU(3)$_c$ was spontaneously broken but later restored was analyzed in Ref. [114]. The attractiveness of such a multi-step phase transition scenario stems from the fact that it may generate the observed baryon asymmetry of the universe [115].

To describe the evolution of energy density $\epsilon(T)$ and entropy density $s(T)$ of the early universe, it is customary to normalize both quantities to their values $\epsilon_0(T)$ and $s_0(T)$ corresponding to an ideal massless Bose gas with a single degree of freedom (DoF) [112,116]

$$g_{\rm eff}(T) = \frac{\epsilon(T)}{\epsilon_0(T)}, \quad \epsilon_0(T) = \frac{\pi^2}{30} T^4, \tag{2}$$

$$h_{\rm eff}(T) \equiv \frac{s(T)}{s_0(T)}, \quad s_0(T) = \frac{2\pi^2}{45} T^3, \tag{3}$$

and call $g_{\rm eff}(T)$ and $h_{\rm eff}(T)$ the effective numbers of DoFs in energy and entropy, respectively. For the particular case of a non-interacting gas consisting of $N_{\rm F}$ Dirac fermions, $N_{\rm V}$ massive vectors, $N_{\rm V0}$ massless vectors and $N_{\rm S}$ neutral scalars, the two functions are identical and read

$$g_{\rm eff}(T) = h_{\rm eff}(T) = \frac{7}{8} 4N_{\rm F} + 3N_{\rm V} + 2N_{\rm V0} + N_{\rm S}, \tag{4}$$

where the prefactors account for the DoF of each of the considered particles.

It is worth mentioning that in a generic case of interacting (non-ideal) gas $g_{\rm eff}(T)$ and $h_{\rm eff}(T)$, they are not identical and depend on temperature. Using the relationship

$p = sT - \epsilon$ between the pressure p, the entropy density s, the energy density ϵ and the generalized EoS parameter w,

$$p = w\epsilon\,. \tag{5}$$

and the speed of sound c_s can be expressed in terms of the effective DoF measures $g_{\rm eff}(T)$ and $h_{\rm eff}(T)$,

$$w(T) = \frac{sT}{\epsilon} - 1 = \frac{4h_{\rm eff}(T)}{3g_{\rm eff}(T)} - 1\,, \tag{6}$$

$$c_s^2(T) = \frac{dp}{d\epsilon} = T\frac{ds}{d\epsilon} + s\frac{dT}{d\epsilon} - 1 = \frac{4}{3}\left[\frac{4h_{\rm eff}(T) + Th'_{\rm eff}(T)}{4g_{\rm eff}(T) + Tg'_{\rm eff}(T)}\right] - 1\,, \tag{7}$$

where the prime indicates differentiation with respect to temperature T [116]. It is worth mentioning that the causality condition between the speed of sound and the speed of light $c_s \leq 1$ induces the inequality

$$\frac{h_{\rm eff}(T)}{g_{\rm eff}(T)} \leq \frac{3}{2}\,, \tag{8}$$

with an upper bound saturated for $w = 1$ corresponding to the case of absolutely stiff fluid.

Taking into account all permissible interactions in the SM, one can calculate either directly [117] or on a lattice the temperature dependence of $\epsilon(T)$ and $s(T)$ and extract corresponding DoFs. This is illustrated on the right panel of Figure 3 showing the temperature dependence of $g_{\rm eff}(T)$ in the SM. For realistic values of both $g_{\rm eff}(T)$ and $h_{\rm eff}(T)$ for a wide temperature interval from 10 keV to 10 TeV, see Ref. [117].

The simultaneous presence of the EW and the QCD matter in thermal equilibrium is one of the remarkable differences between the QGP produced in accelerator experiments and the primordial QGP in the early universe [13,118,119]. To find out which form of matter prevails, let us use Equation (2) and compare the number of DoFs in an ideal massless gas consisting only of EW or QCD matter. Including only the particles which at the temperature $T \lesssim T_{\rm EW}$ can be considered as massless, we obtain $g_{\rm eff}^{\rm EW} = \frac{7}{8}(12+6) + 2 = 17.75$ DoFs for the EW case. The first term in the brackets corresponds to charged leptons e, μ, τ, and the second term corresponds to neutrinos ν_e, ν_μ, ν_τ. The last term corresponds to photons. For non-interacting or weakly interacting QCD matter, Equation (4) reduces to

$$g_{\rm eff}^{\rm QCD} = 2 \times 8 + \tfrac{7}{8}(3 \times N_F \times 2 \times 2)\,, \tag{9}$$

where the first term accounts for the two spin and $N_c^2 - 1 = 8$ color DoFs of the gluons and the second term accounts for $N_c = 3$ colors, N_F flavors, two spin and two particle-antiparticle DoFs of the quarks. Including only the quarks with the mass $m_i/T \simeq 0$, $i = (u,d)$, s, c, b, we obtain successively $g_{\rm eff}^{\rm QCD} = (37, 47.5, 56, 68.5)$ DoFs. In thermal equilibrium at the temperature $T \lesssim T_{\rm EW}$ and for N_F active quark flavors, the QGP contains a factor of $g_{\rm eff}^{\rm QCD}/g_{\rm eff}^{\rm EW} \simeq 2 - 4$ more energy and pressure than those for the EW matter. For temperatures $T \gg T_c^{\rm EW}$, deep inside the EW era, all six quarks u, d, s, c, b, t can be considered to be massless, cf. the left panel of Figure 3, and $g_{\rm eff}^{\rm QCD} = 79$. At the same time, the EW matter acquires $g_{\rm eff}^{\rm EW} = \frac{7}{8}(12+6) + 8 + 4 = 26.75$ DoFs, where $8 = 2 \times (3+1)$ are the DoFs of massless gauge bosons, $W^\pm, W^0, B^0$, and the last term is due to the Higgs scalar doublet. For this case, the QGP has a factor of $g_{\rm eff}^{\rm QCD}/g_{\rm eff}^{\rm EW} \simeq 3$ larger energy density and pressure than those of the EW matter. Hence, we conclude that the QGP was the most dense form of matter filling the early universe during both the QCD and EW epochs.

2.2. *Creation of Primordial Black Holes during the Phase Transitions*

According to inflationary theories, initially, very small inhomogeneities in the matter distribution were produced by the end of the exponential expansion regime. Such inhomogeneities filling the early universe are described by the metric perturbations $\delta g_{\mu\nu}$ which can be decomposed into three irreducible pieces—scalar, vector and tensor ones, see, e.g., Refs. [84,120]. While the scalar part is induced by energy density fluctuations $\delta\epsilon$, the vector and tensor perturbations are related to the rotational motion of the fluid and to the gravitational waves, respectively [84]. Given the scope of this review, in the following, we focus only on one spectacular phenomenon related to metrics fluctuations in the early universe—the matter collapse into primordial black holes (PBHs).

Whereas their existence was proposed already a half-century ago first by Zeldovich and Novikov [121] and later by Hawking [122], it was the detection of gravitational waves from mergers of tens of solar mass $M_\odot$ black hole binaries [123] which has led to a surge of current interest in the PBHs as a Cold Dark Matter (CDM) candidate [73,124–126]. It can be shown that the creation of PBHs due to the gravitational collapse of hot and dense matter occurs for the density contrast $\delta = \delta\epsilon/\epsilon$ exceeding the critical threshold $\delta_c(w[T]) \approx 0.3 - 0.45$, which generally depends on the EoS parameter w [73,125]. Such large values of δ can be generated, e.g., during a period of inflation in the very early universe [126] or during an intermediate period dominated by long-lived massive particles (for recent work, see e.g., Ref. [127] and references therein) or when the universe in the course of the phase transition passes a local minimum in the pressure-to-energy density ratio $w = p/\epsilon$ [125].

For the PBHs forming from Gaussian inhomogeneities with root-mean-square amplitude $\delta_{\rm rms}$, the present CDM fraction for PBHs with a mass around M is found as [125,128]

$$f_{\rm PBH}(M) \approx 2.4\beta(M)\sqrt{\frac{M_{\rm eq}}{M}}, \quad \beta(M) = \frac{2}{\pi}\int_x^\infty e^{-y^2}dy, \quad x = \frac{\delta_c(w[T(M)])}{\sqrt{2}\delta_{\rm rms}(M)}, \tag{10}$$

where $\beta(M)$ is the fraction of horizon patches undergoing collapse to PBHs when the temperature of the universe is T, $M_{\rm eq} = 2.8\times 10^{17} M_\odot$ is the horizon mass at matter-radiation equality, and the numerical factor comes from the ratio of measured baryon $\Omega_{\rm b}$ and CDM $\Omega_{\rm CDM}$ abundances. In Equation (10), we have explicitly taken into account dependence of the critical overdensity δ_c on the EoS parameter $w(T)$. The temperature depends on the PBH mass M as $T \approx 200\sqrt{M_\odot/M}$ MeV. Note that the parameter $\delta_{\rm rms}(M)$ appearing in Equation (10) can be always adjusted to counterbalance the theoretical uncertainties in the value of δ_c so that the current PBH DM fraction is preserved [128]. It is worth mentioning that in the scenarios where PBHs are formed during inflation, their abundance is larger than the Gaussian result by orders of magnitude, but also the mass function has a more pronounced tail at larger masses [126].

In fact, there are a plethora of other mechanisms for PBHs formation (including besides the already mentioned FOPTs, bubble collisions, and the collapse of cosmic strings, necklaces, domain walls, non-standard vacua, etc., see e.g., the recent reviews [73,125]). In the following, in conformity with the topic of our review, we will concentrate on the softest point (SP) mechanism of creation of the PBHs discussed in Ref. [128]. Its virtue stems from the fact that by tracing the origin of PBHs to the SM phase transitions, it is capable of explaining the provenance of part, if not all, of the CDM in the universe [124].

The SP, corresponding to a local minimum in the pressure-to-energy density ratio $w = p/\epsilon$ as given in Equation (6), gives rise to elongation of the expansion time of the hot and dense matter. In HICs, the interest in locating the SP was started by the recognition that the longest-lived fireball might provide a clear signature of the QGP-to-hadron phase transition [129]. Shortly after that, the formation of horizon-size PBHs due to a substantial reduction of pressure during adiabatic collapse in the course of the QCD transition was analyzed in the context of the early universe in Refs. [130,131]. Even though the previously used assumption of the first-order character of the phase transition was later on replaced by a crossover scenario, the lattice calculations have found a local minimum in $w = 0.145(4)$ at $T = 159(5)$ MeV [132].

To become acquainted with the influence of the SPs on the cooling of the universe during its radiation-dominated era, let us follow Ref. [128] and inspect the behavior of the function $g_{\rm eff}(T)$ shown on the right panel of Figure 3. Let us focus on the temperatures when a part of the radiation matter transforms into a non-relativistic matter. Starting from $T \approx 200$ GeV downwards, this happens first to the top quark at $T \approx m_t = 172$ GeV, which is followed by the Higgs boson at 125 GeV and the Z and W bosons at 92 and 81 GeV, respectively. The fact that these particles become non-relativistic at nearly the same time of universe expansion induces a significant drop in the number of relativistic DoFs, from $g_{\rm eff} = 106.75$ down to $g_{\rm eff} = 86.75$. Further changes at the b,c-quark and τ-lepton thresholds are too small to be noticeable. Hence, further on, $g_{\rm eff}$ remains approximately constant until the QCD transition at around 160 MeV. The number of relativistic DoFs then falls abruptly to $g = 17.25$. A little later, pions become non-relativistic, and then muons, yielding $g_{\rm eff} = 10.75$. Thereafter, $g_{\rm eff}$ remains constant until e^+e^- annihilation and neutrino decoupling at around 1 MeV, when it drops down to $g_{\rm eff} = 3.36$ [128].

Provided that total entropy is conserved, an abrupt reduction of $g_{\rm eff}(T)$ leads to a sudden drop in the speed of sound $c_s(T)$, cf. Equation (7), and hence to a drop of pressure, $p = w(T)\epsilon$, cf. Equation (6). The effect is clearly visible on the left panel of Figure 4 showing the four periods in thermal history of the universe when $w(T)$ reaches its local minimum. After each period, w returns back to its relativistic value of 1/3, but each sudden drop modifies the probability of gravitational collapse of any large curvature fluctuations present at that time [128].

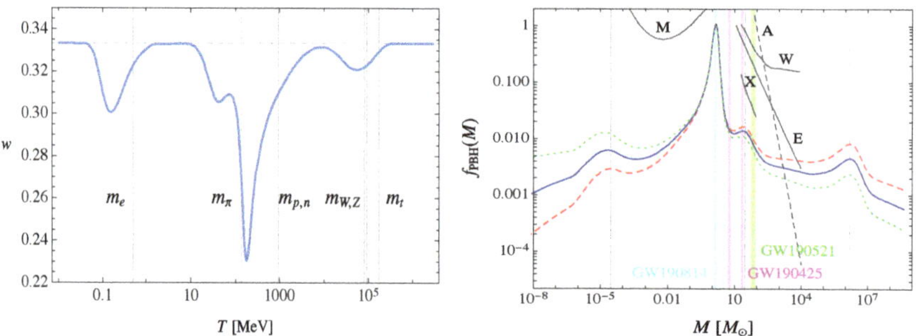

Figure 4. (**Left**) EoS parameter w as a function of temperature T. The gray vertical lines correspond to the masses of the electron, pion, proton/neutron, W, Z bosons and top quark, respectively. The gray dashed horizontal line indicates value of $w = 1/3$. Adapted from Ref. [128]. (**Right**) The mass spectrum of PBHs $f_{\rm PBH}(M)$ in solar mass units $M_\odot$. The gray vertical lines correspond to the EW and QCD phase transitions and e^+e^- annihilation. The vertical colored lines indicate the masses of the three LIGO-Virgo events. Gray curves are constraints from microlensing (M), ultra-faint dwarf galaxies and Eridanus II (E), X-ray/radio counts (X), and halo wide binaries (W). The accretion constraint (A) is shown dashed, as it relies on uncertain astrophysical assumptions. Adapted from Ref. [128].

Consider one cooling period $T_1 < T < T_2$ with $w(T) < w(T_{1,2}) = 1/3$ centered around the local minimum $w(T_{\rm SP})$ and define the quantity,

$$\Delta h_{\rm eff}(T) \equiv g_{\rm eff}(T) - h_{\rm eff}(T)\,, \tag{11}$$

measuring departure from the $w = 1/3$ case; see Equation (6). At the endpoints $\Delta h_{\rm eff}(T_2) = \Delta h_{\rm eff}(T_1) = 0$ but for $w(T) < 1/3$, it is always positive $\Delta h_{\rm eff}(T) > 0$, cf. Equation (7). Hence, the initial drop in the entropy DoFs, $h_{\rm eff}(T)$, always precedes the jump in the energy density DoFs, $g_{\rm eff}(T)$. This leads to the following "coarse-grained" scenario for the PBH formation: the reduction in $h_{\rm eff}(T)$ occurring for $T_2 > T > T_{\rm SP}$ is followed by a fall in $g_{\rm eff}(T)$ for $T_1 > T > T_{\rm SP}$. An excess in entropy $\sim \Delta h_{\rm eff}(T_{\rm SP})$ lost by the radiation during its cooling is dumped into the collapsing matter, emerging eventually in the form of PBHs—the matter with the largest entropy density in the universe [133]. This may explain why even at the present stage of the universe evolution, there is by a huge factor far more entropy in supermassive black holes (BHs) in galactic centers than in all other sources of entropy put together [134].

Assuming that the amplitude of the primordial curvature fluctuations is approximately scale-invariant [128], one obtains from Equation (10) the mass spectrum of PBHs $f_{\rm PBH}(M)$ shown on the right panel of Figure 4. The peaks at $M \simeq 10^{-6}, 2, 30$ and $10^6 M_\odot$ correspond to the EW and QCD phase transitions, to pions and muons becoming non-relativistic and to e^+e^- annihilation, respectively. The latter may also provide seeds for the supermassive BHs' formation in galactic nuclei. The largest contribution to $f_{\rm PBH}(M)$ comes from the PBHs formed at the QCD transition epoch and that would naturally have the Chandrasekhar mass ($1.4\,M_\odot$) [128]. Moreover, the peak in the range 1–10 $M_{rm\odot}$ could explain the LIGO/Virgo observations [123]. The latter favor mergers with low effective spins as expected for PBHs, but it is hard to explain BHs of stellar origin [135].

The simple analytical models that describe the dynamical process of gravitational collapse which may be relevant for PBH formation were studied in Ref. [136]. It is also worth noting that the gravitational collapse of large inhomogeneities during the quark-hadron transition epoch may also explain the baryon asymmetry of the universe [137]. The asymmetry can be generated in local hot spots through the violent process of PBH formation at the QCD transition triggered by a sudden drop in the radiation pressure and the presence of large amplitude curvature fluctuations caused by the axion field—the subject to be discussed in Section 2.6.

2.3. Perturbative and Strongly Coupled Regimes of QCD

An important contribution to the effective number of relativistic DoFs, $g_{\rm eff}$, comes from the hadron-to-QGP phase transition—see a big jump in the interval $10^2 \lesssim T \lesssim 10^3$ MeV in Figure 3 (right panel). At higher temperatures, in the QGP region, the strength of the interactions between the quarks and gluons is set by the QCD coupling $\alpha_S(Q)$ which at the one-loop order of perturbation theory takes the form,

$$\alpha_S(Q) \simeq \frac{2\pi}{b_0 \ln(Q/\Lambda_{\rm QCD})}, \qquad b_0 = 11 - \frac{2}{3} N_F, \tag{12}$$

where Q is the momentum transferred during the interaction, $\Lambda_{\rm QCD} \simeq 200$ MeV is the characteristic QCD scale, and N_F is the number of active quark flavors. The logarithmic decrease of $\alpha_S(Q)$ with increasing Q, i.e., with decreasing distance among the quarks and gluons, is due to the fact that, in contrast to the photon in QED, the force carriers in QCD, the gluons, have color charge. Their exchanges in higher-order processes involving both the quarks and the gluons occur more frequently with increasing Q and lead to a color charge spread (or anti-screening). Indeed, the gluon multiplicity increases at low momentum fractions corresponding to the limit of large energies. Dilution of the initial color charge is responsible for the weakening of α_S at small distances $\ell \ll \Lambda_{\rm QCD}^{-1}$, i.e., when the quark experiences a large momentum transfer Q, see Equation (12). This effect known as asymptotic freedom [31,32,138] is illustrated on the left panel of Figure 5 where the values of the $\alpha_S(Q)$ extracted from proton-(anti)proton and lepton-proton collisions are shown [139]. In agreement with Equation (12), a slow logarithmic decrease from $\alpha_S(Q_{\rm min} = 5\,{\rm GeV}) = 0.22$ to $\alpha_S(Q_{\rm max} = 1500\,{\rm GeV}) = 0.08$ is observed.

Before proceeding further, let us recall that quantum field theory (QFT) at finite temperature T is often considered to be equivalent to Euclidean QFT in a space which is periodic, with period $1/T$ along the "imaginary time" axis (for a recent review of this subject, see e.g., Refs. [26,140] and references therein). Thus, in order to formulate the theory at $T > 0$ using its variant at $T = 0$, one should replace zero components of all 4-momenta k^μ in the Euclidean integrals by the discrete Matsubara frequencies—$2\pi n T$ for bosons and $(2n+1)\pi T$ for fermions, and sum over $n \in \mathbb{Z}$ instead of integrating over k^μ. Consequently, the average momentum transferred during the interactions in the hot medium Q can be related to the temperature as $Q = 2\pi T$. In particular, the maximum value of the momentum transferred $Q_{max} = 1500\,\mathrm{GeV}$, which has been so far measured in pp collisions at the LHC, roughly corresponds to the "temperature" $T_{max} = Q_{max}/(2\pi) \simeq 240\,\mathrm{GeV} \simeq T_c^{EW}$.

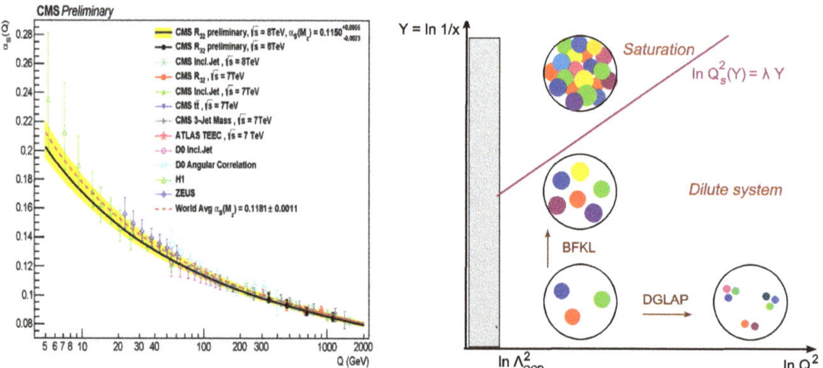

Figure 5. (**Left**) The QCD coupling $\alpha_S(Q)$ as a function of the momentum transfer scale Q obtained by using the MSTW2008 NLO PDF set [141]. Adapted from Ref. [139]. (**Right**) The partonic phase diagram showing evolution of the partons' density and size as a function of their rapidity $Y = \ln(1/x)$ and the logarithm of momentum transfer squared $\ln Q^2$. Adapted from Ref. [142].

The asymptotic freedom formula given by Equation (12) is based on the applicability of QCD perturbation theory to the processes at high momentum transfers, as it is a well-known fact that the perturbative expansion is an example of asymptotic series. It looses its validity with decreasing Q when the perturbative approximation breaks down. Interestingly, by performing the matching of the fundamental theory onto the effective chiral Lagrangian at the infrared scale $Q \simeq 4\pi f_\pi \simeq 1\,\mathrm{GeV}$, at which the ranges of validity of perturbative QCD and chiral perturbation theory (describing interactions among low-momentum hadrons) descriptions meet, one can infer the information about the behavior of $\alpha_S(Q)$ at large distances. Such a matching implies that the QCD coupling in the infrared region is "frozen" at $\langle\alpha_S\rangle_{IR} \simeq 0.56$ [143], incidentally at twice the upper scale value of $\alpha_S(Q)$ shown on the left panel of Figure 5.

Let us note that evolution of the strong coupling parameter $\alpha_S(Q)$ described at the leading order by Equation (12) is valid only for DoFs that dominate the thermodynamical evolution, i.e., for the partons (quarks and gluons) with momenta of order $T = Q/2\pi$, and it does not apply to the long wavelength non-perturbative modes residing at the length scales of $\ell > T^{-1}$. Those modes are occupied by a liquid in which neighboring "unit cells" are tightly coupled to each other [144]. This strongly coupled regime [10] makes the QGP behave as the ideal fluid [145,146]. The fluidity of the QGP was first established in the collisions of ultra-relativistic nuclei at RHIC [53–55] and later confirmed at higher energies of the LHC [147–149]. The most prominent signals of the strong interaction in the deconfined bulk manifest in a collective flow of matter [8,146] and in a spectacular phenomenon of suppression of very energetic partons passing through the QGP medium [8,150,151]. Direct evidence for the non-perturbative character of deconfined matter comes from the low-momentum spectra of direct photons measured in Au+Au collisions at RHIC. The temperatures obtained from the spectra $T \simeq 220$ MeV [152] point to the initial temperatures $T_{ini} \simeq 300$–600 MeV at early times of $\tau_0 = 0.6$–0.15 fm$/c$, which are way below the perturbative regime of QCD.

By definition, plasma is a state of matter in which charged particles interact via long-range (massless) gauge fields [153]. This distinguishes it from neutral gases, liquids, or solids in which the inter-particle interaction is of short range. So, plasmas themselves can be gases, liquids, or solids, depending on the value of the plasma parameter Γ, which is the ratio of interaction energy to kinetic energy of the particles forming the plasma [39]. For a classical plasma of N particles with charge Ze occupying a volume V,

$$\Gamma \equiv \frac{(Ze)^2}{ak_BT}, \qquad a(T) = \left(\frac{3V}{4\pi N}\right)^{1/3} \approx 0.62n(T)^{-1/3}, \tag{13}$$

where $n(T) = N/V$ is the temperature-dependent particle number density. While most plasmas are ideal with $\Gamma < 10^{-3}$, a strongly interacting plasma has $\Gamma \gtrsim 1$. For plasma with $Z \simeq 1$ at the temperature $T \simeq 10^6$ K $\simeq 100$ eV, the number density n must be as high as 10^{26} cm^{-3} to make $\Gamma \simeq 1$ [39]. Ion plasma in a white dwarf has Γ = 10–200, in transient plasmas produced in explosive shock tubes, the values of Γ = 1–5 are found [39]. A more down-to-earth example is table salt (NaCl), which can be considered as a crystalline plasma made of permanently charged ions Na^+ and Cl^- [153]. At temperatures of $T \approx 10^3$ K, still too small to ionize non-valence electrons, the table salt transforms into a molten salt, which is a liquid plasma with $\Gamma \approx 60$.

Generalization of Equation (13) to the QGP case was suggested in Ref. [154]

$$\Gamma \simeq 2\frac{C_{q,g}\alpha_S}{aT}, \qquad C_q = \frac{N_c^2-1}{2N_c} = \frac{4}{3}, \qquad C_g = N_c = 3, \tag{14}$$

where the strong coupling α_S is a slowly varying function of temperature, C_q and C_g are the Casimir invariants of fundamental and adjoint irreducible representations of the color SU(3)$_c$ group corresponding to quarks and gluons, respectively, and $a = a(T)$ is the average inter-parton distance at a given temperature T as follows from Equation (13). The factor 2 in Equation (14) takes into account the equal importance of chromoelectric (CE) and chromomagnetic (CM) interactions in ultra-relativistic systems. For ideal massless QCD gas with N_F active quarks and $d_F = g_{\rm eff}^{\rm QCD}$ degrees of freedom, see Equation (9), the particle number density reads

$$n = d_F\frac{\zeta(3)}{\pi^2}T^3 \approx d_F\left(\frac{T}{2}\right)^3, \qquad d_F = 2\times 8 + \tfrac{7}{8}(3\times N_F\times 2\times 2)\,. \tag{15}$$

From Equations (13) and (15), it follows that $a \simeq 1.24d_F^{-1/3}T^{-1}$, and so, the term aT, appearing in the denominator of Equation (14), depends on T through the temperature-dependent number of active quark flavours $N_F(T)$ only. Consequently, for an ideal massless QCD gas, the temperature-dependence of the plasma coupling parameter Γ is driven by $\alpha_S d_F^{1/3}$. For the QGP created in HICs at RHIC $T \approx 200$ MeV and $\alpha_S = 0.3$–0.5 with $N_F = 2$, Equations (15) and (14) yield $d_F = 37$ and $\Gamma \simeq 2$–8 well inside the strongly coupled regime. At much higher temperatures, say, at $T \simeq T_{\rm EW}$, with $\alpha_S = 0.08$ and $N_F = 5$, we obtain $d_F = 52.5$ and $\Gamma \simeq 0.5$–1.5—the value located in the vicinity of the strongly coupled regime. At even higher temperatures, the number of active quark DoFs saturates at $N_F = 6$ and the evolution of $\Gamma(T)$ becomes solely driven by the (logarithmically decreasing) QCD coupling $\alpha_S(T)$, cf. (12). Let us note that the ideal gas approximation serves only as a lower estimate of Γ because it ignores the interactions in the partonic liquid. The latter will slow down the temperature dependence of the average inter-parton distance a, thus weakening the strong coupling parameter dependence on T.

A more in-depth approach to strongly coupled non-Abelian plasmas [155] was expected to come from the gauge/string duality [156]—a correspondence between d-dimensional conformal QFT and $(d+1)$-dimensional string or gravity theory. In these theories, the graviton needs not live in the same spacetime as the QFT, but due to the holographic principle, the description of gravity within a volume of spacetime can be thought of as encoded on a lower-dimensional boundary to the region in the formalism of conformal field theory [157,158]. However, the inherently conformal character of the gauge QFT used in the duality with anti-de Sitter gravity (AdS) is at variance with QCD where the scale invariance is broken by the confinement scale (for a recent review of confinement dynamics, see Ref. [22]), causing the running of the coupling. This limits the applicability of gauge/string duality [156] to temperatures $T \gg T_c^{QCD}$ and hence to weak couplings.

2.4. QCD at High Parton Densities and Saturation

To proceed further, a different type of analysis of the QCD dynamics of partonic matter is needed. There are two independent paths along which the density of partons can evolve, and these are illustrated in the right panel of Figure 5. The two together form the basis of our current understanding of high-energy scattering in QCD.

The first path follows the development of partonic cascade in variable Q. For partons that occupy a transverse area $1/Q^2$, the increase of Q and hence of the temperature $T \sim Q$ leads to dilution of their density. The process is controlled by the Dokshitzer–Gribov–Lipatov–Altarelli–Parisi (DGLAP) [159–161] equations describing the evolution of partonic density as a function of evolution variable $\ln(Q^2/\Lambda^2_{QCD})$ [17,162].

The second path follows the development of a parton shower in variable $x = k^+/P^+$, which is a fraction of the light cone momentum[1] P^+ of the parent parton, which has radiated a parton emerging with the light-cone momentum k^+. In the $\boldsymbol{x}_\perp$ plane transverse to the direction of the fast-moving primary parton, the partonic cascade initiated by the primary parton can be visualized as a Brownian motion-like trajectory developing from $x = 1$ toward $x \to 0$. The corresponding Gribov diffusion process is controlled by the so-called evolution parameter $Y = \ln(1/x)$, leading to a difference in rapidity between the primary and radiated partons, with a coefficient being the diffusion constant proportional to α_S. Its evolution in the Y variable is described by the Balitsky-Fadin-Kuraev-Lipatov (BFKL) equation that is complementary to the DGLAP evolution realized in $\ln Q^2$ (see standard textbooks, e.g., Refs. [17,162] and references therein).

At fixed Q, the radiated partons (mostly, soft gluons with $x \ll 1$) are typically of the same size. When a parton-parton interaction cross-section $\sim\alpha_S/Q^2$ multiplied by $xG_A(x, Q^2)$—the probability to find at fixed Q a parton carrying a fraction x of the parent parton momentum—becomes comparable to the geometrical cross section πR_A^2 of the object A occupied by the gluons, the partons "overlap". The repulsive interactions among the gluons ensure that their occupation number f_g (the number of gluons with a given x multiplied by the area each gluon fills up divided by the transverse size of the object) saturates at $f_g \sim 1/\alpha_S$. Note that this is a very generic behavior—the same density scaling as the inverse interaction strength α^{-1} is characteristic of a number of condensation phenomena such as the Higgs condensate, see, e.g., Ref. [163], or superconductivity [164].

The phenomenon of saturation [165] is thus important for gluons with transverse momenta $k_\perp \le Q_s$ [166,167], where

$$Q_s^2(x) = \frac{\alpha_S(Q_s)}{2(N_c^2-1)} \frac{xG_A(x, Q_s^2)}{\pi R_A^2} \sim \frac{1}{x^\lambda} \tag{16}$$

is the x-dependent saturation scale representing a fixed point of the parton density evolution in x or, equivalently, the emergent "close packing" scale [167]—see the right panel of Figure 5 where the saturation line $\ln Q_s^2(Y) = \lambda Y$, $Y = \ln x$ is also displayed. Such gluons form a highly coherent configuration called Color Glass Condensate (CGC) [142,167,168], or glasma [169], which due to the high occupation number f_g has properties of QCD in the classical regime [166].

At high temperatures, one usually expects that quantum effects become less important. To show that, for the CGC, we follow Ref. [166] and write the gluonic part of the QCD action in terms of the gauge field potential $\mathcal{A}_\mu^a$ and field strength $\mathcal{F}_{\mu\nu}^a$ which are obtained by rescaling of their equivalents A^a_μ and $F^a_{\mu\nu}$ used in a more traditional approach when the coupling constant g_s multiplies the interaction terms in the Lagrangian,

$$A^a_\mu \to \mathcal{A}^a_\mu \equiv g_s A^a_\mu\,, \quad F^a_{\mu\nu} \to g_s F^a_{\mu\nu} \equiv \mathcal{F}^a_{\mu\nu} = \partial_\mu \mathcal{A}^a_\nu - \partial_\nu \mathcal{A}^a_\mu + f^{abc}\mathcal{A}^b_\mu \mathcal{A}^c_\nu\,, \tag{17}$$

$$S_g = -\frac{1}{4}\int F^a_{\mu\nu}F^{\mu\nu,a}d^4x = -\frac{1}{4g_s^2}\int \mathcal{F}^{\mu\nu,a}\mathcal{F}^a_{\mu\nu}d^4x\,, \tag{18}$$

where f^{abc} with $(a,b,c) \in \{1,\ldots,8\}$ are the SU(3) group structure constants. For a classical configuration of gluon fields, by definition, $\mathcal{F}^a_{\mu\nu}$ does not depend on the coupling, and the action is large, $S_g \gg \hbar$. The number of quanta in such a configuration is then

$$f_g \sim \frac{S_g}{\hbar} \sim \frac{1}{\hbar g_s^2}\rho_4 V_4\,, \tag{19}$$

where we rewrote Equation (18) as a product of four-dimensional gluon condensate density $\rho_4 \sim \langle \mathcal{F}^{\mu\nu,a}\mathcal{F}^a_{\mu\nu}\rangle$ and spacetime volume V_4. The number of quanta f_g in such a configuration depends only on the product of the Planck constant $\hbar$ and the strong coupling squared $g_s^2 = 4\pi\alpha_S$. The classical limit $\hbar \to 0$ is indistinguishable from the weak coupling limit $g_s^2 \to 0$ [166]. Thus, the weak coupling limit of small α_S corresponds to the semi-classical regime.

An equivalent argument employs the fact that the path integral formulation of the quantum theory in Minkowski space sums over all field configurations weighted with $\exp(-iS_g/\hbar)$. Since g_s^2 appears in the exponential in the same place as $\hbar$, cf. Equation (18), this already suggests that for $g_s^2 \to 0$, the path integral is dominated by the classical configurations. Such configurations are believed to describe the matter inside incident nuclei during the initial stage of relativistic HICs at RHIC and the LHC [142,168].

Although in its original formulation, the QCD saturation is used for partons with fixed Q in case of the macroscopic bodies, it is more relevant to consider the partons at a fixed temperature $T = Q/2\pi$, avoiding at the same time the quantum entanglement problem [170], which is inevitably present in the description of microscopic objects. In this generalized setting, an object A filled with gluons may represent not only a fast-moving proton or nucleus but also the interior of the expanding early universe. Moreover, as follows from Equation (16), the saturation phenomenon is not necessary related to the growth of the gluon density at small x. For a big fast-moving domain of space filled with deconfined quarks and gluons, with radius $R \gg 1$ fm, and hence also for the fast-expanding early universe itself characterized by the Hubble horizon $L_H \ggg 1$ fm, the saturation limit can be reached even at $x \simeq 1$ [162][2]. In the extreme case, when the gluonic part of QCD matter completely decouples from the QCD fermionic fields and forms the vacuum condensate, the first term in Equation (9) can be neglected, and we obtain $g^{\rm QCD}_{\rm eff}/g^{\rm EW}_{\rm eff} \simeq 8/3$, making the CGC a prevailing form of matter during both the QCD and EW epochs.

Thus, in the periods of cosmological evolution when $T \gg \Lambda_{\rm QCD}$, including a very hot QCD era, EW era and beyond, it is perfectly conceivable that the universe was dominated by the fully saturated gluonic matter with occupation number $f_g \sim \alpha_S^{-1}$. If during its subsequent cooling, the universe followed a trajectory in the $[\ln(1/x), \ln Q^2]$ plane staying still above the $\ln Q_s^2 = \lambda Y$ line, see the right panel of Figure 5, the CGC phase would be a prevailing form of matter down to the temperatures $T \approx (2-5)\times\Lambda_{\rm QCD}$. For lower temperatures, the glasma is expected to fragment into a strongly interacting QGP.

The issue of emergence of classical behavior in the cosmological history has drawn recently a great deal of attention because of its conceptual as well as practical importance, see e.g., Ref. [171] and references therein. Although the origin of the observed anisotropies in the cosmic microwave background (CMB) radiation is traced back to vacuum fluctuations of quantum fields in the very early universe [84,120,172], there is a general expectation that the main characteristics of the universe can be described in classical terms even in its early history [171,173,174]. This is consistent with the fact that the initial conditions of the Hot Big Bang were determined by cosmic inflation driven by the so-called inflaton field [84,86,175,176]. Apart from the quantum fluctuations of this field, its very emergence may well be a quantum phenomenon, e.g., the axion condensation (for a more thorough discussion, see below).

Let us note that the word "glass" appearing in the acronym CGC is used in condense matter physics to describe a non-equilibrium, disordered state of matter acting like solids on short time scales but liquids on long time scales [177]. In the glasma case, there are two scales present—the light cone time $\tau_{\rm wee}$ of low x (or wee) partons and the light cone time $\tau_{\rm valence}$ of primary (or valence) partons. For partons of transverse momentum $k_\perp$, we obtain

$$\tau_{\rm wee} = \frac{1}{k^-} = \frac{2k^+}{k_\perp^2} = \frac{2xP^+}{k_\perp^2} \ll \frac{2P^+}{k_\perp^2} \approx \tau_{\rm valence}\,, \tag{20}$$

suggesting that the valence parton modes are static over the times scales over which the wee modes are probed [167]. It is quite tempting to identify $\tau_{\rm valence}$ with the quantum break-time discussed in Refs. [178,179] defined as the time-scale after which true quantum evolution of the parton densities departs from the classical mean field evolution.

Glasses are formed when liquids are cooled too fast to form the crystalline equilibrium state. The fast cooling leads to an enormous number of possible configurations $N_{\rm gl}(T)$ into which the glasses can freeze and consequently to their large entropy $S = \ln N_{\rm gl}(T)$, which does not vanish, even at $T = 0$, see e.g., Refs. [177,180]. In case of the CGC, the fast cooling is expected to take place in the Grand Unified Theory (GUT) era (see Section 2.5) when about one-third of all gauge bosons are gluons. By the end of that period, the excess of effective entropy DoF $h_{\rm eff}$ is almost completely absorbed by the saturated gluonic matter.

The gluon condensation into (many domains of) the saturated phase was also facilitated by the fact that the wee partons "see" the color charge of other gluons over very large distances given by their transverse wavelength $\lambda_{\rm wee} \sim 1/k^+ = 1/(xP^+)$. Since the glasma domains were formed in separate and completely different gluonic configurations, the saturated gluonic matter occupying the early universe had a substantial excess in the entropy DoF, $h_{\rm eff}^{\rm QCD}$, over the effective number of DoF in energy, $g_{\rm eff}^{\rm QCD}$. Consequently, the value of the generalized EoS parameter w was higher than that of the ideal massless gas; see Equation (6).

One possibility of how the EoS of a non-equilibrium matter comprising weakly interacting gluons can be approximated by the ideal massless gas of quasi-particles with $w(T) > 1/3$ follows from Equation (48) discussed later in Section 3.1. The glasma with $w(T) \approx 1/D$ may be looked upon as either a two-dimensional sheet ($D = 2$), a one-dimensional string ($D = 1$) or, more generally, a fractal with the Hausdorff dimension $1 \le D < 3$. Recent investigations of the dynamics of expanding glasma show that the spatial asymmetry introduced by the initial geometry is effectively transmitted to the azimuthal distribution of the gluon momentum field, even at very early times [181,182].

2.5. The Running Couplings of the Standard Model and Their Unification

The importance of QCD interactions in the EW era of the universe evolution can be also seen by comparing the corresponding couplings—the strong $\alpha_{\rm S}$, electromagnetic $\alpha_{\rm EM}$ and weak $\alpha_{\rm W}$ ones. This is illustrated on the left panel of Figure 6 showing the RG flow in the scale $\mu = Q$ of the electromagnetic, weak and strong coupling parameters above $Q = 100$ GeV. Note that due to the fact that the gauge group of SM interactions $\mathrm{SU(3)_C} \times \mathrm{SU(2)_L} \times \mathrm{U(1)_Y}$ is not a simple Lie group, the theory has not one but three coupling parameters, which are often denoted as $\alpha_1 \equiv \alpha_{\rm EM}$, $\alpha_2 \equiv \alpha_{\rm W}$ and $\alpha_3 \equiv \alpha_{\rm S}$ [183].

To see how the values of the coupling parameters influence the state of early universe, let us compare the collision time among its constituents $t_c \sim 1/(\sigma n v)$ (σ is the effective cross-section, n is the particle number density and v is their relative velocity) with the characteristic time-scale of the universe expansion $t_H \sim 1/H$ [84]. Let us first restrict ourselves to the temperatures $T \gtrsim T_{\rm EW}$ when all particles of the SM are ultra-relativistic and the gauge bosons are massless. Then, the cross-sections for strong and EW interactions have a similar energy dependence $\sigma \sim \alpha^2/T^2$, where $\alpha \simeq 10^{-1} - 10^{-2}$ are the corresponding dimensionless running couplings $\alpha_{1,2,3}$ varying logarithmically with T; see Figure 6. Taking into account Equations (39) and (42) for $n \sim T^3$, we find

$$t_c \sim \frac{1}{\sigma n v} \sim \frac{1}{\alpha^2 T} \ll t_H \sim \frac{1}{H} \sim \frac{1}{\sqrt{\epsilon}} \sim \frac{1}{T^2}. \quad (21)$$

Thus, for the temperatures $10^{15} - 10^{17}$ GeV $\gtrsim T \gtrsim T_{\rm EW}$, the local equilibrium in the fluid is established before expansion of the universe becomes relevant.

At the lower temperatures down to $T_c^{\rm QCD} \approx 160$ MeV, the strong interactions prevail over the EW ones, and the local equilibrium is controlled by the interactions among the quarks and gluons in the strongly interacting QGP liquid. Even though the number density of the medium may be smaller[3], the big effective cross-section among the particles forming the medium guarantees that the condition for local equilibrium $t_c \ll t_H$ is satisfied. Thus, over that whole range of the temperatures, the early universe is in the local equilibrium. It develops along the maximal possible entropy path, making it amenable to the hydrodynamical description.

In spite of its success, the SM cannot be the ultimate theory of particle physics. Such long-standing problems as the absence of a suitable DM candidate, no explanation of the observed baryon asymmetry in the universe, as well as various hierarchy problems in the underlined mass spectra (such as the flavor problem, the neutrino mass problem and the Higgs hierarchy problem) call for various bottom-up extensions of the SM framework as well as continuous attempts to derive the SM structure from a top-down perspective.

As was earlier discussed, e.g., in Ref. [184], the Higgs boson quartic coupling in the SM turns negative at scales $\gtrsim 10^{10}$ GeV, rendering the vacuum state of the theory unstable at high energies. The current theoretical developments and experimental measurements suggest that the metastability of the Higgs vacuum is favored. This means a vacuum decay may occur with possibly catastrophic consequences for cosmology, since there are many catalysts that could trigger such a decay in the early universe. For a comprehensive review on cosmological implications of the Higgs vacuum metastability, see, e.g., Ref. [185].

Incidentally, almost immediately after the discovery of the asymptotic freedom, it was suggested [186] that at very high energies, the three gauge interactions of the SM are merged into a single force. The model, a first example of the Grand Unified Theory (GUT), is based on the smallest simple Lie group which contains the SM gauge groups SU(5) $\supset$ $\mathrm{SU}_c(3) \times \mathrm{SU}(2)_{\rm L} \times \mathrm{U}(1)_{\rm Y}$. Among its 24 gauge bosons, there are in addition to eight gluons of QCD and four EW gauge bosons $W^{\pm}, Z$ and γ also 12 new ones called X and Y. Their emission or absorption makes it possible to transform a lepton into a quark or vice versa. Hence, the SU(5) GUT does not conserve baryon and lepton numbers separately, making it the first theory providing an explicit mechanism for the proton decay $p \to e^+\pi^0$, with the half-time $\tau_p \simeq M_X^4/m_p^5 \simeq 10^{30} - 10^{31}$ years, where M_X is the mass of SU(5) gauge boson at the scale of Grand Unification and m_p is the mass of the proton. Although it was later found to disagree with experimental lower limit of $\tau_p \geq 8.2 \times 10^{33}$ years [187] SU(5), unification is still considered an important example and a reference point of GUT model-building.

The basic property of the SU(5) theory and its later GUT successors [188–192] is that by virtue of the unification into a single (simple Lie) gauge group at very high energies, strict unification of the SM gauge couplings must take place. This is hinted at but not really achieved in the SM; see the left panel of Figure 6. First, α_1 and α_2 cross each other at $\mu \sim 10^{13}$ GeV; then, α_1 crosses α_3 at $\mu \sim 10^{14}$ GeV, and finally, α_2 and α_3 cross at $\mu \sim 10^{17}$ GeV, providing a hint of unification close to the Planck scale. Thus, as ultraviolet (UV) completions of the SM describing the physics at very high energies, GUTs are sometimes connected to theories of gravity such as string theory.

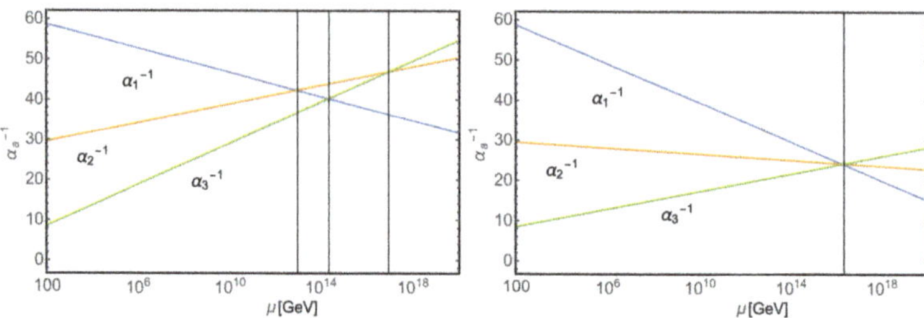

Figure 6. (**Left**) RG flow of the inverse SM gauge couplings α_a^{-1} as functions of the renormalization scale parameter μ. Index $a = 1, 2, 3$ stands for QED ($a = 1$), weak ($a = 2$) and QCD ($a = 3$) couplings. (**Right**) RG flow of the MSSM gauge couplings. Adapted from Ref. [192].

Let us now try to speculate on how the Grand Unification might have influenced the dynamics of the early universe at the temperatures when gluons decouple from quarks, forming a condensate. In SU(5) GUT, with increasing temperatures $T \gg T_c^{\rm EW}$, the gluon exchange between quarks (antiquarks) becomes overshadowed by the exchange of EW massless gauge bosons $W^{\pm}, W^0, B^0$ to be finally superseded by the GUT super-heavy X and Y gauge bosons. The latter also facilitate the conversion of quarks into leptons and vice versa. Since the transformations occur in the thermal and chemical equilibrium, the number of quarks and leptons remains constant on average.

One of the fundamental issues with the GUT models, which remains a challenge today, is the large hierarchy between the mass scale of Grand Unification and the EW scale. The latter is a source of large loop corrections to the Higgs mass. The problem is usually solved by extending the GUT with supersymmetry, the hypothetical symmetry between fermions and bosons (for a recent review on the concepts of supersymmetric GUTs, see e.g., Ref. [192]), which is broken at lower energies. Some of its minimal realizations, such as the Minimal Supersymmetric Standard Model (MSSM) [193], predict the unification of all three gauge couplings at the same scale; see the right panel of Figure 6.

Notwithstanding the drawbacks of GUTs, their cosmological signatures look quite promising. With the critical temperature $T_c^{\rm GUT}$ of the GUT phase transition approximately of the same order of magnitude as the particle masses at that temperature, the phase transitions that take place in GUTs at $T \gtrsim 10^{14}$ GeV, as a rule, prove to be FOPTs [194]. Such transitions proceed via bubble nucleation [195]. While an isolated spherical bubble may produce GWs through sound waves in the plasma and magneto-hydrodynamics turbulence effects (see for example Refs. [196–199] and references therein), the process of bubble collision contributes to the GWs spectrum in the quadrupole approximation [86,200]. This contrasts with the SM phase transitions where the crossovers do not lead to a strong enhancement over the primordial GW spectrum. Moreover, the FOPTs also generate a primordial magnetic field during the turbulence phase of the plasma and bubble collision [86,201], and in some instances, they may generate topological defects such as domain walls and strings [195,202].

Let us add that an FOPT in the EW sector though precluded in the SM is possible in many of its scalar sector extensions [203]. In the most exotic scenario, a very peculiar history of the universe may occur: a first-order QCD phase transition (with six massless quarks) triggers an EW FOPT, which is eventually followed by a low-scale reheating of the universe where hadrons (likely) deconfine again, before a final, conventional crossover QCD transition to the current vacuum [204].

2.6. Axions

A promising avenue connecting the strong interactions with physics beyond the SM having at the same time far-reaching consequences in cosmology is provided by hypothetical ultra-light particles—the axions [183,205–208]. The QCD, unlike the EW interaction, is symmetric under time reversal and hence under a combined charge conjugation C and parity P operation CP. In principle, one can add to the QCD Lagrangian the term:

$$\mathcal{L}_Q = \theta \frac{g_s^2}{32\pi^2} F^a_{\mu\nu} \tilde{F}^{\mu\nu}_a \,, \tag{22}$$

where $F^a_{\mu\nu}$ is the gluon field strength tensor (see Equation (17)), and $\tilde{F}^{\mu\nu}_a = \frac{1}{2}\epsilon^{\mu\nu\alpha\beta} F^a_{\alpha\beta}$ is its dual. For a non-zero value of the parameter θ, called the vacuum angle [209], the strong coupling permits violation of the CP symmetry. However, since $\mathcal{L}_Q$ can be written as a total derivative of K^μ, the Chern–Simons current, $\mathcal{L}_Q = \partial_\mu K^\mu$, the new term does not produce any effects in perturbation theory and is therefore usually neglected. Nevertheless, classical configurations, topological in nature, do exist, one example being the instantons [210], for which this term cannot be ignored. For instance, in the semi-classical dilute instanton gas approximation (DIGA) [211], the QCD vacuum energy density ϵ_0 depends on θ as

$$\epsilon_0(\theta) = -2Ce^{-S_{\text{inst}}} \cos\theta \,, \;\; S_{\text{inst}} = \frac{8\pi^2}{g_s^2} \,. \tag{23}$$

Here, C is a positive constant and S_{inst} is the QCD instanton action [207,208], cf. Equation (18). Moreover, since $\mathcal{L}_Q$ in Equation (22) preserves the charge conjugation C, it contributes directly to the neutron electric dipole moment $d_n \approx e m_q / m_n^2 \theta$, where e is the proton charge, m_q denotes the mass of u, d quarks, and m_n is the neutron mass. Current measurements [212] provide an upper bound on the CP-violation parameter, $|\theta| \lesssim 10^{-10}$.

Within the SM, the smallness of the θ parameter becomes a true fine-tuning problem. Since θ could acquire an $\mathcal{O}(1)$ contribution from the observed CP-violation in the EW sector (via the common quark mass phase, arg $\det(M_q)$, where M_q is the quark mass matrix), its not obvious why it becomes cancelled to a high precision by the (unrelated) gluon term [207]. To solve this problem, the SM is augmented with an extra pseudo-scalar particle called axion A, whose only non-derivative coupling is to the CP-violating topological gluon density $F^a_{\mu\nu}\tilde{F}^{\mu\nu}_a$ that is suppressed by a large scale f_A. With $\theta \to \theta + \phi(x)/f_A$, where ϕ is the angular DoF of spin-zero complex field,

$$\varphi = |\varphi| e^{i\theta} = |\varphi| e^{i\phi/f_A} \,, \tag{24}$$

the minimum of the vacuum energy occurs when the coefficient $\theta + \phi/f_A$ in front of $F^a_{\mu\nu}\tilde{F}^{\mu\nu}_a$ vanishes.

It is worth noting that interactions of the scalar field with ordinary matter is controlled by the factor $\partial_\mu \varphi / f_A$. Thus, even at its originally suggested value of $f_A \sim 250$ GeV at the EW breaking scale [213], the axions interact so weakly that they emerge without attenuation from reactor cores or stellar interiors. Present astrophysical constraints push f_A to substantially higher values, which are somewhere between a few 10^8 GeV and a few 10^{17} GeV.

In cosmology, since their introduction, the axion-like particles were considered to be potentially important candidates if not for all than at least for the main component of the DM—a form of matter accounting for about one-quarter of its total energy density [72]. In order to fulfill their mission, the axions must contribute a non-negligible amount to the energy density of the universe and should have not been in thermal equilibrium with the cosmological plasma at any time in the history of the universe. This, together with the smallness of axion mass, implies their large occupation numbers [207]—the situation already encountered when we have discussed the properties of saturated gluon matter, cf. Equation (19). This implies that the axions over the whole history of the universe can be modeled by solving the classical field equations of a scalar condensate [214].

Starting from the Peccei–Quinn (PQ) scalar field φ introduced in Equation (24), the Lagrangian invariant under the global $U(1)_{PQ}$ transformation reads [215]:

$$\mathcal{L} = \frac{1}{2}|\partial_\mu \varphi|^2 - V_{\rm eff}(\varphi, T), \quad V_{\rm eff}(\varphi, T) = \frac{\lambda}{4}(|\varphi|^2 - f_A^2)^2 + \frac{\lambda}{6}T^2|\varphi|^2. \tag{25}$$

Focusing on the early universe, the evolution of the field φ passes the following milestones. At high temperatures $T \gg T_c = \sqrt{3}f_A$, the effective potential $V_{\rm eff}(\varphi, T)$ depicted on the left panel of Figure 7 has the $U(1)_{PQ}$ symmetric minimum at $\varphi = 0$. With increasing time t, the universe cools, and the vacuum with $\varphi = 0$ becomes unstable. Due to the misalignment mechanism, the field starts to roll down from $\varphi = 0$, and the potential becomes tilted. At $T \lesssim f_A$, the PQ symmetry is spontaneously broken—the field acquires the vacuum expectation value $\langle\varphi\rangle = f_A$. Then, the axion—the Nambu-Goldstone boson of the spontaneously broken $U(1)_{PQ}$ symmetry—becomes a massless angular DoF at the minimum of the potential. The detailed results depend on whether the PQ phase transition occurs before or after inflation. While in the former case, only one θ_0 angle contributes (all other values are inflated away), in a post-inflationary scenario, the initial value of the angle θ takes all values in the interval $\langle -\pi, \pi\rangle$. Eventually, when the universe cools to the temperatures of a few GeV, the axion obtains a mass through the QCD non-perturbative instanton effect known as the axial anomaly [183,216,217].

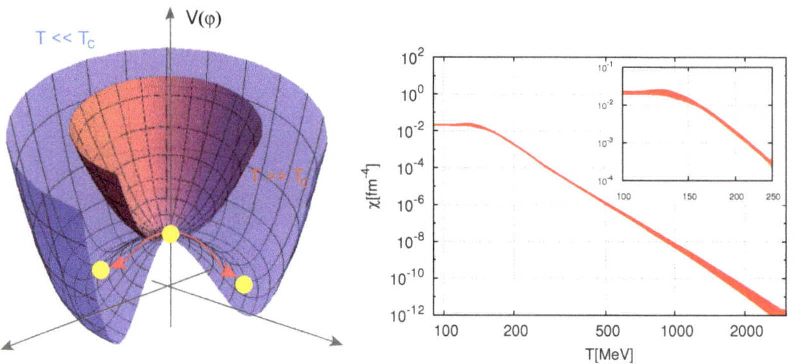

Figure 7. (**Left**) Potential of the PQ scalar $V(\varphi)$ at different temperatures $T \gg T_c$ (pink) and $T \ll T_c$ (violet). The yellow circles show the positions of the minimum. Adapted from Ref. [215]. (**Right**) Continuum limit of $\chi_{\rm top}(T)$ from LQCD. The inserted sub-figure shows the behavior around the QCD phase transition temperature. Adapted from Ref. [89].

At the leading order in f_A^{-1}, the axion mass $m_A(T)$ at some temperature T can be extracted from the QCD-generating functional $\mathcal{Z}(\theta)$ in the presence of a theta term [218],

$$m_A^2(T) = \frac{\delta^2}{\delta\phi^2}\ln\mathcal{Z}\left(\frac{\phi}{f_A}\right)\Big|_{\phi=0} = \frac{1}{f_A^2}\frac{d^2}{d\theta^2}\ln\mathcal{Z}(\theta)\Big|_{\theta=0} = \frac{\chi_{\rm top}(T)}{f_A^2}, \tag{26}$$

where $\chi_{\rm top}(T)$ is the QCD topological susceptibility. This quantity is typically computed using the lattice methods developed by several groups, see e.g., Refs. [89,219–228], but also in the framework of analytical approaches [218,229]. Its temperature dependence can be extracted either from the DIGA or from the LQCD calculations; see the right panel of Figure 7. The lattice simulations performed, for instance, in Ref. [89] have revealed that for $T > T_* = 150$ MeV, the susceptibility falls as $\chi_{\rm top}(T) \sim T^{-b}$ with $b = 8.16$, extending thus the previous DIGA result $\chi_{\rm top}(T) = \chi(0)T^{-8}$ up to $T \approx 3$ GeV. At the temperatures of $T = 100 - 140$ MeV in the vicinity of the QCD chiral phase transition temperature T_c, see Section 2.1, $\chi_{\rm top}(T)$ flattens. Further analysis performed in Ref. [89] exploiting the QCD

EoS obtained therein has revealed that in the post-inflationary scenario, depending on the fraction of DM consisting of axions, $m_A = 50$–1500 μeV. In particular, for the 50% axion content of the DM, one obtains the axion mass scale of $m_A = 50(4)$ μeV. Since at $T < T_c$, the chiral perturbation theory [183] becomes applicable, it is worth making a comparison with the original formula for the axion mass,

$$m_A^2 = \frac{m_u m_d}{(m_u + m_d)^2} \frac{m_\pi^2 f_\pi^2}{f_A^2} \implies m_A \approx 5.7\left(\frac{10^{12}\text{GeV}}{f_A}\right)\mu\text{eV}\,, \tag{27}$$

where f_π is the pion decay constant, and m_d and m_u are the down- and up-quark masses appearing in the QCD Lagrangian [183,208]. Let us note that if $m_A \gtrsim 20$ eV, the axions decay faster than the age of the universe.

For temperatures above the chiral phase transition, the axion potential computed in DIGA reads [211]:

$$V(\phi, T) = \chi_{\text{top}}(T)\left[1 - \cos\left(\frac{\phi}{f_A}\right)\right]. \tag{28}$$

Expanding Equation (28) around $\phi/f_A = 0$, we obtain $V(\phi) = m_A^2\phi^2/2$ at a finite T. Assuming the Friedmann-Lemaître-Robertson-Walker (FLRW) metric and classical axion field with this potential, the corresponding unperturbed energy density $\bar{\epsilon}_A$ and pressure $\bar{p}_A$ due to the axion field read

$$\bar{\epsilon}_A = \frac{1}{2}\dot{\phi}^2 + \frac{1}{2}m_A^2\phi^2 \quad\text{and}\quad \bar{p}_A = \frac{1}{2}\dot{\phi}^2 - \frac{1}{2}m_A^2\phi^2\,, \tag{29}$$

respectively. Substituting $\bar{\epsilon}_A$ and $\bar{p}_A$ from Equation (29) into the fluid Equation (39), we obtain

$$\ddot{\phi} + 3H\dot{\phi} + m_A^2 = 0\,. \tag{30}$$

At early times $H(t) \gg m_A(t)$, we can neglect m_A in Equation (30) to obtain the solution $\phi(t) = \phi_0$—the axion field is frozen at a constant value $\phi_0 \sim f_A$.

With increasing time t, eventually, the oscillating term proportional to $m_A^2(t) \equiv m_A^2(T(t))$ in the equation of motion of the axion field (28) begins to contribute. At the time t_{osc} defined implicitly as $m_A(t_{\text{osc}}) \approx 3H(t_{\text{osc}})$, the universe is sufficiently large to host a sizeable fraction of one oscillation period—the axion field starts to oscillate with an amplitude damped by the expansion rate. A solution of Equation (30) then reads [120],

$$\phi(t) = \phi_1\left(\frac{a(t_1)}{a(t)}\right)^{3/2} \cos\left(\int_0^t m_A(t)dt + \alpha\right), \tag{31}$$

where t_1 is the time at which $H(t_1) = m_A$, i.e., when the temperature drops below the QCD chiral phase transition temperature T_c, $\phi_1 \sim f_A$ is the constant and α is the phase. In particular, for $m_A(t) = m_A$ and a radiation-dominated universe, one obtains $\phi_1 \approx 1.44\phi_0$ and $\alpha = -3\pi/8$ [120].

For the initial conditions at the onset of oscillations $\theta_i \equiv \theta(t_{\text{osc}})$, $\dot{\theta}_i \equiv \dot{\theta}_i(t_{\text{osc}})$, where θ_i is called the initial misalignment angle, we obtain from Equation (30)

$$\theta_i = \theta_{\text{PQ}} + \frac{\dot{\phi}_{\text{PQ}}}{H_{\text{PQ}}} \quad\text{and}\quad \dot{\theta}_i = \dot{\theta}_{\text{PQ}}\left(\frac{H(t_{\text{osc}})}{H_{\text{PQ}}}\right)^{3/2}, \tag{32}$$

where $\theta(t_{\text{PQ}}) \equiv \theta_{\text{PQ}}$, $\dot{\phi}(t_{\text{PQ}}) \equiv \dot{\phi}_{\text{PQ}}$, $H_{\text{PQ}} \equiv H(t_{\text{PQ}})$ and $a_{\text{PQ}} \equiv a(t_{\text{PQ}})$ are the values at the PQ symmetry breaking time $t_{\text{PQ}} \ll t_{\text{osc}}$. In the second equation, we have also used $a \sim 1/T$ and $H \sim T^2G^{1/2}$ [208].

While in the pre-inflationary scenario, inflation selects one patch of the universe within which the spontaneous breaking of the PQ symmetry leads to a homogeneous value of the initial misalignment angle ϕ_i, in the post-inflationary scenario, the PQ symmetry breaks with θ_i, taking different values in patches that are initially out of causal contact; see the left panel of Figure 7. However, today, they populate the volume enclosed by our Hubble horizon. In the post-inflationary scenario, the initial misalignment angle ϕ_i takes all possible values on the unit circle. For a quadratic potential, $V(\phi) = m_A^2\phi^2/2$, this is equivalent to an assumption that the initial condition reads $\phi_i \equiv \sqrt{\langle\phi_i^2\rangle} = \pi/\sqrt{3}$, where the angle brackets represent the value averaged over $\langle -\pi, \pi\rangle$ [208].

Starting from $T_{\rm osc}$, the number of axions in a comoving frame becomes frozen, and their number density evolves as

$$n_A(T_{\rm osc}) \approx m_A(T_{\rm osc}) f_A^2 \langle\phi_i^2\rangle\,. \tag{33}$$

For isotropic evolution, the ratio of the number density n_A to the entropy density s in the comoving frame is conserved, i.e., $n_A(T)/s(T) = n_A(T_{\rm osc})/s(T_{\rm osc})$ leading for $T \ll T_c$ to the expression for the axion energy density,

$$\epsilon_A^{\rm mis} = m_A n_A(T_{\rm osc}) \frac{h_{\rm eff}(T)}{h_{\rm eff}(T_{\rm osc})}\left(\frac{T}{T_{\rm osc}}\right)^3 = m_A f_A^2 \langle\phi_i^2\rangle \frac{h_{\rm eff}(T)}{h_{\rm eff}(T_{\rm osc})}\left(\frac{T}{T_{\rm osc}}\right)^3 T_c\,, \tag{34}$$

where the effective number of DoFs of the entropy $h_{\rm eff}(T)$ is defined in Equation (3). Let us note that in contrast to Ref. [208] in Equation (34), the constancy of the axion mass for $T \lesssim T_c$ is already taken into account.

After the spontaneous $U(1)_{\rm PQ}$ symmetry breaking the axion field, ϕ, being an angular variable, takes values in the interval $\langle 0, 2\pi f_A\rangle$, cf. Equation (24). Consequently, the axion potential $V(\phi)$ given by Equation (28) is periodic in ϕ with period $\Delta\phi = 2\pi f_A/N_{\rm DW}$. In the other words, $V(\phi)$ has an exact $Z_{N_{\rm DW}}$ discrete symmetry. The axion acquires a periodic potential with $N_{\rm DW}$ equivalent minima. The Kibble mechanism [195,230] then dictates that, depending on the homotopy group $\pi(\mathcal{M})$ of the manifold $\mathcal{M}$ of degenerate vacua, the topological defects—domain walls, strings or monopoles—form each time the symmetry is broken [195,202]. With $\mathcal{M} = U(1)_{\rm PQ}$ and $\pi(\mathcal{M} = Z_{N_{\rm DW}})$, the production of axionic strings, which are vortex-like topological defects that form as soon as the symmetry is spontaneously broken, is possible. Those are not important when the PQ symmetry is broken before inflation—they are inflated away—but they play an important role in the post-inflationary scenario. When the Hubble parameter H becomes comparable to the axion mass m_A, the axion starts to roll down to one of the minima. Since the axion field settles into different minima in different places of the universe, domain walls are formed between the different vacua; see the left panel of Figure 7. It is worth mentioning that this phenomenon is similar to ice formation on the surface of a pond or a puddle when the water begins to freeze in many places independently, and the growing plates of ice join up in random fashion, leaving zigzag boundaries between them [195].

As an example, let us consider a planar wall orthogonal to the z-axis $\phi = \phi(z)$. The solution of the classical field equation with potential given by Equation (28) reads [231]:

$$\frac{\phi(z)}{f_A} = \frac{2\pi k}{N_{\rm DW}} + \frac{4}{N_{\rm DW}} \tan^{-1} e^{m_A z}\,. \tag{35}$$

This configuration interpolates between the two allowed vacua, $\phi/f_A = 2\pi k/N_{\rm DW}$ at $z \to -\infty$ and $\phi/f_A = 2\pi(k+1)/N_{\rm DW}$, which are separated by the wall of thickness $1/m_A$.

Astrophysical signatures of the axion can be broadly divided into its couplings to elementary/composite particles, i.e., photons, electrons, protons or neutrons, and to the macroscopic objects in the universe such as BHs [208]. In the latter case, when the Compton length of the axions becomes of order of the BH size, they form gravitational bound states around it. The phenomenon of superradiance [232,233] causes the axion occupation numbers to grow exponentially, providing a way to extract very efficiently energy and angular momentum from the BH. The presence of axions could be inferred by observations of BH masses and angular momenta. Current measurements exclude the region of $6\times 10^{17}\,{\rm GeV} \le f_A \le 10^{19}\,{\rm GeV}$.

In particle physics signatures, the most important is the decay channel of axion into two photons with the decay width $\Gamma_{A\to\gamma\gamma} = g^2_{A\gamma\gamma} m_A^3/(64\pi)$ in terms of the model-dependent coupling constant $g_{A\gamma\gamma}$. The main uncertainty is due to the electromagnetic and color anomalies of the axial current associated with the axion. The most relevant process induced by the $g_{A\gamma\gamma}$ is the Primakoff process—the conversion of thermal photons in the electrostatic field of electrons and nuclei into axions,

$$\gamma + Ze \to A + Ze\,. \tag{36}$$

A strong bound on exotic cooling processes in the sun is provided by the helioseismological considerations [234] giving the following constraint: $g_{A\gamma\gamma} \le 4.1 \times 10^{-10}\,\text{GeV}^{-1}$. For more details, consult Refs. [205–208,215]. Another option is the inverse Primakoff scattering, which allows solar axions to coherently scatter off the atomic electric field and back-convert into photons in the detector volume, $A + Ze \to \gamma Ze$, proceeding through a t-channel photon exchange [235].

The axion besides being a well-recognized DM candidate may provide also an interesting explanation of a wider range of phenomena related to the early universe dynamics. As a first example, let us mention the SMASH model—a minimal extension of the SM with additional particle content comprising three sterile right-handed neutrinos $N_i, i = 1,2,3$, a color triplet Q and a complex SM-singlet scalar σ, whose VEV of $v_\sigma \sim 10^{11}$ GeV breaks the lepton number and the PQ symmetry simultaneously [236]. At low energies, the model reduces to the SM, which is augmented by seesaw-generated neutrino masses and mixing, plus the axion. In this scenario, the inflaton, a scalar field driving cosmic inflation in the very early universe, is a mixture of σ and the SM Higgs fields. The reheating of the universe after inflation occurs via a mechanism known as a Higgs portal [237]—by the DM particles, which interact only through their couplings with the Higgs sector of the theory. The model provides a consistent picture of particle physics from the EW scale to the Planck scale $M_{\rm PL}$ and of cosmology from inflation until today. In particular, in the SMASH model framework, the PQ symmetry is first broken and then restored non-thermally during preheating for $f_A = 4 \times 10^{16}$ GeV.

The second example, the Axion Quark Nugget (AQN) DM model, see, e.g., Ref. [238] and references therein, replaces the commonly accepted baryogenesis scenario with a charge separation process in which the global baryon number of the universe remains zero at all times. Similarly to Witten's idea of stranglets [5], the AQN DM is composed of quarks and anti-quarks but now in a new high-density CSC phase. Initially, nuggets of both matter and antimatter are formed with equal probability as a result of the dynamics of the axion domain walls which at the same time provide the extra pressure needed to stabilize the CSC phase. Later on, due to the global CP violating processes associated with the initial misalignment angle $\theta_0 \neq 0$ during the early formation stage, the populations of the nuggets with the positive and negative baryon number become different. The unobserved antibaryons hidden inside the DM would not participate in nucleosynthesis and, therefore, according to the usual definition would not contribute to the visible matter. However, since antimatter nuggets can interact with regular matter via annihilation leading to electromagnetic radiation, their existence has observational consequences [238].

It is worthwhile mentioning here an interesting generalization of the PQ mechanism where in addition to θ angle, also the strong coupling α_S is promoted to a dynamical quantity. The latter evolves through the VEV of a singlet scalar field that mixes with the Higgs field [239,240]. In the resulting cosmic history, the QCD confinement and EW symmetry breaking initially occur simultaneously close to the weak scale.

To conclude this section, as we have already noticed in several examples above, the non-perturbative dynamics of QCD often exhibits very non-trivial and rather unexpected consequences at cosmological scales. An attractive mechanism proposing that the QCD axion may emerge as a composite state has been discussed very recently in Ref. [241]. In particular, it was suggested that Majorana neutrinos, that combine into Cooper pairs, can form collective low-energy degrees of freedom. This motivates the existence of the

QCD axion as a collective excitation of the neutrino condensate. Such a condensate can be produced after the QCD phase transition epoch in a cold coherent state by means of a misalignment mechanism, thus providing an alternative DM candidate. In this case, a QCD anomalous portal provides the necessary means for a tiny mass gap generation by neutrinos. Furthermore, the Cosmological Constant emerges as a result of the spontaneously broken mirror symmetry of the QCD ground state triggered by the quantum gravity effects as suggested by Refs. [65,66] (see also below). Hence, one concludes that QCD may be responsible for the dynamical generation of both the DM and DE components of the universe such that a complete knowledge of the QCD in the infrared regime may be absolutely critical for understanding of the cosmological evolution since the latest QCD transition epoch and for eventual formation of the current state of the universe.

3. Dynamics of the Early Universe

3.1. Simple Models with Constant Speed of Sound

In accord with the observations, the Standard Cosmological Model (SCM) postulates that the cosmic matter at scales larger than 100 Mpc is homogeneously and isotropically distributed. Consequently, thermodynamic pressure p at early times of its evolution depends on temperature T and various chemical potentials $\mu_i, i = B, Q, L, \ldots$ corresponding to baryon number B, electric charge Q, lepton number L etc., only via the energy density ϵ. Solution of the Einstein equations of general relativity

$$R_{\mu\nu} - \tfrac{1}{2}g_{\mu\nu}(R - 2\Lambda) = -8\pi G T_{\mu\nu}\,, \tag{37}$$

where $R_{\mu\nu}$ is the Ricci tensor, $R = R_{\mu\nu}g^{\mu\nu}$ is the scalar curvature, $g^{\mu\nu}$ is the metric tensor, and Λ and G are the cosmological and gravitational constants; preserving the homogeneity and isotropy of space under its time evolution is a spacetime of constant curvature parameter $k = \{+1, 0, -1\}$. It is described by a single function—the time-dependent scale factor $a(t)$ [84,120,242] which connects the Lagrangian (or comoving) coordinates r with the physical Euler coordinates $\hat{r}(t) = a(t)r$. The metric tensor $g_{\mu\nu}$ in the preferred coordinate system where these symmetries are clearly manifest reads[4]

$$ds^2 = g_{\mu\nu}x^\mu x^\nu = dt^2 - a^2(t)\left[\frac{dr^2}{1-kr^2} + r^2(d\theta^2 + \sin^2\theta d\phi^2)\right], \qquad k = \mathrm{const}\,. \tag{38}$$

Using this metric in Equations (37) and (38) and neglecting the dissipative terms yields the Friedmann equation for the time evolution of $a(t)$ and the fluid equation for the time evolution of $\epsilon(t)$, respectively (see e.g., Refs. [84,175]),

$$H^2(t) \equiv \left(\frac{\dot{a}}{a}\right)^2 = \frac{8\pi G}{3}\epsilon - \frac{k}{a^2} + \frac{\Lambda}{3}, \qquad \dot{\epsilon} + 3(\epsilon + p)H(t) = 0\,, \tag{39}$$

where $H(t)$ is the Hubble parameter. From Equation (39), it follows that the expanding universe is characterized by a natural time-scale $H^{-1} = a/\dot{a}$. Any particle species will remain in thermal equilibrium with the cosmic fluid so long as the mean interaction time t_c allows rapid adjustment to the falling temperature provided that $t_c < H^{-1}$.

In the period of the universe evolution when $\epsilon \gtrsim 1$ GeV fm^{-3}, the terms containing constants Λ and k in Equation (39) can be safely neglected, transforming the above two equations into a single one describing the time evolution of the energy density [6],

$$-\frac{d\epsilon}{3\sqrt{\epsilon}(\epsilon + p)} = \sqrt{\frac{8\pi G}{3}}dt\,. \tag{40}$$

For the time-independent speed of sound c_s and for the energy densities negligible compared to the initial density $\epsilon(t > 0) \ll \epsilon(t = 0)$ integration of Equation (40), the calculations yield [13]

$$\epsilon(t) = \frac{1}{6\pi G(1+c_s^2)^2 t^2}, \qquad c_s^2 \equiv \frac{dp}{d\epsilon} = \text{const}. \tag{41}$$

Substituting this equation into the fluid Equation (39), we obtain the expansion rate of the early universe

$$\dot{a} \sim t^{-\alpha}, \qquad \alpha = \frac{1+3c_s^2}{3(1+c_s^2)}. \tag{42}$$

In particular, for the massless non-interacting gas with $c_s^2 = 1/3$, one obtains $\dot{a} \sim t^{-1/2}$ and hence $a(t) \sim t^{1/2}$.

It is worth mentioning that in the cosmological literature, see, e.g., Ref. [120], it is customary to express the EoS in terms of the parameter w, Equation (5). For the constant speed of sound case, i.e., for $w = \text{const}$, the solution of the Friedmann Equation (39) yields

$$\epsilon \sim a^{-3(1+w)}. \tag{43}$$

For the non-relativistic matter (dust) with $p = 0$, we obtain $\epsilon \sim a^{-3}$, for the massless non-interacting gas, $\epsilon \sim a^{-4}$, for the EoS $p = \epsilon$ corresponding to absolutely stiff fluid [243], $\epsilon \sim a^{-6}$, and for the vacuum energy with the EoS $p = -\epsilon$, Equation (43) gives $\epsilon = \text{const}$.

Let us now follow Ref. [244] and consider an ideal gas of free particles in D-dimensional space. Its particle number density n, energy density ϵ and pressure p expressed in terms of single-particle statistical sum $f(E,T,\mu)$ of particle with energy E, momentum P and spin s reads (see e.g., Ref. [245]):

$$f(E,T,\mu) = \frac{1}{\exp[(E-\mu)/T] \pm 1}, \qquad n = \gamma \int f(E,T,\mu) d^D P, \tag{44}$$

$$\epsilon = \gamma \int f(E,T,\mu) E(P) d^D P = \gamma S(D) \int_0^\infty f(E,T,\mu) E(P) P^{D-1} dP, \tag{45}$$

$$p = -\frac{T}{V} \ln Z = -\gamma T \int \ln f(E,T,\mu) d^D P = \gamma \frac{S(D)}{D} \int_0^\infty f(E,T,\mu) \frac{\partial E(P)}{\partial P} P^D dP, \tag{46}$$

where $d^D P = [D\pi^{D/2}]/[\Gamma(D/2+1)] P^{D-1} dP \equiv S(D) P^{D-1} dP$ is a volume element of the D-dimensional hypersphere and $\gamma \equiv (2s+1)(2\pi)^{-D}$. When evaluating the first integral in Equation (46), we have performed integration by parts assuming that the particle energy $E(P)$ is some generic function of P. The substitution of Equations (45) and (46) into Equation (5) constrains $E(P)$ to satisfy the differential equation

$$\frac{P}{D} \frac{\partial E(P)}{\partial P} = w E(P), \tag{47}$$

whose solution for $w = \text{const}$ reads

$$E(P) = \xi P^{wD}, \tag{48}$$

with ξ some arbitrary constant. Hence, the medium with constant speed of sound squared $c_s^2 = w$ in ordinary three-dimensional space can be equivalently described as an ideal gas of quasi-particles with energy E and momentum P satisfying the dispersion relation (48) in D-dimensional space [244]. It is worth mentioning that at some instances, the dispersion relation (48) can be satisfied by the real particles. The case of $w = 0$ corresponds to non-relativistic particles, while $wD = 1$ corresponds to massless particles in D-dimensional space with the EoS $p = \epsilon/D$ and hence with the sound velocity $c_s = D^{-1/2}$.

Unfortunately, this is not the case of the absolutely stiff fluid, as first discussed by Zeldovich [243]. Notwithstanding that its EoS $p = \epsilon$ can be used to describe a large variety of systems such as phonon-like excitations in a thin channel ($D = 1$) [244], thin film ($D = 2$) of non-relativistic quasi-particles [244], interiors of neutron stars [246,247], Big Bang nucleosynthesis [248] or warm self-interacting DM component [249]. The cosmology and thermodynamics of the FLRW universe with bulk viscous stiff fluid was studied in Ref. [250]. The fact that the stiff fluid saturates the holographic covariant entropy bound was used in Ref. [251] to describe a cosmology of the very early universe. Last but not least, a stiff perfect fluid is energetically equivalent to a time-like massless scalar field ϕ, see, e.g., Ref. [252]. From its energy-momentum tensor,

$$T^{\phi}_{\mu\nu} = \partial_{\mu}\phi\partial_{\nu}\phi - \frac{1}{2}g_{\mu\nu}\partial_{\alpha}\phi\partial^{\alpha}\phi\,, \qquad \partial_{\alpha}\phi\partial^{\alpha}\phi > 0\,, \tag{49}$$

we obtain that $p \equiv T^{\phi}_{kk} = \epsilon \equiv T^{\phi}_{00} = \partial_{\alpha}\phi\partial^{\alpha}\phi$.

3.2. Equation of State of the Early Universe

As already discussed in Section 3.1, the only way how to write the EoS compatible with homogeneity and isotropy of the universe is through the pressure expressed as a function of the energy density. Thus, given the barotropic EoS $p(\epsilon)$ of the expanding matter, we can for instance use Equation (40) to predict the temporal evolution of the energy density $\epsilon(t)$ and the temperature $T(t)$. For the ideal gas EoS, Equation (41) yields the relation between the temperature T of the early universe and the time $t_{\rm sec}$ elapsed from the Big Bang, $T_{\rm MeV} = \mathcal{O}(1)/\sqrt{t_{\rm sec}}$ [84], i.e., the same time-dependence as for the time derivative of the scale factor $\dot{a}(t)$; see Equation (42) and the discussion below it. For the QGP-to-hadronic matter transition, this leads to $t^{\rm QGP}_{\rm sec} \sim 10^{-5}$ s, and for the EW phase transition, this leads to $t^{\rm EW}_{\rm sec} \sim 10^{-11}$ s.

More sophisticated EoS can be based either on the phenomenological models or on the microscopic theory. In the latter case, a quantum physics formulation of statistical mechanics in terms of the S matrix, which describes the scattering processes taking place in the thermodynamical system of interest, is available—see Ref. [253]. It provides a simple prescription for calculating the grand canonical potential $Z(T, \mu)$ of any gaseous system given the free-particle energies and S-matrix elements. The application of S matrix formulation to study the thermal properties of an interacting gas of hadrons can be found in Ref. [254]. It can be also used to show how the hadron resonance gas model emerges from the S-matrix framework [255]. However, this approach is based on the perturbative expansion of the S-matrix and is therefore not applicable in the strong coupling regime of QFT. There, one resorts to direct calculation of the grand canonical potential on the lattice.

Our first example of the EoS used in a description of the evolution of the early universe is the Bag Model (BM) EoS [1,256,257] based on the phenomenological description of the mass spectrum of the hadron states [11,258,259] in terms of gas of massless color objects—quarks and gluons—moving inside the confining potential—the bag,

$$\epsilon_q(T) = \sigma_q T^4 + \mathcal{B}\,, \qquad p_q(T) = \frac{\sigma_q}{3}T^4 - \mathcal{B}\,, \qquad p_q(\epsilon) = \frac{1}{3}(\epsilon - 4\mathcal{B})\,. \tag{50}$$

In Equation (50), $\sigma_q = \frac{\pi^2}{30}g^{\rm QCD}_{\rm eff}$ is the Stefan-Boltzmann constant with $g^{\rm QCD}_{\rm eff}$ given by Equation (9). The BM EoS (50) incorporates color confinement through the bag constant $\mathcal{B} = \epsilon_{\rm bag} - \epsilon_{\rm vac} > 0$, indicating the difference between the energy densities of the physical vacuum and the ground state for quarks and gluons in the medium. The latter can be interpreted as the energy needed to create a bubble in the vacuum in which the non-interacting quarks and gluons are confined. While the fit to hadron masses made in the original BM predict $\mathcal{B}^{1/4} \approx 140$ MeV, the value of $\mathcal{B}^{1/4} \approx 220$ MeV is frequently quoted in works dealing with the vacuum structure of QCD, see, e.g., Ref. [11].

Let us note that BM EoS represents a bare-bones model of hadron-to-QGP phase transition. At small energy densities, hadrons—the bubbles inside the non-perturbative vacuum—occupy only a small fraction of the total considered volume V. An increase of ϵ

leads to the coalescence of several bubbles into larger ones. For $\epsilon \geq 4\mathcal{B}$, the volume V is filled with one large bubble, cf. Equation (50), whose surface coincides with the enclosing walls. Hence, there is no longer free surface against the vacuum, and the new de-confined phase of matter is created [256]. The critical temperature T_c of the phase transition can be estimated using Gibbs criteria. Equating the BM pressure (50) with the pressure p_h of the hadron (pion) gas with 3 DoF and hence with $\sigma_h = 3\pi^2/30$, we obtain

$$p_q(T_c) = \frac{\sigma_q}{3}T_c^4 - \mathcal{B} = p_h(T_c) = \frac{\sigma_h}{3}T_c^4, \qquad T_c = \left(\frac{3\mathcal{B}}{\sigma_q - \sigma_h}\right)^{1/4}. \tag{51}$$

For $N_F = 3$ active quark flavors u, d, and c, Equation (2) yields $T_c \approx 0.67\mathcal{B}^{1/4}$ and $\approx$150 MeV. Let us add that Equation (51) describes the first-order phase transition with energy density discontinuity,

$$\Delta\epsilon = \epsilon_q(T_c) - \epsilon_h(T_c) = 3p_q(T_c) + 4\mathcal{B} - 3p_h(T_c) = 4\mathcal{B}. \tag{52}$$

Last but not least, expressing the BM EoS (50) in terms of the dimensionless interaction measure—the trace anomaly describing the thermal contribution to the trace of the energy-momentum tensor $\mathcal{T}^{\mu\nu} \equiv T^{\mu\nu}$

$$\Theta \equiv \frac{\mathcal{T}^{\mu\mu}(T)}{T^4} = \frac{\epsilon_q - 3p_q}{T^4} = \frac{4\mathcal{B}}{T^4} = \frac{4\sigma_q\mathcal{B}}{\epsilon_q - \mathcal{B}}, \tag{53}$$

we observe a monotonous weakening of the interaction strength with increasing ϵ_q [33], leading ultimately to a Stefan-Boltzmann (SB) value of $\Theta = 0$. At the same time, the entropy density $s_q = (\epsilon_q + p_q)/T = 4/3\sigma_q^{1/4}(\epsilon_q - \mathcal{B})^{3/4}$ converges to its SB limit from below. The speed of sound derived from the BM EoS (50) is energy-density-independent and coincides with that of the ideal gas of massless particles $c_s^2 = dp_q/d\epsilon_q = 1/3$.

More sophisticated EoS can be constructed, e.g., by adding the term $\sim T^2$ to expressions for $\epsilon(T)$ and $p(T)$ in Equation (50) with $\sigma = \sigma_q$

$$\epsilon(T) = \sigma T^4 - CT^2 + \mathcal{B}, \qquad p(T) = \frac{\sigma}{3}T^4 - DT^2 - \mathcal{B}. \tag{54}$$

The motivation for such a modified BM EoS comes from the observation [260,261] that in the case of a pure gauge theory up to temperatures a few times the transition temperature T_c, the dominant power-like correction to the pQCD high-temperature behavior is $\mathcal{O}(T^{-2})$ rather than $\mathcal{O}(T^{-4})$. Moreover, the quadratic thermal terms in the deconfined phase can be also obtained from gauge/string duality [262].

In its original setting with $C = D > 0$, Equation (54) represents the LQCD-motivated "fuzzy" BM EoS of Ref. [263]. On the other hand, for $C = -D < 0$, it represents a particular case of gas of gluonic quasi-particles EoS with a temperature-dependent bag function $\mathcal{B}(T) = -CT^2 + \mathcal{B}$; see e.g., Refs. [264,265]. In the following discussion, we will keep the values of the constants C and D in Equation (54) unrestricted.

By inverting $\epsilon(T)$ in Equation (54) with respect to temperature squared T^2,

$$T^2(\epsilon) = \frac{C + \sqrt{C^2 + 4\sigma(\epsilon - \mathcal{B})}}{2\sigma} > 0, \tag{55}$$

and substituting for $T^2(\epsilon)$ into Equation (54), we obtain the barotropic form of the EoS (54):

$$p(\epsilon) = \frac{1}{3}(\epsilon - 4\mathcal{B}) - \frac{1}{3}\mathrm{sgn}(A)|A|T^2(\epsilon) \quad A = 3D - C. \tag{56}$$

The corresponding sound velocity squared reads

$$c_s^2(\epsilon) = \frac{dp(\epsilon)}{d\epsilon} = \frac{1}{3}\left(1 - \frac{\text{sgn}(A)|A|}{\sqrt{C^2 + 4\sigma(\epsilon - \mathcal{B})}}\right). \tag{57}$$

Note that for $A = 0$, both Equations (56) and (57) degenerate to the corresponding result from the BM EoS; see Equation (50). In the context of Equation (54), the situation with $C = 3D = \frac{3}{2}N_F\mu_B^2$ describes the EoS of ideal QGP with non-zero baryon chemical potential [266]. It is worth mentioning that by tuning the phase transition temperature to $T_c = 160$ MeV, this EoS predicts a six times higher value of the bag constant $\mathcal{B}$ than the original BM [266].

Therefore, in the following, we discuss only the non-trivial case of $A \neq 0$. First, we check which values of the constants C and D in Equation (57) are compatible with the condition $c_s^2 > 0$. While for $A < 0$ (i.e., $C > 0$ and $-C < D < C/3$), Equation (57) is always satisfied for $A > 0$ ($C \geq 0$ and $D > C/3$ or $D > -C > 0$), there exists a lower bound ϵ_0 on energy density

$$\epsilon > \epsilon_0 = \frac{A^2 - C^2}{4\sigma} + \mathcal{B}. \tag{58}$$

Thus, in both cases ($A > 0$ or $A < 0$), Equation (55) represents the genuine non-trivial EoS of non-ideal high-density matter each with its own sound velocity approaching for $\epsilon \to \infty$, either from below or from above, the SB limit of $\sqrt{1/3}$. However, only for $A < 0$, the second term on the right-hand side of Equation (56) with $-\frac{1}{3}AT^2(\epsilon)$ represents independent pressure. It is also worth mentioning that for the latter case ($A < 0$), the trace anomaly

$$\Theta(\epsilon) = \frac{4\mathcal{B}}{T^4(\epsilon)} + \frac{\text{sgn}(A)|A|}{T^2(\epsilon)}, \tag{59}$$

with $T^2(\epsilon)$ defined in Equation (55), acquires a peak at the energy density

$$\epsilon_p = \frac{2\mathcal{B} \pm \sqrt{2\mathcal{B}|A|C}}{|A|\sigma} - \mathcal{B}. \tag{60}$$

Standard explanation of this phenomenon within the SU(3)$_c$ gauge theory, see, e.g., Ref. [267], relies on the fact that in the region around and just above the critical temperature T_c of hadron-to-QGP phase transition, the energy density rises much more rapidly than the pressure, leading to the observed rapid increase of Θ. Since asymptotically ϵ/T^4 and $3p(T)/T^4$ converge to their common Stefan-Boltzmann value of σ, see Equation (54), there must be some temperature T_p (and hence also some energy density $\epsilon_p = \epsilon(T_p)$) at which the growth rates change roles, with the pressure now increasing more rapidly. The further decrease of Θ is in good approximation given by T^{-2}, so that $T^2\Theta(T)$ becomes approximately constant very soon above T_c and up to about $5T_c$ [263].

Note that Equation (40) with the initial condition $\epsilon_0(t_0) = 10^4$ GeV fm^{-3} at $t_{\text{sec}}^{\text{QGP}} \ll t_0 = 10^{-9}$ s $\ll t_{\text{sec}}^{\text{EW}}$ was used in Ref. [13] to study the sensitivity of dilution and cooling of the early universe to the changes in the primordial QGP EoS. The latter might modify the pattern of the emission of GWs or the generation of baryon number fluctuations. No dramatic changes between different EoS, including Equations (50) and (54) (with $C = D$) and others, were found in the whole considered time interval. This finding seems to be supported by the LQCD calculations which show that the transition from primordial QGP matter to hadronic matter proceeded as a continuous crossover [92]. The latter does not introduce any fluctuations on length scales much longer than the natural length scales of QCD $\sim \Lambda_{\text{QCD}}^{-1}$, so it has probably left no imprint in the microseconds-old universe that survived so as to be visible in some way today [7].

A different conclusion was, however, reached in Ref. [74] where a novel mechanism for the production of GWs during the QCD phase transition has been proposed. It was found that while the energy density of the homogeneous gluon condensate is smoothly decaying in cosmological time t_{sec}, the pressure of the condensate undergoes a sequence of violent oscillations at the characteristic QCD time scales—the process called relaxation—the basic feature of the QCD transition. Such relaxation processes generate a very specific multi-peaked GWs signature in the domain of radio frequencies. In particular, gravitational echoes of the QCD transition potentially accessible by the FAST [268] and SKA [269] GW telescopes are sourced by the formation of domain walls in the QCD vacuum (also responsible for color confinement in the infrared regime). More details about the dynamical theory of the YM vacuum will be provided below in Section 4.

Recently, the stochastic GWs induced by the scalar perturbations of the metrics were investigated in Ref. [270] as a cosmological probe of the sound speed c_s during the QCD phase transition. Although the GW propagation itself does not depend on c_s, the sound speed value affects the dynamics of primordial density. Induced stochastic GWs can thus be an indirect probe of both the EoS parameter $w = p/\epsilon$ and $c_s = \sqrt{dp/d\epsilon}$. In particular, similarly to the conclusions of Ref. [74], the GW frequency $\sim 10^{-8}$ Hz corresponding to the Hubble scale during the QCD phase transition appears to be in the range of the planned GW detectors.

Let us now turn to a description based on the fundamental theory. Using the Lagrangian of the SM, one can extract the thermodynamical quantities either directly from the lattice calculations [89], deduce them from lattice simulations using a dimensionally-reduced EFT [90] or use the perturbation theory. With the temperature T depending on the lattice spacing a and the number of lattice points in the temporal direction N_t as $T = (aN_t)^{-1}$, the reliable Monte Carlo simulations of QFT in the high-temperature regime need very fine lattices. At the same time, as the lattice spacing a is reduced, the autocorrelation times for zero temperature simulations rise, and the costs of these simulations explode beyond feasibility [89]. This makes the LQCD simulations in the region of temperatures higher then the few GeVs practically impossible. Alternatively, one can vary the gauge coupling g_s, which leads to changing T as well, although the spacial and temporal dimensions do not [90]. In addition to that, even at the temperatures of the EW phase transition $T_c^{\rm EW}$, the dynamics needs to be treated with lattice methods. This is due to the fact that when the particle momenta in the range $k \sim g^2T/\pi$ are considered, then the dynamics of the system is non-perturbative [194,271,272]. Values much below and far above this value p/T^4 can be determined by a direct perturbative computation [90].

In the SM framework, the basic thermodynamical observable is the pressure [273], which can be formally defined through the grand canonical partition function $\mathcal{Z}$ as

$$\mathcal{Z} \equiv \exp\left[\frac{p_B(T)V}{T}\right], \quad p_B(T) = p_E(T) + p_M(T) + p_G(T), \tag{61}$$

where p_B denotes the "bare" result related to the physical (renormalized) pressure as $p(T) = p_B(T) - p_B(0)$. The pressure terms $p_E(T), p_M(T)$ and $p_G(T)$ appearing in Equation (61) collect the contributions from the momentum scales $k \sim \pi T$, $k \sim gT$, and $k \sim g^2T/\pi$, respectively. The couplings g relevant in the SM are $g \in \{h_t, g_1, g_2, g_3\}$, where h_t denotes the Yukawa coupling between the top quark and the Higgs boson, and g_1, g_2, g_3 are the SM couplings related to U(1)$_Y$, SU(2)$_{\rm L}$ and SU$_c$(3) gauge groups, respectively. Note that in Equation (61), the thermodynamic limit $V \to \infty$ is implied.

Using this technique, the SM calculations of the dimensionless function $p(T)/T^4$ and of the trace anomaly $\Theta(T)$ (53) up to $\mathcal{O}(g^5)$ were performed in Ref. [90]. It was found that similarly to EoS Equation (54), Higgs dynamics induces a peak in heat capacity $c(T) = d\epsilon/dT$ occurring around $T_c^{\rm EW} \approx 160$ GeV. This leads to a short period of slower temperature change, and correspondingly, a mildly increased abundance of produced particles. However, in general, the largest radiative corrections originate from QCD effects, reducing the energy density by a couple of percent from the free value even at $T > 160$ GeV [90]. It is worth emphasizing that the above-mentioned effects do not exhaust all possible phenomena relevant for dynamics of the early universe. In particular, as already discussed in Section 2.4, at high temperatures, the saturation phenomena become ubiquitous. In this case, the gluon

DoFs are frozen in the classical field (condensate) and do not contribute to the density of the QGP. To shed more light on the microscopic theory in the canonical picture, the dynamical aspects of relativistic gluon and hot meson plasmas, both coupled to the homogeneous condensate, are discussed in detail in Refs. [274–277], respectively.

In Ref. [278], the results from Refs. [89,90,279] were used to analyze the early universe EoS $p(\epsilon)$. The extracted EoS depicted on the left panel of Figure 8 covers the broad interval in energy density $10^{-2} \leq \epsilon \leq 10^{16}$ GeV fm^{-3} and corresponds to the evolution periods down from the GUT era, through the EW era and the QGP era, into the hadron era. The apparent smoothness of the function $p(\epsilon)$ hides in the complicated temperature dependence $\epsilon(T) = g_{\text{eff}}(T)\epsilon_0(T)$ depicted on right panel of Figure 3.

While the GUT EoS $p_{\text{GUT}} = (0.330 \pm 0.024)\epsilon$ valid for $10^8 \lesssim \epsilon \leq 10^{16}$ GeV·fm^{-3} (the red triangles in Figure 8) appears only slightly below the ideal gas limit, the hadronic-era EoS $p_h = (0.003 \pm 0.002) + (0.199 \pm 0.002)\epsilon$ characterizes the region $\epsilon \lesssim 1$ GeV·fm^{-3}. Most interesting is the intermediate region containing both the QCD and the EW epochs, which can be described by a single function $p_{\text{SM}} = a + b\epsilon + c\epsilon^d$ with $a = 0.048 \pm 0.016, b = 0.316 \pm 0.031, c = -0.21 \pm 0.014, d = -0.576 \pm 0.034$ [278]. It is worth noting that the critical energy density ϵ_c defined implicitly by the equality of the pressures in hadronic and QGP phases $p_h(\epsilon_c) = p_{\text{SM}}(\epsilon_c)$ reads: $\epsilon_c \simeq (1.2 \pm 0.2)$ GeV·fm^{-3}.

As can be seen from the parametrization of the fit,

$$p_{\text{SM}} = p_1(\epsilon) + p_2(\epsilon), \quad p_1(\epsilon) = b\epsilon, \quad b > 0, \quad p_2(\epsilon) = a + c\epsilon^d, \quad a > 0, \quad c < 0, \quad d < 0, \tag{62}$$

throughout the whole QCD and EW eras, there are two independent contributions $p_1(\epsilon)$ and $p_2(\epsilon)$ to the overall pressure of the universe. While p_1 is always positive, the second pressure p_2 is negative up to $\epsilon \lesssim (7-13)$ GeV·fm^{-3}. Although the corresponding value of the trace anomaly can not be directly deduced from the fitted EoS (62)

$$\Theta = \frac{\epsilon - 3p}{T^4} = \frac{\epsilon(1-3b) - 3a - 3c\epsilon^d}{T^4}, \tag{63}$$

it is positive for $\epsilon \gtrsim (3-4)$ GeV·fm^{-3}, and for very large energy densities, it falls as $\Theta \sim g_{\text{eff}}\epsilon^{-3/2}$, cf. Equation (2).

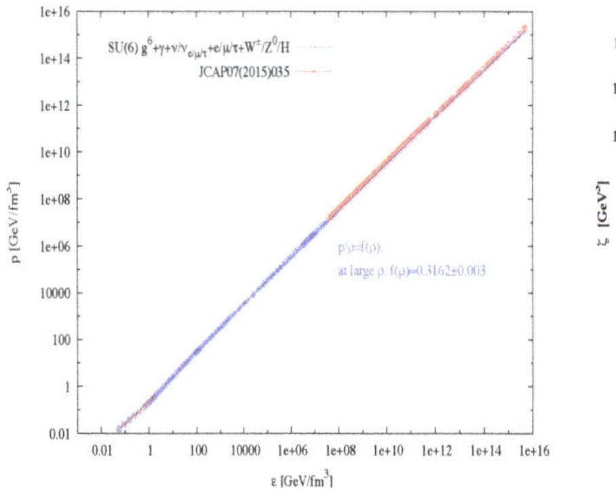

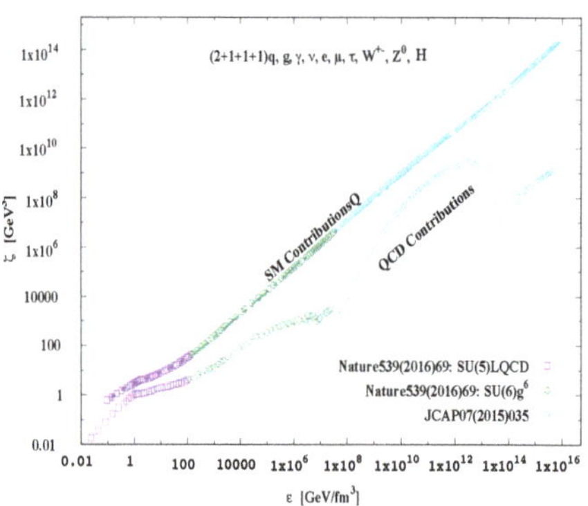

Figure 8. (**Left**) The combined EoS $p(\epsilon), \epsilon \equiv \rho$ of QCD and EW matter, using non-perturbative results [89] extended to include other DoFs such as γ, neutrinos, leptons, EW, and Higgs bosons as well as perturbative results [90,279]. Adapted from Ref. [278]. (**Right**) Bulk viscosity ζ at $\mu_B = 0$ as a function of the energy density ϵ. The top symbols stand for the SM contributions, while the bottom ones stand for the QCD contributions, only. Adapted from Ref. [280].

Note that the sound velocities

$$c_{s,1}^2(\epsilon) = \frac{dp_1}{d\epsilon} = b > 0, \qquad c_{s,2}^2(\epsilon) = \frac{dp_2}{d\epsilon} = cd\epsilon^{d-1} > 0, \tag{64}$$

are well defined, making it possible for each of two components to represent the EoS of some substance. Let us add that analytical expressions for the scale factor of the universe $a(t)$ and Hubble parameter $H(t)$ deduced from the EoS (62) were recently discussed in Ref. [281].

While the first component of pressure (8) corresponds roughly to the EoS of the massless gas of non-interacting particles, the second one, in agreement with the asymptotic freedom of QCD [33], dies out with increasing ϵ. Interestingly, $p_2(\epsilon)$, up to the constant additive term a, coincides with the generalized Chaplygin EoS used in Ref. [282] to describe the evolution from a phase dominated by non-relativistic matter to a phase dominated by the Cosmological Constant (or DE)[5]. For $d = -1$ in Equation (62), we obtain the ordinary Chaplygin gas [283,284] with EoS $p = c/\epsilon$.

Another possibility of how to read the term $p_2(\epsilon)$ in Equation (62) is, in analogy with Refs. [264,265] and Equation (54), to interpret it as the density-dependent bag function $\mathcal{B}(\epsilon) = -(a + c\epsilon^d)$, cf. Equation (62). The latter may account for the density-dependent character of the physical vacuum due to, e.g., the instanton liquid [210,285]. Instantons [272], classical solutions to the Euclidean equations of motion, are localized in all the four dimensions and correspond to tunneling events between degenerate classical vacua in Minkowski space. Since tunneling lowers the ground-state energy, the instantons provide a simple understanding of the negative non-perturbative vacuum energy density. Yet, the Euclidean-based instanton model is not the only solution representing the QCD vacuum. It remains, in fact, questionable to what extent it represents the reality due to non-analyticity of the gluonic field operators and the associated color confinement property. Below, in Section 4, we elaborate on a recently proposed alternative picture and new gluonic vacuum solutions, which are readily formulated in Minkowski and FLRW spacetimes.

3.3. *Hydrodynamical Description of Dissipative Effects and the Early Universe*

According to the currently accepted scenario, see, e.g., Refs. [84,286], the evolution of the early universe must include a number of dissipative processes in order to explain the current large value of the entropy per baryon. Some of them, such as the decoupling of neutrinos during the radiation era [120] or different cooling rates of the fluid components in the expanding universe [287], can result from the conventional physics; others, involving more exotic mechanisms, assume entropy production via string creation [288] or the GUT phase transitions [289]. The hydrodynamical description of dissipative effects is summarized in Appendix B.

Let us now follow the evolution of the early universe in terms of the EoS including bulk viscosity ζ defined in Equation (A11). In Ref. [280], data from non-perturbative [89] and perturbative [90,279] SM simulations were used to study behavior ζ over a wide range of temperatures T, entropy densities s and energy densities ε. It was found that $\zeta/(Ts)$ decreases exponentially with T increasing. The bulk viscosity dependence on the energy density at zero baryon chemical potential $\mu_B = 0$ is displayed on the right panel of Figure 8. Looking first on the QCD contributions only, it is apparent that the non-monotonic dependence of $\zeta(\epsilon)$ can be divided into four regions. The first, spanning $\epsilon \lesssim 100$ GeV/fm^3, corresponds to the hadron-QGP phase. The second region, up to $\sim 5 \times 10^7$ GeV/fm^3, contains both the QCD and the EW phases of matter. The third one seems to form an asymmetric parabola with focus at the critical energy density of the universe, $\rho_c \simeq 10^{12}$ GeV/fm^3 [280]. The fourth region shows a rapid increase in ϵ emerging from a non-continuous point.

In the SM contributions, as shown in the upper curve of Figure 8, besides gluons and $(2+1+1+1)$ quarks, the contributions of the gauge bosons: photons, $W^{\pm}$, and Z^0, the charged leptons: neutrino, electron, muon, and tau, and the Higgs bosons: scalar Higgs particle, were also taken into account. An overall conclusion drawn in Ref. [280] is that over the entire range of energy densities, the SM contributions are very significant. It is worth mentioning that the characteristic structures observed with the QCD contributions only are almost removed when adding also the EW contributions.

It is worth noting that the bulk viscosity could be important even outside the realm of the Hot Big Bang. In Ref. [290], the theory of the inflationary epoch, covering cold and warm inflation, as well as the models of late universe expansion in the presence of bulk viscosity, were analyzed. Assuming that the viscous effects during the inflationary epoch can be represented by a generalized and inhomogeneous EoS of the form:

$$p = -\epsilon + A\epsilon^{\beta} + \zeta(H) = -\epsilon + A\epsilon^{\beta} + \bar{\zeta}\left(\frac{8\pi G\epsilon}{3}\right)^{\gamma/2}, \tag{65}$$

where $A, \beta, \bar{\zeta}$, and γ are positive constants and $\zeta(H) = \bar{\zeta}H^{\gamma}$ is the Hubble parameter-dependent bulk viscosity, the authors have studied the behavior of various inflationary observables. Let us turn to important implications of non-perturbative QCD vacuum dynamics in cosmological evolution.

3.4. *Theory of Hot Meson Plasma Interacting with the QCD Vacuum*

Shortly after the confinement phase transition, the universe enters the state of hot meson plasma whose thermal evolution has been thoroughly explored in the framework of the Linear Sigma Model (LσM) in Ref. [277]. This analysis exploits the non-perturbative method of a generating functional derived from the effective Lagrangian at finite temperatures (for further references on this method, see Refs. [291–294]) and accounting for the quartic self-interactions of the σ-meson only. The latter approximation corresponds to a realistic well-motivated configuration of the "hadron gas" interacting with the non-linear σ-field and reproduces some of the basic thermodynamical characteristics of the hot meson plasma observed also in other approaches, in particular, in LσM-based scenarios [293,295] and Polyakov-loop extended Nambu–Jona-Lasinio (PNJL) models [296–298], but also features additional properties such as a possibility for $\sigma \to \pi\pi$ decays in the plasma above the critical temperature.

The effective LσM chiral Lagrangian accounting for the lightest scalar and pseudoscalar degrees of freedom $\pi^{\pm}, \pi^0, K^{\pm}, K^0, \bar{K}^0, \eta, \eta'$, and the σ-meson, in the hadron gas approximation at $T = 0$, reads [277],

$$\begin{aligned}
\mathcal{L}_{\rm eff} =& \frac{1}{2}\partial_\mu\sigma\partial^\mu\sigma + 2g^2v_0^2\sigma^2 - g^4\sigma^4 + \\
& \frac{1}{2}(\partial_\mu\pi_\alpha\partial^\mu\pi_\alpha + \partial_\mu\eta\partial^\mu\eta + \partial_\mu\eta'\partial^\mu\eta') + \partial_\mu\bar{K}\partial^\mu K - \\
& \frac{1}{2}\Big[2\kappa g^2(m_u+m_d)\sigma^2\pi_\alpha\pi_\alpha + \frac{2}{3}\kappa g^2(m_u+m_d+4m_s)\sigma^2\eta^2 + \\
& \quad \frac{4}{3}\kappa g^2(m_u+m_d+m_s+\Lambda_{\rm an})\sigma^2\eta'^2\Big] - \kappa g^2(m_u+m_d+2m_s)\sigma^2\bar{K}K,
\end{aligned} \tag{66}$$

where $m_{u,d,s}$ are the constituent up, down and strange quark masses, respectively, g is the σ quartic coupling, and $\Lambda_{\rm an} \simeq 0.5$ GeV is the gluon anomaly term that provides an explicit breaking of $U(1)_L \times U(1)_R$ symmetry. The QCD order parameter of the hadron mater $v_0 = 265 \pm 15\,\text{MeV}$ represents the amplitude of the quark-gluon (quantum-topological) condensate, such that

$$\epsilon_{\rm top}(T=0) = -\left(\frac{b}{32} + \frac{(m_u+m_d+m_s)l_g}{4}\right)\langle 0|\frac{\alpha_s}{\pi}G^a_{\mu\nu}G_a^{\mu\nu}|0\rangle \equiv -v_0^4, \tag{67}$$

given in terms of the gluon correlation length $l_g \simeq (1.2\,\text{GeV})^{-1}$ [210,299], and the coefficient of the one-loop β-function of three-flavor QCD, $b = 9$. The quark-gluon condensate contributes together with the perturbative hadronic vacuum $\epsilon_{\rm vac}^{\rm had}$ emerging due to regularized contributions from meson fluctuations to the net QCD ground-state energy density,

$$\begin{aligned} \epsilon_{\rm vac}^{\rm QCD} &\equiv \epsilon(T=0) = \epsilon_{\rm top} + \epsilon_{\rm vac}^{\rm had} = -v_0^4 - \frac{1}{128\pi^2}\big(m^4_{\sigma({\rm vac})} + 3m^4_{\pi({\rm vac})} \\ &+ m^4_{\eta({\rm vac})} + 4m^4_{K({\rm vac})} + m^4_{\eta'({\rm vac})}\big) \simeq -7\times 10^9\,{\rm MeV}^4\,. \end{aligned} \tag{68}$$

that satisfies the vacuum equation of state in the zero-temperature limit, $\epsilon(T=0) = -p(T=0)$. The hadronic vacuum term is negative $\epsilon_{\rm vac}^{\rm had} < 0$ and appears to have a relatively small magnitude, i.e., $\epsilon_{\rm vac}^{\rm had}/\epsilon_{\rm vac}^{\rm QCD} \approx 0.15$ [277]. The current u, d, s quark masses break the global chiral $SU(3)_L \times SU(3)_R$ symmetry explicitly, while it is also broken spontaneously by means of a σ-field expectation value,

$$\sigma = \langle\sigma\rangle + \tilde\sigma\,, \qquad \langle\sigma\rangle \equiv \frac{v}{g}\,, \tag{69}$$

In order to generalize the effective Lagrangian approach to finite temperatures, following theoretical foundations laid out in, e.g., Refs. [292,300–302], one should resume the daisy and superdaisy contributions. This is effectively achieved by utilizing an approximation that the expectation values of different fields are independent of each other, while omitting odd-point correlation functions and factorizing the four-point correlation functions into a product of two-point ones, for instance, $\langle\eta^2\tilde\sigma^2\rangle = \langle\eta^2\rangle\langle\tilde\sigma^2\rangle$, etc. Then, as a result of the minimization procedure of the non-equilibrium vacuum potential, one obtains the equation of state for the σ-condensate,

$$\begin{aligned} v^2 &= v_0^2 - 3g^2\langle\tilde\sigma^2\rangle - \frac{1}{2}\kappa(m_u+m_d)\langle\pi_\alpha\pi_\alpha\rangle \\ &- \frac{1}{6}\kappa(m_u+m_d+4m_s)\langle\eta^2\rangle - \frac{1}{3}\kappa(m_u+m_d+m_s+\Lambda_{\rm an})\langle\eta'^2\rangle \\ &- \frac{1}{2}\kappa(m_u+m_d+2m_s)\langle\bar K K\rangle\,, \end{aligned} \tag{70}$$

as well as the equations of motion for the (pseudo)scalar field fluctuations about the evolving non-trivial ground state, for example, $\partial_\mu\partial^\mu\tilde\sigma + m_\sigma^2\tilde\sigma = 0$, etc. Those fluctuations after quantization correspond to physical mesons with masses $m_\sigma^2 = 8g^2v^2$, etc. that are, in general, dependent on temperature. The vacuum values of the pseudo-scalar pseudo-Goldstone meson masses are found in terms of the light quark condensates via the Gell-Mann–Oakes–Renner relation [303–305], while the σ-meson has been identified phenomenologically with the scalar $f_0(500)$ state with mass, $m_{\sigma({\rm vac})} \simeq 400-500$ MeV.

The generating functional in the case of zeroth chemical potential is the free energy density of the considered meson plasma found in terms of the spatial part of the energy-momentum tensor as follows,

$$\mathcal{F}(T, v, m_\sigma^2, \mathcal{M}^2) \equiv \frac{1}{3}\langle T_i^i\rangle\,,$$

whose minimization over the independent variables m_σ^2 and $\mathcal{M}^2 \equiv v^2 + g^2\langle\tilde\sigma^2\rangle$ provides the physical meson masses and the equation of state for the condensate v^2. Substituting the meson masses expressed in terms of the temperature T and the order parameter v into $\mathcal{F}$, one obtains the non-equilibrium Landau functional,

$$\mathcal{F}_{\rm NE}(T,v) \equiv \mathcal{F}(T, v, m_\sigma(T,v), \mathcal{M}(T,v))\,, \tag{71}$$

that provides the critical temperature of the chiral phase transition, where the finite-condensate phase becomes unstable, by means of

$$\left.\frac{d^2\mathcal{F}_{\rm NE}}{dv^2}\right|_{T=T_c} = 0\,, \tag{72}$$

while the stability condition of the low-symmetry phase reads $d^2\mathcal{F}_{\rm NE}/dv^2 \geq 0$. Finally, resolving the order parameter as a function of temperature, i.e., $v = v(T)$, one finds the so-called equilibrium Landau functional as,

$$\mathcal{F}_{\rm E}(T) \equiv \mathcal{F}_{\rm NE}(T, v(T))\,, \tag{73}$$

that matches the usual free energy definition. A variety of thermodynamic observables of the meson plasma are then derived in terms of $\mathcal{F}_{\rm E}(T)$ such as pressure, entropy density, energy density, heat capacity and the speed of sound squared,

$$p(T) = -\mathcal{F}_{\rm E}(T)\,, \quad \sigma(T) = -\frac{d}{dT}\mathcal{F}_{\rm E}(T)\,, \quad \epsilon = \mathcal{F}_{\rm E} + T\sigma\,, \quad c_V = T\frac{d\sigma}{dT} = \frac{d\epsilon}{dT}\,, \quad u^2 = \frac{\sigma}{c_V}\,, \tag{74}$$

respectively. While at $T = 0$, the QCD vacuum equation of state, $\epsilon = -p$, is satisfied, both pressure and energy density grow with T due to positive particle contributions. The total energy density of the "plasma + condensate" system vanishes at $T_{\epsilon=0} \simeq 237$ MeV for $m_{\sigma({\rm vac})} \simeq 500$ MeV.

Using the above formalism, in Ref. [277], the properties of different phases and transitions between them are explored in detail. For instance, at temperatures $T > T_c$, hadrons deconfine into quarks and gluons, while the condensates melts away, yielding a deconfined (or zero-condensate) phase of the QCD matter. Such a phase becomes metastable at temperature $T_0 < T_c$, when the σ-meson fluctuation becomes massless $m_\sigma(T_0) = 0$, while $v(T_0) = 0$ is still valid, and T_0 is then found by resolving the extremum conditions on the generating functional. At another T_1, two minima of $F_{\rm NE}(T, v)$ become equal such that the zero-condensate phase stabilizes and the corresponding temperature is found from the following equation, $F_{\rm NE}(T_1, 0) = F_{\rm E}(T_1)$. The first-order chiral phase transition to the zero-condensate phase then occurs at some temperature between T_1 and T_c, and its strength increases with the σ-meson mass in the vacuum. Note, as expected from the first-order nature of this transition, the entropy density of the meson plasma grows with T and remains finite at all values of T, while the heat capacity appears to have a singularity, and the speed of sound squared vanishes at the critical temperature $T = T_c$.

The thermal evolution of the meson mass spectrum and the condensate of the LσM is shown in Figure 9. Interestingly enough, both the condensate and the masses of all the mesons decrease with temperature for $T < T_c = 438\,$MeV. The critical temperature becomes reduced if the fermions (quarks and baryons) are introduced [293]. As a result of the "hadron gas" approximation, the σ-mass rapidly falls to zero at $T_0 = 402$ MeV and then grows much faster than the masses of other mesons (such that $m_\sigma > 2m_\pi$ almost for any values of T) in contrast with the corresponding predictions of other existing approaches such as PNJL. As a result, at low temperatures, the hadronic plasma is dominated by pions. There is also a significant phase co-existence domain of size $(T_c - T_0)/T_c \simeq 0.1$. Pressure $p(T)$, energy density $\epsilon(T)$ and the EoS $(\epsilon(T) - 3p(T) - A)/T^4$, where $A = \epsilon(T = 0) - 3p(T = 0)$ is the net vacuum contribution, are shown in Figure 10 from left to right, respectively. Due to the positive "hadron gas" contribution, both $p(T)$ and $\epsilon(T)$ rise with temperature. Their profiles can be approximately reconstructed as a sum of the negatively-definite constant QCD vacuum term $\epsilon(T = 0)$ and the contribution of the relativistic hadron plasma $\propto T^4$ (dashed lines), with the numerical coefficient $\alpha \simeq 3.5$ and the effective number of DoFs, $g_i \simeq 9$.

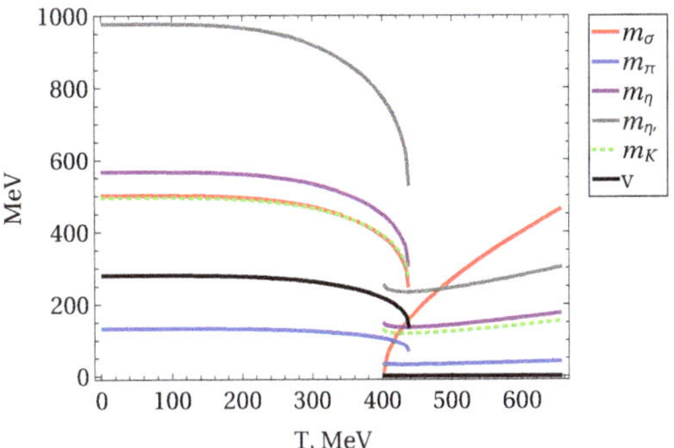

Figure 9. The condensate and meson masses as function of T. Here, $m_{\sigma(\mathrm{vac})} = 500\,\mathrm{MeV}$, $g^2 = 0.4$, $T_c = 438\,\mathrm{MeV}$, $T_0 = 402\,\mathrm{MeV}$, $T_1 = 430\,\mathrm{MeV}$.

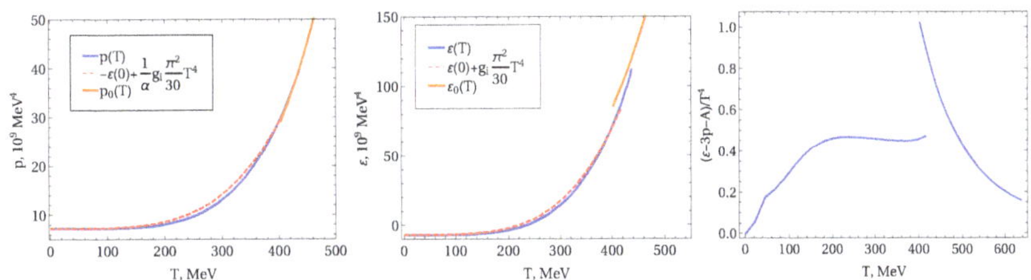

Figure 10. (**Left**) Pressure $p(T)$ as a function of temperature for the finite-condensate $v(T) \neq 0$ phase compared to that in the zero-condensate $v(T) = 0$ phase, $p_0(T)$ (solid lines), and to the approximated result (dashed line). (**Middle**) The same but for the energy density. (**Right**) The normalized EoS $(\epsilon(T) - 3p(T) - A)/T^4$ as a function of temperature, where $A \equiv \epsilon(T = 0) - 3p(T = 0)$ is the net vacuum contribution.

3.5. Cosmological Constant and Vacuum Catastrophe

The tight observational constraint on the DE EoS

$$w_{\mathrm{DE}} = -1.03 \pm 0.03\,, \tag{75}$$

comes through a combination of the cosmological data from various sources such as the Type Ia supernovae, the baryon acoustic oscillations, the CMB anisotropies, and the weak gravitational lensing, etc.; for details on the confidence level and datasets, see Ref. [306]. These constraints are consistent with the standard cosmological model known as the ΛCDM (Λ term plus DM in the form of CDM, as two dominant components of the universe). Specifically, the DE is considered to be in the form of Cosmological Constant or Λ-term density,

$$\epsilon_{\mathrm{DE}} = \epsilon_\Lambda\,, \qquad \epsilon_\Lambda \equiv \frac{\Lambda}{\kappa}\,, \qquad \kappa = 8\pi G\,, \tag{76}$$

expressed in terms of Λ-term in conventional normalization, Λ; see Equation (37). The latter satisfies the EoS $w_{\mathrm{DE}} = -1$ exactly. For a detailed recent review on achievements and challenges of the ΛCDM, see, e.g., Ref. [307].

Classically, an arbitrary Λ-term density ϵ_0 can be readily added to the right-hand side of the Einstein equations of GR that determine the macroscopic evolution of the universe,

$$R_{\mu\nu} - \frac{1}{2}g_{\mu\nu}R = \kappa(\epsilon_0 g_{\mu\nu} + T_{\mu\nu})\,, \qquad T_{\mu\nu} = -\frac{2}{\sqrt{-g}}\frac{\delta S_m}{\delta g^{\mu\nu}}\,, \qquad S_m = S_m[\phi,\psi,A_\mu,g_{\mu\nu}]\,, \tag{77}$$

in terms of action of matter fields S_m and their energy-momentum tensor $T_{\mu\nu}$. In quantum theory, a non-trivial contribution to the ground-state energy density emerges as an average of the energy-momentum tensor over the Heisenberg vacuum state [308,309]

$$\langle 0|T_{\mu\nu}|0\rangle = \epsilon_{\rm vac} g_{\mu\nu}\,, \qquad \epsilon_{\rm vac} \neq 0\,. \tag{78}$$

The latter is proportional to the trace of the energy-momentum tensor; hence, it represents an effect of conformal symmetry breaking in a given fundamental QFT through either the formation of a Bose-Einstein condensate in a massless theory or through nonzero mass-dimensional terms in the original Lagrangian. It is generically ill-defined and should be renormalized, with the classical ("bare") ϵ_0 being treated as a counter-term in the initial Lagrangian, such that the divergences are cancelled between the two yielding a finite, but renormalization scale μ dependent, vacuum energy density, $\epsilon_\Lambda(\mu)$. This is the physical vacuum energy density that emerges in cosmological measurements performed at some fixed scale $\mu = \mu_{IR}$ in the present universe, such that

$$\epsilon_\Lambda(\mu_{IR}) \equiv \epsilon_0 + \epsilon_{\rm vac}\,. \tag{79}$$

A macroscopic Cosmological Constant effect causing the universe to expand with acceleration (de-Sitter phase) is usually identified with energy density of the quantum vacuum that acquires contributions from all the incident vacuum subsystems existing in the SM and beyond. These would correspond to all quantum fields existing at energy scales ranging from the quantum gravity (Planck) scale, $M_{\rm PL} \sim 10^{19}$ GeV, down to the QCD confinement scale, $M_{\rm QCD} \sim 1$ GeV—the maximal and minimal energy scales of particle physics, respectively. The current vacuum state of the universe is considered to be produced in the aftermath of the latest QCD phase transition associated with hadronization of the cosmological plasma.

In the framework of SM, besides the zero-point energy contributions to the ground state coming from each elementary particle, there are two major vacuum condensates whose characteristics are well established in particle physics—the weakly coupled classical Higgs condensate responsible for spontaneous EW symmetry breaking giving masses to the SM vector bosons and fermions and the strongly-coupled quantum-topological quark-gluon condensate in QCD.

One of the important aspects of the cosmological QCD transition epoch concerns the formation of the negatively-definite (CM) contribution to the ground-state of the universe that has received little attention in the literature so far. For illustration of this effect, let us consider the conformal anomaly term in the trace of the effective QCD energy-momentum tensor [310–312],

$$T_\mu^{\mu,\rm QCD} = \frac{\beta(g_s^2)}{2} F_{\mu\nu}^a F_a^{\mu\nu} + \sum_{q=u,d,s} m_q \bar{q} q\,, \tag{80}$$

where m_q are the light (sea) quark masses $q = u, d, s$, g_s and β are the QCD coupling constant and the β-function, respectively, and $F^a_{\mu\nu}$ is the gluon field stress tensor. Taking the vacuum average, we obtain

$$\begin{aligned}\langle 0|T_\mu^{\mu,\rm QCD}|0\rangle &= -\frac{9}{32}\langle 0| : \frac{\alpha_S}{\pi} F_{\mu\nu}^a(x) F_a^{\mu\nu}(x) : |0\rangle + \frac{1}{4}\Big[\langle 0| : m_u \bar{u} u : |0\rangle + \langle 0| : m_d \bar{d} d : |0\rangle \\ &+ \langle 0| : m_s \bar{s} s : |0\rangle\Big] \simeq -(5\pm 1)\times 10^{-3}\,{\rm GeV}^4\,, \qquad \alpha_S = \frac{g_s^2}{4\pi}\,,\end{aligned} \tag{81}$$

representing the maximal value of the averaged quantum-topological QCD contribution to the physical vacuum energy density, whose spacetime dynamics is not fully understood and yet to be established. This contribution, also known as the quark-gluon condensate, is predicted by the theory of QCD instantons [210,299] and plays an important role in the chiral symmetry breaking and in dynamics of color confinement as well as in generation of mass of light mesons in hadron physics as suggested by the Gell-Mann–Oakes–Renner relation [303–305].

The so-called "Vacuum Catastrophe" reflects the fundamental problem of consistent matching between the macroscopic observable ϵ_Λ value, close to the critical energy density of the universe ρ_c,

$$\epsilon_\Lambda \simeq 0.7\rho_c \simeq 2.5 \times 10^{-47}\,\mathrm{GeV}^4 > 0\,, \qquad \rho_c \equiv \frac{3H_0^2}{\kappa}\,, \tag{82}$$

and the characteristic sizes of microscopic QFT predictions for the energy scale of each separate vacuum condensate [313,314]. Indeed, considering the topological QCD vacuum energy density $\epsilon_{\rm vac}^{\rm QCD}$ alone at macroscopic time and space separations typical for cosmological measurements, see Equation (81), its value is off by over forty orders of magnitude and has a wrong sign compared the observable ϵ_Λ.

Indeed, such a large and negative contribution to the vacuum density creates a big problem for existence of the spatially flat universe, as the right-hand side of the corresponding Friedmann equation must be positive at all times (see, e.g., Refs. [67,307,315,316]). The presence of a negative cosmological constant of the QCD scale would necessarily trigger a fast collapse of the universe at the time scale of a microsecond, as the other components of the cosmological plasma energy-density decay as $\propto 1/a^n$ (with $n = 4$ for relativistic and $n = 3$ for non-relativistic media). This would prevent the universe from traversing the QCD horizon scale such that no macroscopic evolution would be possible. For a recent review on the status of this problem and the existing approaches, see, e.g., Ref. [67] and references therein.

A consistent resolution of the so-called "old" Cosmological Constant problem (why is ϵ_Λ small and positive?) and the "new" Cosmological Constant problem (why is ϵ_Λ non-zeroth and exists at all?) [307] may require a dynamical mechanism for compensation of different short-distance vacuum configurations in the infrared limit of the corresponding QFT. Such a vacuum self-alignment in the non-perturbative regime would be desired in order to avoid a major fine tuning of different parameters of the fundamental theory [317], and it may be considered as a new physical phenomenon [65,67,315,318].

As has been pointed out in Ref. [27], in the thermal SU(2)/SU(3) YM theories, the confining phase at low temperatures is expected to be void of energy density and pressure. Along these lines, one introduces a natural hypothesis about a heterogeneous structure of the non-perturbative QCD vacuum in the infrared limit of QCD [67]. Such a structure is characterized by the presence of at least two distinct vacuum subsystems which contribute with opposite signs to the net QCD vacuum energy density and mutually eliminate each other on average at large space and time separations, $\Delta x \sim \Delta t \gg 1/\Lambda_{\rm QCD}$. Then, it is reasonable to assume that a phase transition in the QCD vacuum has occurred in the course of cosmological evolution. Such a transition has led to a dramatic drop in the net vacuum energy density due to an (almost) exact cancellation between the different vacuum subsystems in the IR limit of QCD.

Generically, an ultimate goal would be to develop a universal framework that consistently describes quantum vacua (condensates) dynamics in real cosmological time at both macroscopic (IR limit) and microscopic (UV limit) separations and then identify the phase transitions between those. This remains one of the major unsolved problems of fundamental physics [319,320] (see also Refs. [67,314]).

Since the effect of the negative QCD condensate term must be eliminated somehow beyond the Fermi scale of QCD, Ref. [277] explores the simplest scenario for cosmological evolution by invoking an additional positively-definite cosmological constant in the framework of the effective meson plasma model, and also using the Bag-like model of the QCD crossover transition [321] for comparison. The basic working assumption adopted in this work was that a positive cosmological constant has been formed (stochastically) at the QCD scale together with the negative (topological) term, which means that the QCD vacua effects are dynamically eliminated at distances beyond the typical hadron scale. In Figure 11 (left), the net QCD vacuum energy density ϵ is shown as a function of the normalized scale factor in both scenarios with and without an additional positive Λ-term "compensator". In this scenario, as the QCD vacuum evolves with temperature and the universe eventually collapses, a "backward" QCD transition from the meson plasma to QGP may occur above the critical QCD temperature. Provided that there exists a mechanism for a bounce from the singularity (for possible scenarios for such a bounce, see, e.g., Refs. [322–325] and references therein), a possible series of such sequential "direct" and "backward" QCD transitions implies that the universe may, in principle, oscillate around the QCD epoch for some time. Eventually, the negative QCD vacuum effect is eliminated, the universe enters the phase of unbound expansion, and the standard cosmological evolution takes off [277] (see Figure 11, right).

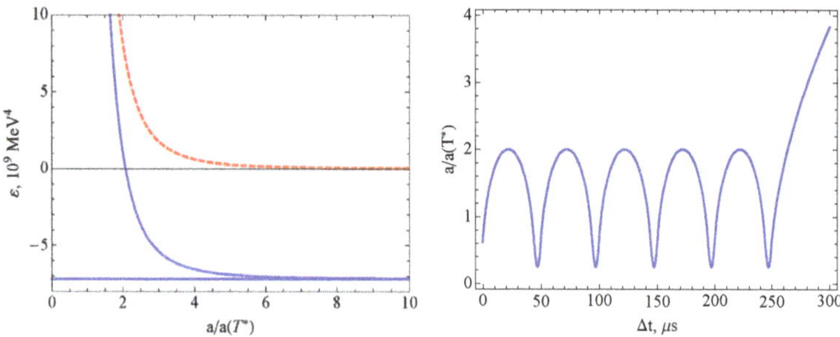

Figure 11. (**Left**) The QCD vacuum energy density as a function of the (normalized) scale factor a in the meson plasma model (solid line) and the same quantity but with an extra positive Λ-term (dashed line) that exactly compensates the negative (topological) QCD term at large time-scales. (**Right**) A scenario of the universe oscillating during the QCD phase transition epoch with stochastic generation of a positively-definite QCD-scale Λ-term "compensator".

4. Dynamics of Ground State in YM Theories

Let us discuss the properties of quantized YM theories and their major implications in cosmology while focusing primarily on QCD-like strongly-coupled dynamics and its connections to confinement and to the nearly vanishing value of the cosmological constant.

4.1. YM Ground State as a Time Crystal

While Gross, Wilczek and Politzer proved the asymptotic freedom of non-Abelian gauge theories at large momentum transfers [31,32,138], Savvidy showed that the perturbative QCD vacuum at zero field strength is unstable [326] and thereby demonstrated the existence of the vacuum condensate (see also Batalin, Matinyan and Savvidy [327]). Nielsen and Olesen worked out an argument for why the explanation of vacuum condensates in terms of a homogeneous field filling the vacuum is problematic: such a field mode is unstable, at least on a static Minkowski background [328]. The ground state of QCD is a non-perturbative quantum-topological state of the YM theory that is of primary importance for the understanding of color confinement dynamics [329] as well as of hadronic and effective quark masses. For a thorough discussion on the QCD vacuum and its implications, see, e.g., Ref. [330] and references therein.

Despite the possible theoretical frameworks that are available for the description of the quantum ground state in YM theories, an interesting case of the formation of a metastable condensate in the Savvidy approach, with gravitational back-reaction providing a stationary stabilization, was studied in Ref. [74]. Intriguingly, the authors showed that the relaxation process induced by the QCD phase transition provides a novel mechanism for the production of GWs in the early universe. Such production is enabled through the SSB of time translation invariance that is reminiscent of what happens in the time-crystals that were theoretically predicted by Wilczek in Refs. [331,332] and observed by the Monroe group [333] (for a detailed review, see Ref. [334]). Within the setting of early cosmology, as discussed in Ref. [74] that included the perturbative all-order effective action, it was derived that the energy density of the quark-gluon mean-field decays monotonically in time, while the pressure density undergoes violent oscillations at the characteristic QCD scale. This mechanism entails the generation of a primordial multi-peaked GW signal that eventually is shifted into the radio frequencies' domain. If detected, such a signal would represent an unprecedented echo of the QCD phase transition and it is, in principle, observable through forthcoming measurements at the FAST and SKA telescopes.

The scenario depicted in Ref. [74] requires the emergence of a quark-gluon condensate characterized, as mentioned above, by violent oscillations of the pressure density. Such oscillations would be periodic in multiples of the inverse characteristic scale $\Lambda_{\mathrm{QCD}} \simeq 0.1$ GeV. A very similar result that is consistent with this analysis was derived in Ref. [103] in which the authors deployed holography with the aim of analyzing relativistic collisions in a one-parameter family of strongly coupled gauge theories that were undergoing thermal phase transitions. An oscillating behavior of the pressure density was discovered also in this latter work, and again, the period of the oscillations was found to be a multiple of $\Lambda_{\mathrm{QCD}}^{-1}$. It was concluded that out-of-equilibrium physics smoothes out the details of the transition.

Most parts of analyses developed that involve lattice QCD tacitly assume an analytical continuation of the results obtained on the Euclidean space to the Minkowski spacetime. The underlying argument, as has been commonly advocated, relies on the idea that locally, on any FLRW spacetime, the dynamics of QCD can be studied on a "frozen" Minkowski-like background and, thus, results may be analytically continued to the Euclidean space. Relying on this assumption, a notable theorem due to Maiani and Testa [335] showed that the scattering of asymptotic states can be counter-Wick rotated only in the infinite volume limit (with the scale being the threshold amplitude) and for time scales much smaller than the level spacing due to momentum discretization.

Specifically, the authors of [335] started out with the Osterwalder-Schrader theorem that ensures that the Euclidean correlation functions can be analytically continued back to the Minkowski spacetime. However, this theorem heavily relies on the so-called reflexion positivity condition [336], and this condition is not fulfilled in the aforementioned cases when the FLRW dynamics are studied at time scales that exceed the Hubble time by one order of magnitude. Indeed, for the cases discussed, for instance in Refs. [74,103], non-perturbative effects that are originating from violent oscillations of the pressure density definitely spoil the time reflexion positivity condition. This insight suggests that the Maiani-Testa theorem is inapplicable in the context of the current discussion.

In fact, recent analyses developed according to different frameworks surprisingly confirm a quite different picture than the one on which the lattice QCD analyses are based: for instance, the determined period of oscillations in [74,103] that exceeds the Hubble time by one order of magnitude (in Planck units), as was mentioned above. Thus, effects of the spacetime curvature cannot be neglected due to the influence of non-trivial non-perturbative effects that are recovered beyond the characteristic-time of QCD, $\Lambda_{\mathrm{QCD}}^{-1}$. This provides an argument that should influence the confidence of the application of lattice QCD methods in cosmology; specifically when such methods aim to determine the order of the phase transition.

Finally, we mention, as a further approach to the problem of determining the order of the QCD phase transition in the early universe, a method provided by conformal field theories prescribed by the foundation of the modern understanding of quantum field theory and particle physics (for a comprehensive overview of the basic concepts, see Ref. [337]). Having been hitherto deepened so as to unveil the universal properties of scale invariant critical points, these frameworks were applied to describe continuous phase transitions in fluids and magnets as well as in many other materials. Substantial efforts were devoted to the study of non-perturbative strongly coupled conformal field theories, especially concerning their symmetries and theoretical constraints. Such work has opened the pathway to the development of the so-called conformal bootstrap [338,339], which has proven to be a successful framework in two dimensions and which finally was extended within the last decade so as to account for higher dimensionality including the physically relevant cases of three and four dimensions (for a detailed pedagogical review on the conformal bootstrap approach in d dimensions, see, e.g., Refs. [340,341]). Notably, significant progress in analytical methods has been achieved, shedding light on the possibilities of how to cast the bootstrap equations and, in parallel, more powerful numerical techniques were developed in attempts to find their solution [342]. This has brought about ground-breaking results including the determination of critical exponents and correlation function coefficients in the Ising O(N) models in three dimensions [343].

4.2. Effective Action Approach

In this section, the effective action approach is outlined in order to provide a background to the discussion of the contribution to the vacuum energy density through the trace of the energy-momentum tensor (EMT). In 1977, Matinyan and Savvidy [344] and Savvidy [326] investigated the asymptotic behavior of the effective Lagrangian density in gauge theories building on earlier work by Heisenberg and Euler and by Schwinger (see references therein). The behavior was studied using RG methods in order to relate the effective picture of strong fields to the short-range properties of gauge theories. The quantum corrections to the classical action as found by Schwinger were discussed both in these two publications and further summarized in a recent review [345].

The investigation begins with an examination of corrections to the classical action as suggested by Schwinger, i.e., corrections that allow for an expansion of the effective YM action in the gauge fields $\bar{A}^a_\mu$ (also known as connections):

$$\begin{aligned}\Gamma[A] &= \int \mathrm{d}x\, \mathcal{L}_{\text{eff}} \\ &= \sum_n \int \mathrm{d}x_1 \cdots \mathrm{d}x_n\, \Gamma^{(n)}{}^{a_1\cdots a_n}_{\mu_1\cdots\mu_n}\, \bar{A}^{a_1}_{\mu_1}(x_1)\cdots \bar{A}^{a_n}_{\mu_n}(x_n) \\ &= S_{\text{cl}} + W^{(1)} + W^{(2)} + \ldots . \end{aligned} \tag{83}$$

Here, the effective Lagrangian density $\mathcal{L}_{\text{eff}}$ undergoes a perturbative expansion so that the n-loop corrections provide a deviation from the classical action S_{cl}. The charged vector connection $\bar{A}^a_\mu(x) \equiv \langle 0|A^a_\mu(x)|0\rangle$ is the vacuum expectation value of the field operator and $\Gamma^{(n)}$ is the one-particle irreducible (1PI) vertex function. To each order, $W^{(n)}$ provides the n-loop correction to the classical action.

The effective Lagrangian at all-loop order in an SU(N) YM theory can be defined in terms of an order parameter $\mathcal{J}$ and a running coupling $\bar{g}(\mathcal{J})$. It should be noted that the latter is different from the bare coupling g_{YM} of the classical theory [66,316]. In a non-stationary cosmological background characterized by the FLRW metric, the conventional effective (quantum) YM Lagrangian can be written, through a rescaling of the fields according to Equation (17), as

$$\mathcal{L}_{\text{eff}} = \frac{\mathcal{J}}{4\bar{g}^2}, \quad \bar{g}^2 = \bar{g}^2(\mathcal{J}), \quad \mathcal{J} = -\frac{\mathcal{F}^a_{\mu\nu}\mathcal{F}^{a\,\mu\nu}}{\sqrt{-g}}, \tag{84}$$

where $\mathcal{A}^a_\mu$ are the rescaled SU(N) connections, $\mathcal{F}^a_{\mu\nu}$ are the rescaled field-strength components, and the equality entering the covariant field-strength $\mathcal{F}^a_{\mu\nu}$ in a curved background $\nabla_\mu \mathcal{A}^a_\nu - \nabla_\nu \mathcal{A}^a_\mu = \partial_\mu \mathcal{A}^a_\nu - \partial_\nu \mathcal{A}^a_\mu$ has been accounted for. Furthermore, $g \equiv \det(g_{\mu\nu})$, where $g_{\mu\nu} = a(\eta)^2\,\mathrm{diag}(1,-1,-1,-1)$ is the FLRW metric as a function of the conformal time η. Now, $\mathcal{J}$ simplifies to

$$\mathcal{J} = \frac{2}{\sqrt{-g}}\sum_a (\vec{E}_a\cdot\vec{E}_a - \vec{B}_a\cdot\vec{B}_a) \equiv \frac{2}{\sqrt{-g}}(\vec{E}^2 - \vec{B}^2), \tag{85}$$

which is an expression that emphasizes the dependence on the components of the CE and CM fields: $\vec{E}_a$, $\vec{B}_a$. The running of the coupling $\bar{g}$ as a function of the order parameter $\mathcal{J}$ in Equation (84) fully determines the dynamics of the effective YM theory and is given by the solution of the RG evolution equation [66,345]

$$2\mathcal{J}\frac{\mathrm{d}\bar{g}^2}{\mathrm{d}\mathcal{J}} = \bar{g}^2\beta(\bar{g}^2)\,, \tag{86}$$

where β is the standard beta-function of the YM theory [346,347].

As an aside, it should be noted that apart from $\mathcal{J}$, a second, independent, invariant may be constructed using the dual field strength [344]:

$$\mathcal{G} = -\frac{\mathcal{F}^a_{\mu\nu}\mathcal{F}^{*a\,\mu\nu}}{\sqrt{-g}} = \frac{4}{\sqrt{-g}}\vec{E}\cdot\vec{B}\,, \qquad \mathcal{F}^{*a\,\mu\nu} = \frac{1}{2}\epsilon^{\mu\nu\rho\sigma}\mathcal{F}^a_{\rho\sigma}\,. \tag{87}$$

This quantity is usually disregarded in the effective YM approach, since it vanishes for all fields that have orthogonal electric and magnetic components but it may as well, in principle, be incorporated into the effective Lagrangian.

In order to study the vacuum dynamics on cosmological scales, the spatially averaged quantity $\langle\mathcal{J}\rangle$ should be considered, and two cases are distinguished in which: (i) $\langle\mathcal{J}\rangle$ is positive, meaning that the averaged CE components $\langle\vec{E}^2\rangle$ dominate over the averaged CM terms $\langle\vec{B}^2\rangle$; (ii) vice versa, that is the case of a CM-dominated state with $\langle\mathcal{J}\rangle < 0$ that corresponds to a CM condensate. For the purpose of studying the basic features of the cosmological evolution of the CM and CE condensates in pure gluodynamics, it is sufficient to consider the effective SU(2) YM theory, since SU(2) subgroups can always be picked out of the SU(N) YM theory, and such a subgroup is the part that accounts for the cosmological application [66]. The explicit brackets $\langle\ldots\rangle$ will be dropped for the remainder of the discussion.

Applying the variational principle to the effective YM action as in the classical field theory, one straightforwardly obtains the effective YM equations of motion as described in Appendix C. Similarly, the EMT of the effective YM theory can be found as

$$T^{\nu}{}_{\mu} = \frac{1}{\bar{g}^2}\left[\frac{\beta(\bar{g}^2)}{2} - 1\right]\left(\frac{\mathcal{F}^a_{\mu\lambda}\mathcal{F}^{a\,\nu\lambda}}{\sqrt{-g}} + \frac{1}{4}\delta^{\nu}{}_{\mu}\mathcal{J}\right) - \delta^{\nu}{}_{\mu}\frac{\beta(\bar{g}^2)}{8\bar{g}^2}\mathcal{J}\,, \tag{88}$$

which is particularly useful for our discussion of cosmological evolution of the quantum YM system in what follows.

4.3. Mirror Symmetry of the Ground-State Solutions

The one-loop ground-state solution behaves differently depending on the sign of $\mathcal{J}$ and of the running coupling $\bar{g}_1$. In the case of the CM solution, it is again noted that $\mathcal{J} < 0$ and, hence, $\mathcal{F} > 0$. In addition, one refers to a positive $\bar{g}^2 > 0$ in this case so that an absolute value in the one-loop effective Lagrangian, see Equation (A32), may be removed. Considering the CM branch to one-loop order, which corresponds to a choice of the initial condition in the RG equation such that $\bar{g}_1^2(\mu_0^4) > 0$, the RG solution as derived in Equation (A29) can be written on the compact form

$$\bar{g}_1^2(\mathcal{J}) = \frac{96\pi^2}{bN\ln(-\mathcal{J}/\lambda^4)}\,. \tag{89}$$

Here, note that $\bar{g}_1^2(\mathcal{J}) > 0$ when $-\mathcal{J} > \lambda^4$, with

$$\lambda^4 \equiv \mu_0^4\exp\left[-\frac{96\pi^2}{bN\bar{g}_1^2(\mu_0^4)}\right], \tag{90}$$

and this gives yet another frequently used representation for the CM SU(N) Lagrangian [66]

$$\mathcal{L}^{(1)}_{\text{eff, CM}} = \frac{bN}{384\pi^2}\mathcal{J}\ln\left(\frac{-\mathcal{J}}{\lambda^2}\right). \tag{91}$$

The minimum of the effective Lagrangian at $\mathcal{J}_*$ is taken to be the physical scale of the quantum YM theory, which in the case of a CM vacuum reads

$$-\mathcal{J}_* = \mu_0^4, \tag{92}$$

which is the well-known phenomenon of dimensional transmutation. Readily from Equation (91), the minimal value of the effective CM Lagrangian reads

$$\mathcal{L}^{(1)}_{\text{eff, CM}}(\mathcal{J}_*) = \frac{\mathcal{J}_*}{4\bar{g}_1^2(\mathcal{J}_*)} < 0, \qquad \text{with } \bar{g}_1^2(\mathcal{J}_*) > 0, \tag{93}$$

and this is a negative value. In the standard notation, the CM minimum corresponds to

$$2g_{\text{YM}}^2\mathcal{F}^* = e\mu^4 \equiv \Lambda_{\text{QCD}}^4 \quad\to\quad -\mathcal{J}_* \equiv 2\Lambda_{\text{QCD}}^4, \qquad \mathcal{L}^{(1)}_{\text{eff, CM}}(\mathcal{J}_*) = \frac{-\Lambda_{\text{QCD}}^4}{2g_{\text{YM}}^2}, \tag{94}$$

expressed in terms of the conventional scale, Λ_{QCD}.

The exact ground-state solution for positive $\mathcal{J}_* > 0$ may be found immediately by inspection of the all-loop YM equation of motion; see Equation (A25). This is the CE condensate characterized by

$$\beta(\bar{g}_*^2) = 2, \quad \bar{g}_*^2 = \bar{g}^2(\mathcal{J}_*), \quad \mathcal{J}_* > 0, \tag{95}$$

from which it follows that the equation of motion is trivially satisfied. It should be pointed out that such a special solution is universal for any SU(N) symmetry, i.e., it is independent of N.

The contribution to the vacuum by the CE condensate comes from the trace of the EMT (see e.g., Ref. [348]). The trace at the minimum as given by Equation (88) is

$$T^{\mu}{}_{\mu} = -\frac{\beta(\bar{g}_*^2)}{2\bar{g}_*^2}\mathcal{J}_* = -\frac{1}{\bar{g}_*^2}\mathcal{J}_*. \tag{96}$$

The CE condensate hence contributes positively to the vacuum energy density, and this observation is intriguing when remembering that the CM condensate of Savvidy theory comes as a negative-definite contribution [345].

A striking and very interesting property of the YM effective Lagrangian will be discussed here, namely a mirror symmetry. It is apparent that the Lagrangian of Equation (84) is $\mathbb{Z}_2$-symmetric under simultaneous sign changes of $\mathcal{J}$ and $\bar{g}^2$. The invariance of the Lagrangian under this $\mathbb{Z}_2$ symmetry results in that the two condensates (CE and CM) are associated with two, apart from the overall sign, equal minima of the Lagrangian. Since the running of the coupling is a non-linear function of $\mathcal{J}$ in general, this symmetry may only be realized close to the ground state given in Equation (95) [66]. Therefore, in the vicinity of the ground state, the action is symmetric under the simultaneous transformation

$$\mathbb{Z}_2: \qquad \mathcal{J}_* \longleftrightarrow -\mathcal{J}_*, \qquad \bar{g}_*^2 \longleftrightarrow -\bar{g}_*^2. \tag{97}$$

As implicated by the form of the β-function, see Equation (A28), the imposed symmetry forces an additional change of sign in precisely β: $\beta(\bar{g}_*^2) \longleftrightarrow -\beta(-\bar{g}_*^2)$. There are important consequences of this symmetry of the action. Mainly, the conventional CE condensate effectively becomes mapped onto the CM gluon condensate with $\mathcal{J}_* < 0$ and $\bar{g}_*^2 > 0$ [345].

Taking the order parameter at the ground state to be the physical scale of the YM theory, one has, for the two condensates $\mu_0^4 = |\mathcal{J}_*|$. Then, for the CM branch, with $\bar{g}_*^2 > 0$, the running coupling for the one-loop solution may be rewritten as

$$\bar{g}_1^2(\mathcal{J}) = \frac{96\pi^2}{bN\ln(|\mathcal{J}|/\lambda^4)}, \qquad \lambda^4 = |\mathcal{J}_*|\exp\left[-\frac{96\pi^2}{bN\bar{g}_1^2(\mathcal{J}_*)}\right], \tag{98}$$

c.f. Equation (89) and Equation (A29). Insertion of this expression back into the effective Lagrangian results in

$$\mathcal{L}^{(1)}_{\text{eff, CM}} = \frac{bN}{384\pi^2}\mathcal{J}\ln\left(\frac{|\mathcal{J}|}{\lambda^2}\right), \tag{99}$$

c.f. Equation (91).

Due to the $\mathbb{Z}_2$ mirror symmetry, the minima of the effective Lagrangian on the two symmetry branches where $\bar{g}_*^2$ is either positive or negative come with the same value but with different scales: for the two condensates; the scale is modified by the exponential in Equation (98), so that

$$\lambda_\pm = |\mathcal{J}_*|\exp\left[\mp\frac{96\pi^2}{bN|\bar{g}_1^2(\mathcal{J}_*)|}\right]. \tag{100}$$

The upper sign stands for the CM condensate while the lower sign is the CE branch with $\bar{g}_1^2(\mathcal{J}_*) < 0$.

As is clear from Equation (100) and from the discussion in Ref. [66], the minimum for which $\mathcal{J}_* > 0$ appears in the non-perturbative region defined by $0 < \mathcal{J}_* < \lambda^4$. Therefore, this minimum corresponds to the CE condensate found in Equation (95). The mirror minimum is then swiftly found by applying the $\mathbb{Z}_2$-symmetry transformation. Note that the physical scale for the mirror minimum associated with the CM condensate is exponentially suppressed relative to the minimum point $|\mathcal{J}_*|$ with the consequence that the CM condensate, with $\mathcal{J}_* < 0$, appears in the perturbative region where $|\mathcal{J}_*| > \lambda^4$.

Finally, we may return to the contribution of the CM condensate to the vacuum energy density. Applying the mirror symmetry to the CE minimum gives $\beta \to -2$, and the equations of motion, as presented in Equation (A25), are no longer trivially satisfied. However, as pointed out in Ref. [66], the dynamical equation in the vicinity of the CM condensate becomes

$$\hat{\mathcal{D}}^{ab}_\nu\left[\frac{\mathcal{F}^{b\,\mu\nu}}{\bar{g}^2\sqrt{-g}}\right] = 0. \tag{101}$$

This expression bears a close resemblance to the classical YM equations of motion that are valid in the vicinity of the ground state. The EMT for the CM minimum involves extra terms in comparison to Equation (96):

$$T^\nu{}_\mu(\beta = -2) = \frac{-2}{\bar{g}_*^2}\left(\frac{\mathcal{F}^a_{\mu\lambda}\mathcal{F}^{a\,\nu\lambda}}{\sqrt{-g}} + \frac{1}{4}\delta^\nu{}_\mu\mathcal{J}_*\right) + \delta^\nu{}_\mu\frac{1}{4\bar{g}_*^2}\mathcal{J}_*. \tag{102}$$

However, in the trace of this expression, the terms within the parenthesis cancel out, yielding

$$T^\mu{}_\mu = +\frac{1}{\bar{g}_*^2}\mathcal{J}_*. \tag{103}$$

The conclusion of Ref. [66] is therefore that the contribution of the two condensates to the vacuum energy density cancel each other with equal magnitude and opposite signs as long as the averaging over macroscopic volumes that contains many CE and CM vacuum "pockets" of typical microscopic length-scales of $\sim \lambda_\pm^{-1} \sim \Lambda_{\text{QCD}}^{-1}$ is considered:

$$\epsilon_{\text{vac}} = \frac{1}{4}\langle T^\mu{}_\mu\rangle = \mp\mathcal{L}_{\text{eff}}(\mathcal{J}_*). \tag{104}$$

Provided that the energy scales of "electric gluon" and "magnetic gluon" condensation are not the same, as was elaborated above, the condensates are formed at different spacetime separations. However, they evolve toward the same absolute value of the energy density, but with opposite signs, due to their cosmological attractor nature (see Section 4.4 below). This effectively causes the cancellation of "electric gluon" and "magnetic gluon" contributions to the QCD ground state at sufficiently large separations, i.e., in the deep infrared limit of the theory, although at the expense of a loss of homogeneity at typical length (Fermi) scales of QCD $\sim \Lambda_{\rm QCD}^{-1}$. As the QCD vacuum appears to be locally inhomogeneous at these length-scales, gravity is expected to react to its electric and magnetic pockets in opposite ways such that the local metric fluctuations become averaged out beyond the length-scale of QCD to a small net effect compatible to that of the global observed cosmological Λ-term [65].

In the framework of the YM effective action approach, one may reach an intriguing conclusion that the exact compensation of the CE and CM gluon condensate components in the QCD vacuum (averaged over spacetime volumes above the QCD Fermi scale) is the necessary and sufficient condition for confinement in QCD. Indeed, as will be discussed below, the CE and CM vacuum pockets are always separated by non-analytic domain walls effectively blocking the gluon field from propagating over length-scales beyond the Fermi scale of each such pocket. The domain walls that separate different CM and CE pockets of the QCD vacuum prevent the color fields from propagating over macroscopic distances and, thus, effectively confine them within such pockets. An exact cancellation of their contributions to the net vacuum energy-density emerges in experimental observations as a complete disappearance of the gluon DoFs in the IR limit of the theory, i.e., beyond the QCD Fermi scale. In this picture of confinement, it would be natural to consider such pockets (with quarks and gluons being locked inside of them) as hadronic vacuum excitations. This is fully compatible with the classical limit of the YM theory where the conformal anomaly is absent and where only the radiation-like medium (hadron gas) remains at large separations. It remains to be seen exactly how the domain-wall picture of confinement readily formulated in Minkowski spacetime relates to a more standard center-vortex mechanism of confinement [349–355] strongly supported by lattice simulations in Euclidean spacetime (for a recent review of the current status of this research field, see, e.g., Ref. [22,356]).

Note, with respect to the exact CM/CE cancellation and, hence, color confinement, the restoration of a discrete (mirror) symmetry between "electric gluon" and "magnetic gluon" contributions at the level of the ground state is an intrinsic property of the pure YM theory and the RG flow equations. This generic property is intricately connected to the fact that the QCD-induced component of the cosmological constant term vanishes for averages over macroscopic volumes of physical spacetime. A residual effect of the CE/CM cancellation, emerging due to an effective dynamical breaking of the mirror symmetry by gravitational interactions in the QCD vacuum, rigorously matches (in both the sign and an order of magnitude) the observed value of the cosmological constant [65].

Thus, the color confinement phenomenon and the tiny value of the cosmological constant are the direct and closely connected consequences of the mirror symmetry of the QCD vacuum in the infrared regime. An important implication of the domain walls in the QCD vacuum is that no analyticity of the scattering amplitudes can be assumed in such a case, causing potential problems with the standard imaginary time and Euclidean formulations such as lattice QCD that is relying on the analyticity properties and vacuum triviality of the theory.

4.4. YM Cosmological Attractors

The temporal evolution of the condensate(s) discussed in the subsections above may be described on cosmological scales where short-distance fluctuations are averaged (integrated) out. A simple background can be obtained by splitting the full gauge field into a background field component $\bar{A}_\mu$ and a fluctuating field a_μ: $\mathcal{A}_\mu = \bar{A}_\mu + a_\mu$. This scheme was developed in general by Wetterich in, e.g., Ref. [357,358], first for scalar fields in $SO(N)$ and later for gauge fields, and it was further studied by, e.g., Gies [359] and Eichhorn et al. [360].

The (up to a rescaling) unique SU(2) pure YM theory will be parameterized in terms of a scalar time-dependent but spatially homogeneous field component (the background) on large scales due to the local isomorphism of the isotropic SU(2) gauge group and the SO(3) group of spatial 3-rotations [361–365]. One may therefore obtain a unique decomposition of the gauge field into this spatially homogeneous isotropic part (describing the YM condensate) and a non-isotropic/inhomogeneous component (accounting for the YM waves), the latter being the fluctuations, according to

$$\mathcal{A}^{a}{}_{k} = U(t)\delta^{a}{}_{k} + \tilde{A}^{a}{}_{k}(t,\vec{x})\,. \tag{105}$$

Here, the decomposition has been performed using the gauge condition $\mathcal{A}^{a}{}_{0} = 0$ [66]. It should be stressed that the fluctuations that parameterize the inhomogeneous YM wave modes interpreted as gluons in the field-theoretical framework average out over large distances by definition in the sense that $\langle \tilde{A}^{a}{}_{k}(t,\vec{x})\rangle \equiv \int \mathrm{d}^3\vec{x}\,\tilde{A}^{a}{}_{k}(t,\vec{x}) = 0$. Further, the homogeneous YM condensate itself can be considered positively definite, $U(t) > 0$, and it contributes to the ground state of the theory. The parameterization of the gauge field in SU(2) as a spatially homogeneous isotropic condensate and wave modes may be generalized to SU(3) for an application to QCD.

A quasi-classical theory of SU(2) YM quantum-wave excitations of the classical homogeneous condensate (i.e., without accounting for the vacuum polarization effects) has been thoroughly discussed in Ref. [276]. The formalism enables a proper extension to an arbitrary gauge and symmetry group with at least one SU(2) subgroup. Among the key results: an excitation of longitudinal wave (plasma) modes as well as an energy swap between the evolving homogeneous condensate and waves have been established in the linear and next-to-linear approximations. As is shown in Figure 12, the condensate tends to loose its energy, leading to the growth of YM wave amplitudes denoted as "particles". This represents a possible mechanism of particle production due to the dynamical vacuum decay which can be particularly relevant for cosmology and also in QGP production in heavy ion collisions. The effect has further been observed in the maximally supersymmetric $\mathcal{N} = 4$ YM theory and in the more complicated two-condensate SU(4) gauge theory. As the next step, it would be important to perform an analogical study of quantum-wave dynamics in the effective action approach, i.e., in the case of a quantum YM vacuum, in order to study the impact of vacuum polarization phenomena on the energy balance in the "condensate + waves" system and hence on the growth of the wave modes.

Now, the dynamical behavior of the homogeneous YM condensate $U(t)$ introduced in Equation (105) will be discussed. The Einstein equations for the pure YM theory in a non-trivial spacetime are obtained through the principle of variation starting from the effective action [315,316] and read as follows:

$$\frac{1}{\kappa}\left(R^{\nu}{}_{\mu} - \frac{1}{2}\delta^{\nu}{}_{\mu}R\right) = \tag{106}$$

$$\bar{\epsilon}\delta^{\nu}{}_{\mu} + \frac{b}{32\pi^2}\frac{1}{\sqrt{-g}}\left[\left(- \mathcal{F}^{a}_{\mu\lambda}\mathcal{F}^{a\,\nu\lambda} + \frac{1}{4}\delta^{\nu}{}_{\mu}\mathcal{F}^{a}_{\sigma\lambda}\mathcal{F}^{a\,\sigma\lambda}\right)\ln\frac{e|\mathcal{F}^{a}_{\alpha\beta}\mathcal{F}^{a\,\alpha\beta}|}{\sqrt{-g}\lambda^4} - \frac{1}{4}\delta^{\nu}{}_{\mu}\mathcal{F}^{a}_{\sigma\lambda}\mathcal{F}^{a\,\sigma\lambda}\right],$$

$$\left(\frac{\delta^{ab}}{\sqrt{-g}}\partial_{\nu}\sqrt{-g} - f^{abc}\mathcal{A}^{c}_{\nu}\right)\left(\frac{\mathcal{F}^{b\,\mu\nu}}{\sqrt{-g}}\ln\frac{e|\mathcal{F}^{a}_{\alpha\beta}\mathcal{F}^{a\,\alpha\beta}|}{\sqrt{-g}\lambda^4}\right) = 0\,. \tag{107}$$

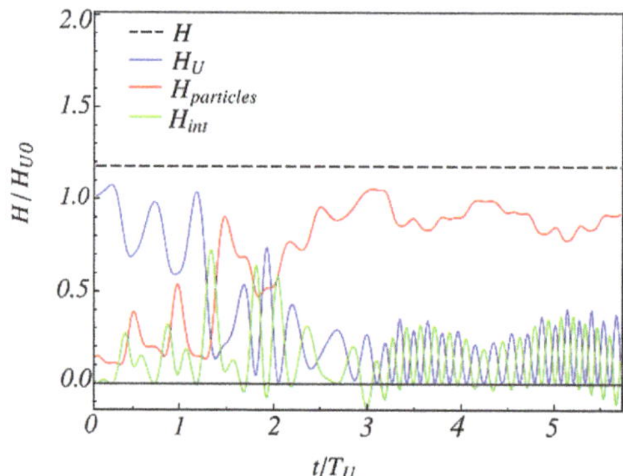

Figure 12. The time dependence of the Hamiltonian corresponding to the YM condensate, H_U, YM wave modes (or particles), $H_{\rm particles}$, as well as the interaction term between them, $H_{\rm int}$ of the YM "condensate + waves" system (with total energy H) in the quasi-classical approximation of small wave amplitudes. Adapted from Ref. [276].

Here, $\lambda = \xi\Lambda_{\rm QCD}$, with $\Lambda_{\rm QCD} \sim 0.1$ GeV being the QCD scale and where ξ has been introduced for scaling purposes, e is the base of the natural logarithm and κ is the gravitational constant. Finally, $\bar{\epsilon}$ describes the ground-state energy density s.t. $\bar{\epsilon} = \epsilon_{\rm top}^{\rm QCD} + \epsilon_{\rm CC}$ in terms of the quantum-topological contribution from QCD and the contribution from the Cosmological Constant. For confined QCD, $\epsilon_{\rm top}^{\rm QCD} \sim -5 \times 10^9$ MeV^4 for an SU(3) color symmetric theory, and this value may be extracted from the evaluation of non-perturbative quantum-topological fluctuations of the quark and gluon fields. The contribution from the Cosmological Constant as obtained from astrophysical measurements is $\epsilon_{\rm CC} \sim 3 \times 10^{-35}$ MeV^4, which is a comparatively minuscule and positive value. The fact that $\epsilon_{\rm top}^{\rm QCD}$ contributes to the ground-state energy of the universe with a large negative value is a severe problem for all existing cosmological paradigms, or more accurately, for the particle physics theories from which it arises. This is because the large negative contribution must be compensated for, in any first-order approximation, to a remarkable precision, resulting in the observed cosmological constant value to an accuracy of a few tens of decimal digits.

The conformal dynamics of the ground state (the condensate) $U = U(\eta)$ and of the scale factor $a = a(\eta)$ are described by the following equations of motion as derived from Equations (106) and (107):

$$\frac{6}{\kappa}\frac{a''}{a^3} = 4\bar{\epsilon} + T^{\mu,U}_{\mu}\,, \qquad T^{\mu,U}_{\mu} = \frac{3b}{16\pi^2 a^4}\left[(U')^2 - \frac{1}{4}U^4\right], \tag{108}$$

$$\frac{\partial}{\partial\eta}\left(U' \ln\frac{6e|(U')^2 - \frac{1}{4}U^4|}{a^4\lambda^4}\right) + \frac{1}{2}U^3 \ln\frac{6e|(U')^2 - \frac{1}{4}U^4|}{a^4\lambda^4} = 0\,. \tag{109}$$

It should be noted that a particular exact solution to Equation (109) can be obtained if the logarithm evaluates to zero at all times: that is, if $|Q| = 1$ for

$$Q \equiv 6e\left[(U')^2 - \frac{1}{4}U^4\right]a^{-4}(\xi\Lambda_{\rm QCD})^{-4}\,. \tag{110}$$

This may be solved for the two special cases $Q = \pm 1$, and the solutions are shown in Figure 13. The homogeneous background $U = U(t)$ displays quasi-periodic singularities in physical time in both cases. It should be stressed that the exact compensation of the CE and CM gluon condensate contributions to the QCD ground-state energy density, as discussed earlier, is realized in particular if the two components $Q = \pm 1$ co-exist in the universe. The cancellation happens over macroscopic distances as the average of the background vanishes in the large-time limit, and importantly, this occurs without any fine tuning. Crucially, the cancellation will arise due to the time-attractor nature of the contributions coming from the two minima; a property that is demonstrated in the following.

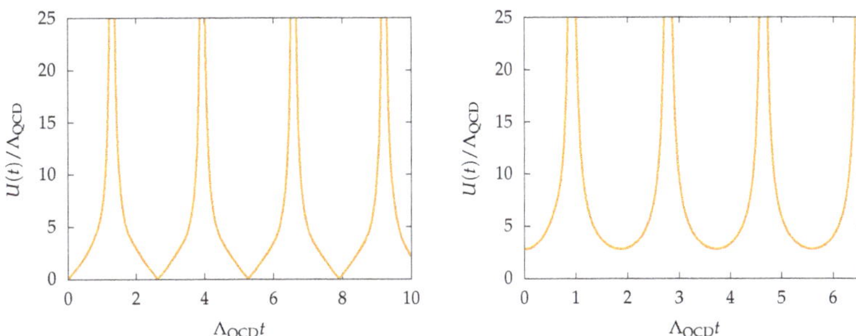

Figure 13. The homogeneous QCD condensate amplitude oscillations. The homogeneous component $U(t)$ displays quasi-periodic singularities in the physical time $t = \int a\,\mathrm{d}\eta$, plotted here in units of the characteristic time scale Λ_{QCD}^{-1}. To the left, the CE vacuum solution of Equation (110) with $Q = 1$ is shown, and to the right, the CM ditto with $Q = -1$ is displayed. $\xi = 4.0$ has been used along with initial conditions $U = 0$ and $U' = 0$, respectively. These results are compatible with those of Ref. [66] up to the scaling of the figure on the right-hand side.

The conformal integral of Equation (108) is

$$\frac{3}{\kappa}\frac{(a')^2}{a^4} = \bar{\epsilon} + T^{0,U}_{\ 0}\,, \qquad T^{0,U}_{\ 0} = \frac{3b}{64\pi^2 a^4}\left\{\left[(U')^2 + \frac{1}{4}U^4\right]\ln\frac{6e|(U')^2 - \frac{1}{4}U^4|}{a^4\lambda^4} + (U')^2 - \frac{1}{4}U^4\right\}, \tag{111}$$

which may be verified by differentiating this expression with respect to η and making use of Equation (109).

Equations (108) and (111) can now be combined in order to find a solution for the scale factor a, the trace of the EMT $T^{\mu}_{\ \mu}$ and the total energy density $T^{0}_{\ 0}$. This is possible since the latter equation incorporates the constraint of Equation (109). Solving this set of equations provides the benefit of obtaining solutions for observable quantities that must necessarily be smooth functions in time. Hence, the quasi-periodic singularities of $U(t)$ may be avoided. Since $t = \int \mathrm{d}\eta\, a(\eta)$, Equations (108)–(111) may be recast in terms of the physical time as

$$\frac{6}{\kappa}\left[\frac{\ddot{a}}{a} + \frac{\dot{a}^2}{a^2}\right] = 4\bar{\epsilon} + T^{\mu,U}_{\ \mu} \equiv T^{\mu}_{\ \mu}(t)\,, \qquad \frac{3}{\kappa}\frac{\dot{a}^2}{a^2} = \bar{\epsilon} + T^{0,U}_{\ 0} \equiv T^{0}_{\ 0}(t)\,, \tag{112}$$

where the energy density of the gluon condensate and the trace in the one-loop effective YM theory read

$$T^{\mu,U}_{\ \mu} = \frac{3b}{16\pi^2 a^4}\left[a^2\dot{U}^2 - \frac{1}{4}U^4\right], \tag{113}$$

$$T^{0,U}_{\ 0} = \frac{3b}{64\pi^2 a^4}\left\{\left[a^2\dot{U}^2 + \frac{1}{4}U^4\right]\ln\frac{6e|a^2\dot{U}^2 - \frac{1}{4}U^4|}{a^4(\xi\Lambda_{QCD})^4} + a^2\dot{U}^2 - \frac{1}{4}U^4\right\}. \tag{114}$$

In order to eliminate the explicit dependence on $U(t)$ and, hence, the obstructing singularities in the two equations, [66] introduced a universal analytic function $g(t)$ that parameterizes the relation between the trace of the EMT and the total energy density. The defining equation of this function is

$$T^{\mu,U}_{\mu} - C = \big(g(t)+1\big)\left[T^{0,U}_{0} - \frac{C}{4}\right], \qquad C \equiv -4\epsilon^{\rm QCD}_{\rm top} = \frac{3b}{16\pi^2}\frac{(\xi\Lambda_{\rm QCD})^4}{6e}. \tag{115}$$

Using $g(t)$, Equation (112) can be written entirely in terms of continuous functions:

$$\frac{6}{\kappa}\left[\frac{\ddot{a}}{a} + \frac{\dot{a}^2}{a^2}\right] = 4\epsilon_{\rm CC} + \big(g(t)+1\big)\left[T^{0,U}_{0} - \frac{C}{4}\right], \tag{116}$$

$$\frac{3}{\kappa}\frac{\dot{a}^2}{a^2} = \epsilon_{\rm CC} - \frac{C}{4} + T^{0,U}_{0}. \tag{117}$$

Note here that $T^{0,U}_{0} = T^{0,U}_{0}(U,\dot{U},a)$. After excluding $T^{0,U}_{0}$ above, the resulting equation for the scale factor that is left to be solved is

$$6\frac{\ddot{a}}{a} + 3(1-g(t))\frac{\dot{a}^2}{a^2} + \kappa\epsilon_{\rm CC}(g(t)-3) = 0. \tag{118}$$

The general solution is[6]

$$a(t) = a^* \exp\left[\sqrt{\frac{\kappa\epsilon_{\rm CC}}{3}}\int_{t_0}^{t} \frac{1+\sqrt{\frac{\epsilon_{\rm CC}}{\epsilon_0}} + \Big(1-\sqrt{\frac{\epsilon_{\rm CC}}{\epsilon_0}}\Big)\exp\Big\{\sqrt{\frac{\kappa\epsilon_{\rm CC}}{3}}\Big(-3(t'-t_0) + \int_{t_0}^{t'} g(\tau)\,{\rm d}\tau\Big)\Big\}}{1+\sqrt{\frac{\epsilon_{\rm CC}}{\epsilon_0}} - \Big(1-\sqrt{\frac{\epsilon_{\rm CC}}{\epsilon_0}}\Big)\exp\Big\{\sqrt{\frac{\kappa\epsilon_{\rm CC}}{3}}\Big(-3(t'-t_0) + \int_{t_0}^{t'} g(\tau)\,{\rm d}\tau\Big)\Big\}}\,{\rm d}t'\right], \tag{119}$$

in terms of the initial values of the scale factor $a^* \equiv a(t=t_0)$ and the total energy density $\epsilon_0 \equiv T^0{}_0(t=t_0)$, respectively. Note that $\epsilon_{\rm CC} \ll \epsilon_0$.

The total energy density, $T^0{}_0(t)$, and the trace of the EMT, $T^\mu{}_\mu(t)$, both explicitly defined in Equation (112), can be found by insertion of the solution above into Equation (117) together with a manipulation of Equation (115). The result is

$$\frac{T^0{}_0(t)}{\epsilon_{\rm CC}} = \left[\frac{1+\sqrt{\frac{\epsilon_{\rm CC}}{\epsilon_0}} + \Big(1-\sqrt{\frac{\epsilon_{\rm CC}}{\epsilon_0}}\Big)\exp\Big\{\sqrt{\frac{\kappa\epsilon_{\rm CC}}{3}}\Big(-3(t-t_0) + \int_{t_0}^{t} g(\tau)\,{\rm d}\tau\Big)\Big\}}{1+\sqrt{\frac{\epsilon_{\rm CC}}{\epsilon_0}} - \Big(1-\sqrt{\frac{\epsilon_{\rm CC}}{\epsilon_0}}\Big)\exp\Big\{\sqrt{\frac{\kappa\epsilon_{\rm CC}}{3}}\Big(-3(t-t_0) + \int_{t_0}^{t} g(\tau)\,{\rm d}\tau\Big)\Big\}}\right]^2, \tag{120}$$

$$\frac{T^\mu{}_\mu(t)}{\epsilon_{\rm CC}} = 4 + \frac{4\big(g(t)+1\big)\Big(1-\frac{\epsilon_{\rm CC}}{\epsilon_0}\Big)\exp\Big\{\sqrt{\frac{\kappa\epsilon_{\rm CC}}{3}}\Big(-3(t-t_0) + \int_{t_0}^{t} g(\tau)\,{\rm d}\tau\Big)\Big\}}{\Big[1+\sqrt{\frac{\epsilon_{\rm CC}}{\epsilon_0}} - \Big(1-\sqrt{\frac{\epsilon_{\rm CC}}{\epsilon_0}}\Big)\exp\Big\{\sqrt{\frac{\kappa\epsilon_{\rm CC}}{3}}\Big(-3(t-t_0) + \int_{t_0}^{t} g(\tau)\,{\rm d}\tau\Big)\Big\}\Big]^2}. \tag{121}$$

It shall be pointed out here that the above solutions for the scale factor, the energy density, and the trace of the EMT do not rely on any approximations but are the general solutions that can be obtained from Equation (112). These cosmological observables may therefore be studied on the full range from t_0 to t, provided that $g(t)$ is known.

For practical analyses, the auxiliary function g may be studied in the vicinity of the exact, large-time cancellation point where $Q(t) \sim 1$. This is done by introducing an expansion of the YM energy density around the asymptotic value of the exact solution, where $T^{0,U*}_{0} = C/4$, such that

$$T^{0,U}_{0}(t) \simeq C/4 + \delta\epsilon(t), \qquad \delta\epsilon \ll C. \tag{122}$$

Depending on the relation between the expansion parameter $\delta\epsilon$ and the remaining scale ϵ_{CC}, the time derivatives $\dot{a}$ and $\dot{T}^{0,U}_{0}(t)$ take two different asymptotic forms. Firstly, in the case of large $\delta\epsilon(t) \gg \epsilon_{CC}$, these are

$$\begin{aligned} \dot{a} &\simeq \sqrt{\frac{\kappa}{3}} a \sqrt{\delta\epsilon}, \\ \dot{T}^{0,U}_{0} &\simeq \sqrt{\frac{\kappa}{3}} (g(t) - 3) (\delta\epsilon)^{3/2}, \end{aligned} \tag{123}$$

when keeping only the leading terms in $\delta\epsilon(t) \ll C$. Secondly, in the opposite case when $\delta\epsilon(t) \ll \epsilon_{CC}$, the same quantities instead become

$$\begin{aligned} \dot{a} &\simeq \sqrt{\frac{\kappa\epsilon_{CC}}{3}} a \left(1 + \sum_{n=1}^{\infty} \binom{\frac{1}{2}}{n} \left(\frac{\delta\epsilon}{\epsilon_{CC}}\right)^{n}\right), \\ \dot{T}^{0,U}_{0} &\simeq \sqrt{\frac{\kappa\epsilon_{CC}}{3}} (g(t) - 3)\delta\epsilon \left(1 + \sum_{n=1}^{\infty} \binom{\frac{1}{2}}{n} \left(\frac{\delta\epsilon}{\epsilon_{CC}}\right)^{n}\right). \end{aligned} \tag{124}$$

The difference between Equations (123) and (124) introduces only a very small correction to the period of $g(t)$, and this correction can safely be neglected, as will be shown later in this section.

It is now possible to study g in the vicinity of the asymptote (Equation (122)) and in the two limits of $\delta\epsilon(t)$ relative to ϵ_{CC}. First, solve Equations (113) and (114) for U, $\dot{U}$ in terms of $T^{0,U}_{0}$, $T^{\mu,U}_{\mu}$. Then, insert the expansion of the energy density from Equation (122) in the resulting expressions as well as in Equation (115). Explicit expressions for U, $\dot{U}$ in terms of the EMT components and the scale factor allows for the formulation of

$$\partial_t U(t) - \dot{U} \equiv 0\,. \tag{125}$$

Explicit computation of the first term results in a differential equation for $g(t)$ when the expansions in Equation (123) are inserted. The final form of such a differential equation is

$$\dot{g}^4 - \frac{8(\xi\Lambda_{QCD})^4}{3e}(1 - g^2)^3 = 0\,. \tag{126}$$

Its implicit analytic solution can be found for the inverse function $t(g)$ over half a period of oscillation of $g(t)$ as

$$t(g) = -\frac{(6e)^{1/4}}{2\xi\Lambda_{QCD}}\left[{}_2F_1\left(\frac{1}{2}, \frac{3}{4}, \frac{3}{2}; g^2\right) g - k\right], \qquad 0 < t(g) < T_g/2\,. \tag{127}$$

The constant $k \approx 2.622$ is defined through the above equation, and the initial condition $g(t_0) = 1$ is adopted for simplicity. These conditions determine $g = g(t)$ as a periodic quasi-harmonic function with unit amplitude. The period of oscillation may be found from Equation (126) as

$$T_g = \frac{2(6e)^{1/4}}{\xi\Lambda_{QCD}} \int_0^1 \frac{dg}{(1 - g^2)^{3/4}} = \frac{2k(6e)^{1/4}}{\xi\Lambda_{QCD}}\,. \tag{128}$$

Using instead the expansions in Equation (124), the above calculation can be repeated for the case of $\delta\epsilon(t) \ll \epsilon_{CC}$. The analogue of Equation (126), now with an additional term proportional to $\alpha \ll 1$, is

$$\frac{dg}{d\tau} \pm 2(1-g^2)^{3/4} - \alpha(1-g^2) = 0, \qquad \tau = t\frac{\xi\Lambda_{QCD}}{(6e)^{3/4}},$$
$$T_g' = T_g\left[1 + \frac{\pi}{4k^2}\alpha^2\right],$$
$$\alpha = \frac{2(6e)^{1/4}}{\xi\Lambda_{QCD}}\sqrt{\frac{\kappa\epsilon_{CC}}{3}} \approx 4.8\times 10^{-24}, \qquad \text{for } \Lambda_{QCD}\sim 210\ \text{MeV}. \tag{129}$$

Hence, the additional term may safely be neglected.

Note that the function $g(t)$ satisfies the following integral constraint

$$\int_0^t d\tau\, g(\tau) = \pm\frac{(6e)^{1/4}}{\xi\Lambda_{QCD}}(1-g^2)^{1/4}, \qquad \frac{nT_g}{2} < t < \frac{(n+1)T_g}{2}, \tag{130}$$

where the upper sign corresponds to even n and the lower corresponds to odd n. It should be noted that this constraint can be used in Equations (119)–(121) in order to explicitly express the general solutions for $a(t)$, $T^0{}_0(t)$ and $T^\mu{}_\mu(t)$ in terms of g. A very good analytic approximation to the exact $g(t)$ may be constructed when keeping only the first two non-vanishing terms of the harmonic Fourier expansion s.t.

$$g(t) \approx A\cos\left(\frac{2\pi t}{T_g}\right) + (1+A)\cos\left(\frac{6\pi t}{T_g}\right). \tag{131}$$

The amplitude A is found through

$$A = \frac{2}{k}\int_0^1 dg\,\frac{g}{(1-g^2)^{3/4}}\cos\left(\frac{\pi}{2k}\int_g^1 \frac{dx}{(1-x^2)^{3/4}}\right) \approx 1.14. \tag{132}$$

A comparison between this approximation and the exact solution $g(t)$ is provided in Figure 14 from which it is clear that the approximation is indeed capturing the universal function to a very good accuracy. For a particular approximation to $g(t)$ discussed above, the physical observables have been plotted in Figure 15 for illustration.

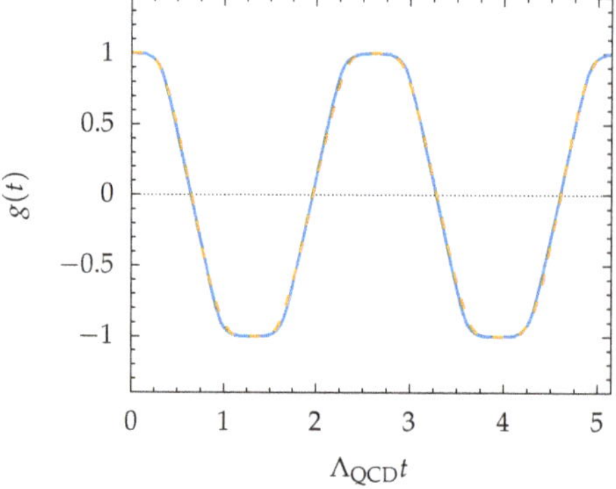

Figure 14. The time dependence of the quasi-harmonic universal function $g = g(t)$. The exact solution in Equation (127) (solid line) has been extrapolated from the solution over a single period $T_g/2$. A harmonic approximation in Equation (131) (dashed line) captures the behavior of $g(t)$ well.

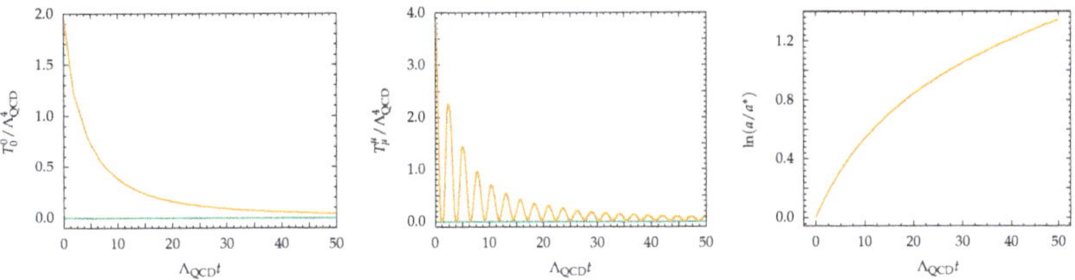

Figure 15. Solutions for the total energy density $T^0_{\ 0}(t)$ (**left**), the trace of the total QCD EMT $T^\mu_{\ \mu}(t)$ (**middle**) and the scale factor $a(t)$ (**right**). The asymptotic values for which $Q \to 1$ are indicated by horizontal lines in the left and middle panels, respectively. The initial conditions have been chosen as $U_0 = 0$, $\dot{U}_0 = (\xi\Lambda_{QCD})^2/\sqrt{3e}$, $Q_0 > 1$, $\xi = 4.0\ \Lambda_{QCD} = 332$ MeV and $\kappa = 10^{-7}$ MeV^{-2}. The energy density and the trace are plotted in dimensionless units, rescaled by Λ^4_{QCD} and for illustrative purposes, ϵ_{CC} was set to $\sim 0.5\,\%$ of $\bar{\epsilon}$. These results are compatible with the qualitative picture in Ref. [66].

4.5. SU(N) and the Functional RG Approach

So far, we have addressed gluodynamics resorting to an all-order effective perturbative approach. Nonetheless, we can extend our investigation so as to include non-perturbative all-order analyses, resorting to the Functional Renormalization Group (FRG) approach [357,366–375]. This latter is a Wilsonian momentum-shell-wise integration method for the path integral, which was developed to delve into the dynamics of interacting quantum field theories and statistical systems in a non-perturbative way when couplings cannot be dealt with using perturbative techniques. A regulator function R_k is taken into account so as to suppress quantum fluctuations at momenta lower than some physical scale, i.e., at an IR cut-off scale k. This scale is in principle different to the one defined in the subsections above, namely μ_0. The former denotes the scale above which all quantum fluctuations are integrated out, while the latter captures instead the one-loop renormalization scale and may further be extended to all-loop accuracy. The regulator function has been discussed by, e.g., Gies in Ref. [359]. A scale-dependent effective action that flows with k is then recovered, i.e., Γ_k, which encodes quantum fluctuations effects at momenta larger than the IR cut-off k. Varying k then allows to smoothly interpolate among the microscopic/short-scale action and the full quantum effective action $\Gamma_{k\to 0}$.

This elucidates why this procedure looks to be tailored ad hoc for cosmological applications, at the IR scale. The Wetterich equation for a non-zero background that fixes the running dependence on the cut-off scale in the FRG approach reads [357,366]

$$\partial_t \Gamma_k = \frac{1}{2}\mathrm{STr}\big(\Gamma^{(2)}_k + R_k\big)^{-1}\partial_t R_k\,, \tag{133}$$

where $\Gamma^{(2)}_k$, a matrix in the field space, denotes the second functional variation of the effective (running) action with respect to the field content of the theory. Above, STr is the supertrace, including a summation over all field components and discrete indices, as well as all the eigenvalues of the Laplacian in the kinetic term. The FRG equation retains a dependence on the full (field-dependent) non-perturbative regularized propagator [367], namely, $(\Gamma^{(2)}_k + R_k)^{-1}$.

The FRG approach has been then adapted to YM SU(N) theories [358–360]. Specifically, in Ref. [360], a numerical extrapolation among low and high energy scales for the full propagators was deployed in order to derive the gluon condensate. Refining these results in Ref. [376], the FRG approach was extended, within a cosmological setting, to the SU(2) case.

In Ref. [376], resorting to approximations that are necessary to solve the FRG equation, which is otherwise too complicated to provide analytic results, the authors considered replacing Γ_k in Equation (133) with the bare action S, allowing for integration on both sides of the FRG equation, namely,

$$\Gamma_k = -\int L_{\rm eff} = \int dk \frac{1}{2}{\rm STr}(S^{(2)} + R_k)^{-1}\partial_t R_k = \frac{1}{2}{\rm STr}\,\log(S^{(2)} + R_k) + {\rm const.}. \quad (134)$$

Setting the bare action to the standard expression $S = \frac{1}{4}\int dx\, F^a_{\mu\nu}F^{a\,\mu\nu}$, which here corresponds to the UV limit of the effective theory, the integration constant can be fixed by requiring that Γ_k vanishes for a vanishing field strength.

The inversion of the regularized propagator then requires the use of a harmonic gauge fixing, entering the action $S_{\rm gf}$ and depending on the α-parameter, with associated Faddeev-Popov ghosts, in $S_{\rm gh}$, namely,

$$S_{\rm gf} = \frac{1}{2\alpha}\int {\rm d}x\, \bar{D}^\mu \bar{a}^a_\nu \bar{D}^\nu a^a_\mu\,, \qquad S_{\rm gh} = \int {\rm d}x\, \bar{D}_\mu \bar{c}_\nu D^\mu c^\nu\,, \quad (135)$$

where the background methods have been deployed, and barred quantities are calculated with respect to background fields. Then, in the Landau gauge where $\alpha \to 0$, the super-trace recasts along the transverse sector as

$$\frac{1}{2}{\rm STr}\,\log(S^{(2)} + R_k) = \frac{1}{2}{\rm Tr}_{\rm trans}\log\left[\bar{\mathcal{D}}^{\mu\nu}_T + R_k(\bar{\mathcal{D}}^{\mu\nu}_T)\right] - \frac{1}{2}{\rm Tr}_{\rm gh}\log\left[\bar{\mathcal{D}}^{\mu\nu}_{\rm gh} + R_k(\bar{\mathcal{D}}^{\mu\nu}_{\rm gh})\right], \quad (136)$$

where the differential operators can be expressed in terms of the SU(N) structure constant f^{abc} and the YM coupling constant $g_{\rm YM}$ as

$$\bar{\mathcal{D}}^{\mu\nu}_{\rm T} = \bar{\Box}\,\delta^{cb}\delta^{\mu\nu} + g_{\rm YM}\,\bar{F}^{a\,\mu\nu}f^{abc}\,, \qquad \bar{\mathcal{D}}^{\mu\nu}_{\rm gh} = \eta^{\mu\nu}\bar{\Box}\,. \quad (137)$$

In order to disentangle the emergence of a condensate as a solution to the FRG equation, one restricts the consideration to the case of SU(2). Although this could seem to be limiting in the wider theoretical perspective, nonetheless, the restriction to SU(2) shall be considered as a selection of a subgroup of SU(N).

Bearing the discussion above in mind, one may select a straightforward expression for the regulator function in terms of a cut-off scale, $R_k(\mathcal{D}) = k^2$. The super-trace STr may then be evaluated using the Schwinger formula

$$\ln A \equiv \int_0^\infty \frac{ds}{s}\left[e^{-sA} - e^{-s}\right], \qquad A > 0\,, \quad (138)$$

and opting for a convenient choice of the background, the self-dual one, as investigated in Ref. [360]. These choices allow for the evaluation of the super-trace as a sum over the eigenvalues of the kinematic terms. In general, the eigenvalues of $\bar{\mathcal{D}}_{\rm T}$ are not known for an arbitrary background. However, those were found in the case of a self-dual background considered in Ref. [360], and a general discussion on the trace technology used in this approach can be found, e.g., in Refs. [358,359]. Applying the results of Ref. [360] in the case of a self-dual background dominated by the color-magnetic field B, Ref. [376] obtained an expression for the regularized effective action in the following form:

$$L_{\rm eff} = \frac{g^2_{\rm YM}\theta}{4\pi^2}\int_0^\infty \frac{{\rm d}s}{s}\left[e^{-As} - e^{-s}\right]\left(\frac{1}{4\sinh^2(s)} + 1 - \frac{1}{4s^2}\right), \quad A = \sqrt{\frac{k^4}{g^2_{\rm YM}\theta}}, \quad \theta \equiv B^2 > 0\,. \quad (139)$$

This expression can be recast in terms of the order parameter in the CM domain, $\mathcal{J} = 2g^2_{\rm YM}\theta > 0$, which was introduced earlier and then analytically continued to the CE branch $\mathcal{J} < 0$ as follows [66],

$$\mathcal{L}_{\rm eff} = \frac{2\mathcal{J}}{16\pi^2}\int_0^\infty \frac{{\rm d}s}{s}\left[e^{-sA[\mathcal{J}]} - e^{-s}\right]\left[\frac{1}{4\sinh^2 s} - \frac{1}{4s^2} + 1\right], \qquad A = \sqrt{\frac{\lambda^4}{\mathcal{J}\tanh(\mathcal{J}/\lambda^4\varepsilon)}}, \quad (140)$$

for $\varepsilon \ll 1$. Indeed, due to

$$\mathcal{J} \tanh\left(\frac{\mathcal{J}}{\varepsilon\lambda^4}\right)\Big|_{\varepsilon\to 0} \to |\mathcal{J}|, \tag{141}$$

the transition between Equation (139) and Equation (140) becomes apparent. Performing the explicit expansion of the rightmost parenthesis above, the form matches that of the one-loop effective Lagrangian in Equation (91) for SU(2).

The results of Appendix D, valid at one-loop order, may now be compared to the all-loop order results. In Figure 16 (left), we show a direct comparison of the one-loop with the all-loop effective Lagrangian for $\mathcal{J} > 0$ (CE branch only). A unique non-trivial minimum is found in the non-perturbative region, $0 < \mathcal{J}^* < \lambda^4$, and it is therefore identified with the CE condensate [66] whose values for one-loop and all-loop cases differ at a permille level and thus express a remarkable consistency of the one-loop approximation. The all-loop running coupling may be straightforwardly extracted from the all-loop effective Lagrangian as

$$(\bar{g}^2)^{-1} = \frac{2}{4\pi^2}\int_0^\infty \frac{\mathrm{d}s}{s}\left[e^{-sA[\mathcal{J}]} - e^{-s}\right]\left[\frac{1}{4\sinh^2 s} - \frac{1}{4s^2} + 1\right], \tag{142}$$

and this is shown in Figure 16 (middle) for the one-loop and all-loop cases. To find an expression for the β-function, one may study the RG equation given by Equation (86). Using

$$\frac{\mathrm{d}\bar{g}^2}{\mathrm{d}\mathcal{J}} = -(\bar{g}^2)^2 \frac{\mathrm{d}}{\mathrm{d}\mathcal{J}}\left(\frac{1}{\bar{g}^2}\right). \tag{143}$$

We may express β as

$$\begin{aligned}\frac{\beta}{\bar{g}^2} &= -2\mathcal{J}\frac{\mathrm{d}}{\mathrm{d}\mathcal{J}}\left(\frac{1}{\bar{g}^2}\right)\\ &= -2\mathcal{J}\frac{2}{4\pi^2}\left(-\frac{\mathrm{d}A}{\mathrm{d}\mathcal{J}}\right)\int_0^\infty \mathrm{d}s\, e^{-sA[\mathcal{J}]}\left[\frac{1}{4\sinh^2 s} - \frac{1}{4s^2} + 1\right],\end{aligned} \tag{144}$$

where

$$\frac{\mathrm{d}A}{\mathrm{d}\mathcal{J}} = -\frac{1}{2\mathcal{J}}\sqrt{\frac{\lambda^4}{\mathcal{J}\tanh(\mathcal{J}/\lambda^4\varepsilon)}}\left(1 + \frac{\mathcal{J}}{\lambda^2\varepsilon}\frac{1-\tanh^2(\mathcal{J}/\lambda^2\varepsilon)}{\tanh(\mathcal{J}/\lambda^2\varepsilon)}\right). \tag{145}$$

The β over the running coupling at one-loop order compared to the corresponding all-loop quantity for SU(2) is displayed in Figure 16 (right). It is clear from the figures that the one-loop approximation accurately captures the main features of the pure YM theory as the differences from the corresponding all-loop quantities are negligible.

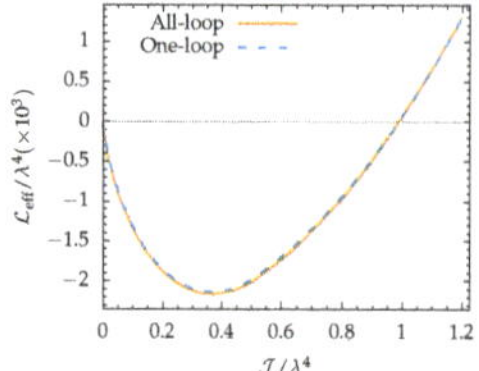

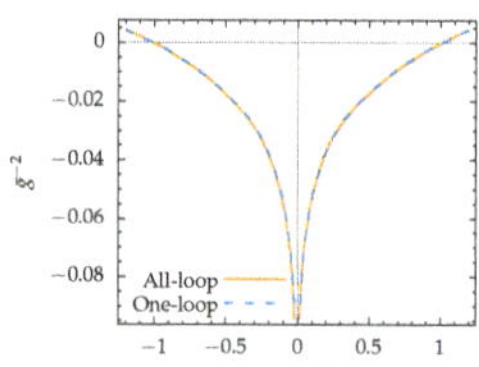

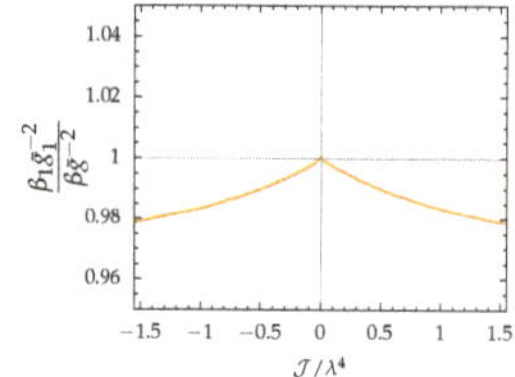

Figure 16. (**Left**) All- and one-loop effective Lagrangian of SU(2) as dependent on $\mathcal{J}/\lambda^4$. Plotted here for the CE branch with $\mathcal{J} > 0$, it is clear that the one-loop result captures the main features of pure YM theory. In the non-perturbative regime shown ($\mathcal{J} < \lambda^4$), the two minima in the vicinity of $\mathcal{J}_* > 0$ coincide. Here, $\epsilon = 0.01$ has been used. (**Middle**) The inverse running coupling $\bar{g}^{-2}$ of SU(2). The (dashed) one-loop result captures well the overall behavior of the running coupling in comparison to the all-loop coupling (solid). Here, $\epsilon = 0.01$ has been used. (**Right**) β over the running coupling at one-loop order compared to the all-loop quantity for SU(2). The ratio of the β-function to the coupling coincides with the all-loop result at one-loop level for small $\mathcal{J}$, while a discrepancy is seen away from zero. Here, $\epsilon = 10^{-5}$ has been used.

As outlined in Ref. [376], when considering the formation of the non-perturbative CE condensate as the YM system rolls down toward the minimum of the effective Lagrangian depicted in Figure 16 (left), the equation that dictates the evolution of this system in cosmological time is found from the continuity equation [376] (see also Ref. [377] for a more recent discussion)

$$\dot{\rho}_{\rm YM} + 3\frac{\dot{a}}{a}\left(\rho_{\rm YM} + p_{\rm YM}\right) = 0\,. \tag{146}$$

Accounting for only homogeneous CE YM fields (i.e., for $\theta = E^2$) in the EMT, for simplicity, the expressions of the energy and pressure densities of the YM system $\rho_{\rm YM}$, $p_{\rm YM}$ can be written as functionals of the effective action [376]:

$$\rho_{\rm YM} = -\mathcal{L}_{\rm eff}(\theta) + 2\theta\mathcal{L}'_{\rm eff}(\theta)\,, \qquad p_{\rm YM} = \mathcal{L}_{\rm eff}(\theta) - \frac{2}{3}\theta\mathcal{L}'_{\rm eff}(\theta)\,, \tag{147}$$

where prime denotes the functional variation with respect to θ. In this case, Equation (146) transforms to

$$\dot{\theta}\left(\mathcal{L}'_{\rm eff} + 2\theta\mathcal{L}''_{\rm eff}\right) + 4\frac{\dot{a}}{a}\theta\mathcal{L}'_{\rm eff} = 0\,, \tag{148}$$

in comoving coordinates. Equation (148) may be integrated, given that $\mathcal{L}_{\rm eff}(\theta)$ is sufficiently well behaved, such that

$$\sqrt{\theta}\mathcal{L}'_{\rm eff}(\theta) = \mathcal{C}\,a^{-2}\,, \tag{149}$$

where $\mathcal{C}$ denotes a coefficient of proportionality that is fixed by the initial conditions. Equation (149) may be solved by inversion, providing the dynamics of a SU(2) YM condensate in cosmological time. Note that Equation (147) is only valid for the dominant electric-field configurations. An analogical analysis of magnetic-field quantum configurations and their cosmological evolution in the effective Lagrangian approach at the one-loop level has been performed in Ref. [377]. Note, generic YM configurations contain both magnetic and electric components. We refer the reader to Section 4.4 above for a thorough discussion of those more generic configurations and their cosmological (real-time) dynamics.

5. Cosmological Implications of Gauge-Fields Driven Inflation

Gauge-fields driven inflation has been considered in several models, starting form the paper by Ford [378], in which a hypercharge cosmological inflationary scenario was envisaged. The theory that was taken into account included, besides the Einstein-Hilbert action, the action for a massive hypercharge field, individuated by the Lagrangian

$$L = \frac{1}{4}F_{\mu\nu}F^{\mu\nu} + V(A^{\alpha}A_{\alpha})\,. \tag{150}$$

Anisotropies that are naturally present in the Maxwell tensor impose to consider a Bianchi type-I metric of the form

$$ds^2 = dt^2 - a^2(t)(dx^2 + dy^2) - b^2(t)dz^2\,. \tag{151}$$

Provided the expression for the energy-momentum tensor of the massive vector field

$$T_{\mu\nu} = F_{\mu\beta}F_{\nu}{}^{\beta} - \frac{1}{4}g_{\mu\nu}F_{\alpha\beta}F^{\alpha\beta} - g_{\mu\nu}V + 2V'A_{\mu}A_{\nu}\,, \tag{152}$$

with prime denoting the functional derivative in $A^{\alpha}A_{\alpha}$, the Einstein equations were recast as

$$2\frac{\dot{a}\dot{b}}{ab} + \frac{\dot{a}^2}{a^2} = 8\pi\epsilon\,, \qquad 2\frac{\ddot{a}}{a} + \frac{\dot{a}^2}{a^2} = -8\pi p_z\,, \tag{153}$$

with ϵ as the energy density and p_z as the pressure density component along the z-axis. The conservation law, derived assuming that the pressure density components along the x and y axes equal $p_x = p_y$, encodes

$$\dot{\epsilon} + \left(2\frac{\dot{a}}{a} + \frac{\dot{b}}{b}\right)\epsilon + 2\frac{\dot{a}}{a}p_x + \frac{\dot{b}}{b}p_z = 0\,, \tag{154}$$

and finally, the only non-vanishing component of the vector field fulfills the equation

$$\ddot{A}_z + \left[2\frac{\dot{a}}{a} - \frac{\dot{b}}{b}\right]\dot{A}_z + 2V'A_z = 0\,. \tag{155}$$

The solutions to this system of equations, Equations (153)–(155), immediately outlined the impossibility to suppress anisotropies over the dynamical evolution of the background and the vector field. Thus, already with the seminal attempt of Ref. [378], it was evident that the over-production of anisotropies at the cosmological level would have provided a main issue, eventually ruling out this class of models.

A possible way to overcome the abundance of anisotropies, severely constrained by the CMB observations by WMAP and Planck, was imagined by Golovnev, Mukhanov and Vanchurin [379], who considered a stochastic distribution of $N \simeq 10^{12}$ vector fields, randomly spanning the space directions. Each vector field was assumed to be massive and regulated by the action

$$S = \int d^4x\sqrt{-g}\left(-\frac{R}{16\pi} - \frac{1}{4}F_{\mu\nu}F^{\mu\nu} + \frac{1}{2}(m^2 + \frac{R}{6})A_{\mu}A^{\mu}\right), \tag{156}$$

where m denotes the mass of the hypercharge vector field considered, such that $F_{\mu\nu} = \nabla_{\mu}A_{\nu} - \nabla_{\nu}A_{\mu} = \partial_{\mu}A_{\nu} - \partial_{\nu}A_{\mu}$.

The equations of motion for the gauge field, which in a covariant fashion recast

$$\frac{1}{\sqrt{-g}}\frac{\partial}{\partial x^{\mu}}(\sqrt{-g}F^{\mu\nu}) + \left(m^2 + \frac{R}{6}\right)A^{\nu} = 0\,, \tag{157}$$

split into a temporal "0" component, implying $A_0 = 0$, and into space components, which finally provide the equation fro the "field strength" $B_i \equiv A_i/a$, i.e.,

$$\ddot{B}_i + 3H\dot{B}_i + m^2B_i = 0\,, \tag{158}$$

having introduced comoving coordinates such that $ds^2 = dt^2 - a^2(t)d\vec{x}^2$, and these were denoted with dot derivative with respect to the cosmological time, so that $H = \dot{a}/a$. For a homogeneous vector field, it was then noticed that the energy-momentum tensor can be expressed by

$$\begin{aligned} T^0_0 &= \frac{1}{2}\left(\dot{B}^2_k + m^2 B^2_k\right), \\ T^i_j &= \left[-\frac{5}{6}\left(\dot{B}^2_k - m^2 B^2_k\right) - \frac{2}{3}H\dot{B}_k B_k - \frac{1}{3}(\dot{H} + 3H^2)B^2_k\right]\delta^i_j + \dot{B}^i\dot{B}_j + H(\dot{B}^i B_j \\ &+ \dot{B}_j B^i) + (\dot{H} + 3H^2 - m^2)B^i B_j, \end{aligned} \tag{159}$$

where the summation over the index k is meant. Summing up now the contributions of a triplet of mutually orthogonal fields $B_i^{(a)}$ with the same magnitude $|B|$, the total energy momentum tensor is averaged to the quantity

$$T^0_0 = \epsilon = \frac{3}{2}\left(\dot{B}^2_k - m^2 B^2_k\right), \qquad T^i_j = -p\delta^i_j = -\frac{3}{2}\left(\dot{B}^2_k - m^2 B^2_k\right)\delta^i_j, \tag{160}$$

where B_k satisfies

$$\ddot{B}_i + 3H\dot{B}_i + m^2 B_i = 0. \tag{161}$$

These latter relations for the energy-momentum tensor can be proved by the fact that for $B_i^{(a)}$, it holds:

$$\sum_i B_i^{(a)} B_{(b)\,i} = |B|^2\delta^{(a)}_{(b)} \to \sum_a B_j^{(a)} B^{(a)\,i} = |B|^2\delta^i_j. \tag{162}$$

Summing up over a large amount of N triple of fields, the energy-momentum tensor finally acquires the expression

$$T^0_0 = \epsilon \simeq \frac{N}{2}\left(\dot{B}^2_k + m^2 B^2_k\right) \qquad T_{ij} \propto \sum_{a=1}^{N} B_i^{(a)} B_j^{(a)} \simeq \frac{N}{3}B^2\delta^i_j + O(1)\sqrt{N}B^2. \tag{163}$$

Within this scenario, anisotropies are then proven to fall off as $1/\sqrt{N}$, hence justifying consistency with the experimental data.

Finally, accounting for an inflationary slow-roll phase, with $\dot{B}_i \simeq 0$, one finally finds for the first Friedmann equation,

$$H^2 = \frac{8\pi}{3}\epsilon \simeq \frac{4\pi}{3}Nm^2B^2. \tag{164}$$

A different perspective was suggested by Maleknejad and Sheikh-Jabbari, who imagined in Refs. [380–382] that it could become relevant at the level of cosmic perturbations, while at the same time driving the inflationary background evolution of the universe. Isotropy could be guaranteed here by aligning the internal indices (of the adjoint representation) of a SU(2) subgroup of SU(N) to the space directions, namely, for the space component of the connection A

$$A^a_i = \psi(t)e^a_i = \psi(t)a(t)\delta^a_i, \qquad e^a = a(t)\delta^a_i, \tag{165}$$

where $\psi(t)$ is a scalar field, e^a_i denotes the triads, which recombine into the space metric $h_{ij} = e^a_i e^a_j$ and are expressed here in the comoving coordinates. For compactness of notation, the authors reshuffled $\phi(t) = \psi(t)a(t)$.

Specifically, the authors considered a YM action provided with the square of a Pontriagin term, i.e.,

$$S = \int d^4x\sqrt{-g}\left[-\frac{R}{2} - \frac{1}{4}F^a_{\mu\nu}F_a^{\mu\nu} + \frac{\kappa}{384}\left(\varepsilon^{\mu\nu\lambda\sigma}F^a_{\mu\nu}F^a_{\lambda\sigma}\right)^2\right], \tag{166}$$

having set $8\pi G = 1$, denoted with $\varepsilon^{\mu\nu\lambda\sigma}$ the totally antisymmetric tensor, and chosen the real parameter κ to be positive.

Labeling with a subscript "YM" the contributions arising from the YM terms, and with κ the ones related to the Pontriagin terms, it was found that

$$\epsilon_{\rm YM} = \frac{3}{2}\left(\frac{\dot\phi^2}{a^2} + \frac{g^2\phi^4}{a^4}\right), \qquad \epsilon_\kappa = \frac{3}{2}\frac{\kappa g^2\phi^4\dot\phi^2}{a^6}, \tag{167}$$

where g is the YM coupling constant entering the definition of the field strength of the SU(2) connection

$$F^a_{\mu\nu} = \partial_\mu A^a_\nu - \partial_\nu A^a_\mu - g\epsilon^{abc}\, A^b_\mu A^c_\nu, \tag{168}$$

and where $\epsilon = \epsilon_{\rm YM} + \epsilon_\kappa$ and $p = \frac{1}{3}\epsilon_{\rm YM} - \epsilon_\kappa$. Having provided these definitions, it was possible to show the emergence of a slow-roll phase of inflation, which was dominated by the κ-term contribution ϵ_κ over the YM contribution $\epsilon_{\rm YM}$, namely $\epsilon_\kappa \gg \epsilon_{\rm YM}$. Furthermore, within the scenario introduced in Refs. [380–382], it was conceivable that sizable effects could be turned on, so as to source primordial magnetic fields at the cosmic perturbation level, with specific possible observational features on the CMB and primordial cosmic magnetic fields outlined.

A novel framework with respect to the ones so far discussed was elaborated by Alexander, Marcianò and Spergel in Ref. [383], in which the authors considered the interaction a model of inflation with as a new ingredient the interaction of an Abelian gauge field with a fermionic charge. This scenario then dramatically differs from only adopting gauge fields in order to generate a realistic inflationary epoch. As a by-product of this approach, researchers considered the possibility of generating the a net-lepton asymmetry. The Sakharov conditions are realized in the model presented in Ref. [383] during the inflationary epoch, due to a dynamical inter-change of the gauge field fluctuations into the lepton asymmetry of the universe.

The action the authors moved from involved, as well as a U(1) hypercharge field A_μ, also a massive scalar field θ, interacting with A_μ through a Chern-Simons term, i.e.,

$$\begin{aligned} S &= S_D + \int d^4x\sqrt{-g}\left[\frac{M^2_{\rm PL}R}{8\pi} - \frac{1}{2}\partial_\mu\theta\partial^\mu\theta - \frac{1}{4}F_{\alpha\beta}F^{\alpha\beta} + \frac{\theta}{4M_*}F_{\alpha\beta}\tilde F^{\alpha\beta}\right], \\ S_D &= \int d^4x\sqrt{-g}\left(\imath\bar\psi\gamma^\mu\nabla_\mu\psi + cc. + M\bar\psi\psi + q\bar\psi\gamma^\mu\psi A_\mu\right), \end{aligned} \tag{169}$$

where $M_{\rm PL}$ denotes the Planck mass, M_* is the mass-scale of the pseudo-scalar decay constant, regulating with *theta* the CP-violating Chern-Simons such as term, $F_{\mu\nu} = \partial_{[\mu}A_{\nu]}$ is the field strength of the U(1) connection A, $\tilde F^{\alpha\beta} = \varepsilon^{\alpha\beta\lambda\sigma}\tilde F_{\lambda\sigma}$ is the gravitational Hodge-dual, and $\gamma^\mu = e^\mu_I\gamma^I$, with γ^I Dirac matrices, e^μ_I inverse tetrad and internal indices $I = 0,\dots 3$.

The model, which is then based on the interaction between a homogeneous and isotropic configuration of a U(1) gauge field and a fermionic charge density $\mathcal{J}_0$, relies on the regulated fermionic charge density as generated from a Bunch-Davies vacuum state, using the procedure outlined by Koksma and Prokopec in Ref. [384]. In conformal coordinates, this was found to redshift as $1/a(\eta)$. Then, within the scenario of Ref. [383], the time-like component of the hypercharge gauge field was found to be sourced by the fermionic charge, consistently with a growth in the gauge field proportional to the scale factor, namely,

$$A_0(\eta) \sim a(\eta)\,. \tag{170}$$

This motivates results with inflation dominated by the energy density stored within the interaction among the gauge field and the fermionic charge, namely $A_0\mathcal{J}_0$, which is approximately constant over the inflationary epoch. The appealing feature of Ref. [383] stands in the possibility to obtain an epoch of cosmic inflation involving the physical description fields already existing in nature, specifically the time-like U(1) gauge field interacting with a fermionic charge density. Nonetheless, the role of the scalar field cannot be underestimated, retaining a certain relevance in producing baryogenesis, and providing a graceful exit from inflation. Indeed, the mechanism that accounted for the graceful exit is strictly interconnected to the one advocated for reproducing the baryogenesis, with the right baryon asymmetry index. The Chern-Simons term, through the coupling to the pseudo-scalar field, converts gauge field fluctuations into lepton number, while the rapid oscillation of the pseudo-scalar field near its minimum allows achieving thermalization of the gauge field and thus to end inflation. The relevance of the coupling between scalar modes, there interpreted as axions, was further investigated in Ref. [385].

An improvement of the scheme first addressed in Ref. [383] was provided in Ref. [386], where the authors analyzed the consistency of the model via the Stückelberg mechanism [387]. This provided an incorporation of the longitudinal scalar DoFs into the hypercharge field, which is now massive. This could be thought again as a further step-forward, toward the realization of an inflationary mechanism relying on the YM dynamics, the hypercharge sector being eventually recognized as an Abelian subgroup of the SU(N) gauge sector.

The action of the theory was then considered to be

$$S = \int d^4x\sqrt{-g}\left[\frac{M_{\rm PL}^2 R}{8\pi} - \frac{1}{4}G_{\alpha\beta}G^{\alpha\beta} - \frac{1}{2}m^2 C_\mu C^\mu + C_\mu \mathcal{J}^\mu + L_{\rm D}\right]$$
$$L_{\rm D} = -\imath\bar{\psi}\gamma^\mu\nabla_\mu\psi + c.c. + M\bar{\psi}\psi\,, \tag{171}$$

having introduced the massive Stückelberg field $C_\mu = A_\mu - \frac{1}{m}\partial_\mu\theta$ and its field strength $G_{\mu\nu} = \partial_{[\mu}C_{\nu]} = \partial_{[\mu}A_{\nu]} = F_{\mu\nu}$, and the fermionic vector current $\mathcal{J}^\mu = q\bar{\psi}\gamma^\mu\psi$. Gauge invariance is ensured in this framework by the transformations

$$A_\mu \to A'_\mu = A_\mu + \partial_\mu\Lambda\,, \qquad \theta \to \theta' = \theta + m\Lambda\,. \tag{172}$$

Within this framework, the authors could prove the existence and the stability of dynamical attractor solutions for the cosmological inflation epoch, which is again driven by the coupling among the fermions and a (massive) gauge field. Numerical analyses then showed that stability is attained for a large basin of the initial conditions, making this inflationary scenario almost independent on these latter: inflation arises without fine tuning and without the need of postulating any effective potential or any non-standard coupling.

An alternative scenario featuring a coupling between an axion-like field and an SU(2) gauge field is known as the chromo-natural inflation [388]. The rotationally invariant homogeneous condensate of the gauge field satisfies an attractor solution that enables it to drive cosmic inflation for the axion decay constant having a natural value at a sub-Planckian scale. Interestingly enough, this scenario features a possibility for termination of inflation as soon as the axion potential vanishes, simultaneously providing a small tensor-to-scalar perturbation ratio.

An inflationary scenario, taking into account non-trivial topological features deployed in Refs. [389–391], was extended in Ref. [392] so as to account, over the universe expansion, for a sector of a strongly coupled QCD-like gauge theory. The idea at the base of investigations in Refs. [389–391] is to perform a periodic (between the Σ and Σ' surfaces) path integral over Euclidean geometries,

$$e^{-\Gamma} = \int_{g,\phi|_\Sigma = g,\phi_{\Sigma'}} D[g,\phi]e^{-S_E g,\phi}\,, \tag{173}$$

with g and ϕ, respectively, representing the metric and matter field and S_E denoting the Euclidean action, so as to extract the 'density matrix' of the universe,

$$\rho[\varphi, \varphi'] = e^{\Gamma} \int_{g,\phi|_{\Sigma,\Sigma'}} D[g, \phi] e^{-S_E[g,\phi]} \,, \tag{174}$$

which describes a microcanonical ensemble, φ denoting field configurations that encode both gravitational and matter variables.

The uniform distribution over the Euclidean spaces actually corresponds, over Lorentzian spaces, to a distribution that is peaked about complex saddle points of the path integral. The latter can be then represented by cosmological instantons, entailing a bounded range values for the cosmological constant.

On the other hand, inflationary cosmologies can be engendered by the very same instantonic solutions [389–391]. The low energy of the accelerated expansion can be then attained at its late stage, resorting to the dynamical evolution of extra dimensions specifically postulated in string theory framework [391]. This results in a bounded range for the very early (inflationary) cosmological constant, which provides a constraint on the available landscape of the string vacua. Finally, the same mechanism can be advocated to give rise to a possible DE candidate, accounting for the quasi-equilibrium decay of the microcanonical state of the universe. Within this scenario, Barvinsky and Zhitnitsky promoted a new picture for the emergence of an inflationary spacetime [392], resorting to considerations developed in Refs. [14,393,394] on the generation, in a strongly coupled QCD-like gauge theory, of the vacuum energy from non-trivial topological features.

The limits of the usual semi-classical expansion were overcome by the dominant contribution of the numerous conformal modes. Integration over the modes then provides the quantum effective action of the conformal field theory $\Gamma_{\rm CFT}[g_{\mu\nu}]$, which can be calculated with methods similar to those ones implemented in determining the conformal anomaly. Starting from the FLRW background, accounting for a periodic factor $a(\tau)$—this is due to the fact that functions of the Euclidean time are supported to the circle $\mathbb{S}^1$—and finally using a local conformal transformation to the static Einstein universe and the very same well-known trace anomaly, one finds

$$g_{\mu\nu} = \frac{\delta\Gamma_{\rm CFT}}{\delta g_{\mu\nu}} = \frac{1}{4(4\pi)^2} g^{1/2} \left(\alpha\Box R + \beta E\gamma C^2_{\mu\nu\rho\sigma}\right), \tag{175}$$

where we introduced the Gauss-Bonnet term $E = R^2_{\mu\nu\alpha\gamma} - 4R^2_{\mu\nu} + R^2$ and the Weyl tensor $C_{\mu\nu\rho\sigma}$.

Considering a spacetime with topology $\mathbb{S}^3 \times \mathbb{S}^1$, and moving from the expression of the energy of the gauge field holonomy, winding across the compactified coordinate of the length $\mathcal{T}$, Barvinsky and Zhitnitsky found that

$$\rho = \rho_{\rm vac}[\mathbb{S}^3 \times \mathbb{S}^1] - \rho_{\rm vac}[\mathbb{R}^4] = \frac{\bar{c}_{\mathcal{T}} \Lambda^3_{\overline{QCD}}}{\mathcal{T}} \,, \tag{176}$$

with $\Lambda_{\overline{QCD}}$ being the scale of the QCD-like gauge theory, $\bar{c}_{\mathcal{T}}$ being a dimensionless constant of order one, and similarly, the full period of the proper Euclidean time on these periodic m-fold garland instantons is given by the analogous integral,

$$\mathcal{T} = \oint_{\mathbb{S}_1} d\tau \,, \tag{177}$$

which in an FLRW metric background reduces to the $2m$-multiple result

$$\mathcal{T} = 2m \int_{a_-}^{a^+} \frac{da}{a} \,, \tag{178}$$

where the integral is between the two neighboring turning points of $a(\tau)$ such that $\dot{a}(\tau_{\pm}) = 0$.

6. Summary

In this review, we have made a brief outlook of the current status of confined and de-confined QCD dynamics in the early universe as well as the key methodology for studies QCD in the strongly coupled regime relevant for cosmological evolution. The covered research areas are broadly inter-disciplinary, and our discussion may not be fully exhaustive. Still, we have identified a few quite unexpected and intriguing connections between currently pursued research in particle physics and possible dynamics of the early universe. Such fundamental questions as the gauge-fields-driven inflation, cyclic universe, particle production mechanisms, non-perturbative real-time dynamics of the QCD ground state, a rather challenging problem of dynamical generation of cosmological DE and DM, the structure of the QCD vacuum, the QCD phase transitions and the role of QCD matter in late-time universe evolution are among the key points of this review. Such a wide breadth of topics, with deep roots into QCD or, more generically, quantum YM field theories, exhibits enormous and critical significance of microscopic dynamics of particle physics and confined field theories for understanding of the macroscopic cosmic evolution. The picture is far from its final shape though, and many more pillars of such connections and possible interplay are yet to be established. We believe that our review can be useful for both young researchers and for more senior experts specialized in both particle physics and cosmology research areas.

Author Contributions: Conceptualization, A.A., A.M., T.L., R.P. and M.Š.; validation, R.P., T.L. and M.Š.; formal analysis, R.P., T.L. and M.Š.; investigation, R.P., T.L. and M.Š.; resources, R.P. and M.Š.; writing—original draft preparation, A.M., T.L., R.P. and M.Š.; writing—review and editing, A.A., A.M., T.L., R.P. and M.Š.; supervision, R.P.; project administration, R.P.; funding acquisition, R.P. and M.Š. All authors have read and agreed to the published version of the manuscript.

Funding: Work of A.A. is supported by the Talent Scientific Research Program of College of Physics, Sichuan University, Grant No.1082204112427 & the Fostering Program in Disciplines Possessing Novel Features for Natural Science of Sichuan University, Grant No. 2020SCUNL209 and 1000 Talent program of Sichuan province 2021. A.M. wishes to acknowledge support by the Shanghai Municipality, through the grant No. KBH1512299, by Fudan University, through the grant No. JJH1512105, the Natural Science Foundation of China, through the grant No. 11875113, and by the Department of Physics at Fudan University, through the grant No. IDH1512092/001. R.P. is supported in part by the Swedish Research Council grant, contract number 2016-05996, as well as by the European Research Council (ERC) under the European Union's Horizon 2020 research and innovation programme (grant agreement No 668679). M.Š. is partially supported by the grants LTT17018 and LTT18002 of the Ministry of Education of the Czech Republic.

Institutional Review Board Statement: Not applicable.

Informed Consent Statement: Not applicable.

Data Availability Statement: Not applicable.

Conflicts of Interest: The authors declare no conflict of interest. The funders had no role in the design of the study; in the collection, analyses, or interpretation of data; in the writing of the manuscript, or in the decision to publish the results.

Appendix A. Elements of Relativistic Hydrodynamics

Equations of relativistic hydrodynamics are based on conservation of the energy-momentum and the current

$$\partial_{\mu} T^{\mu\nu} = 0, \quad \partial_{\mu} j_i^{\mu} = 0, \tag{A1}$$

where $j_i^{\mu}, i = B, Q, L, \ldots$ denotes the conserved currents corresponding to baryon number B, electric charge Q, lepton number L, etc. Both $T^{\mu\nu}$ and j_i^{μ} can be decomposed into time-like and space-like components using natural projection operators, the local flow four-velocity u^μ, and the second-rank tensor perpendicular to it $\Delta^{\mu\nu} = g^{\mu\nu} - u^\mu u^\nu$ [11,25,395,396]:

$$T^{\mu\nu} = \epsilon u^\mu u^\nu - p\Delta^{\mu\nu} + W^\mu u^\nu + W^\nu u^\mu + \pi^{\mu\nu}, \tag{A2}$$

$$j_i^\mu = n_i u^\mu + V_i^\mu, \tag{A3}$$

where $\epsilon = u_\mu T^{\mu\nu} u_\nu$ is the energy density, $p \equiv p_s + \Pi = -\frac{1}{3}\Delta_{\mu\nu}T^{\mu\nu}$ is the total (hydrostatic p_s + bulk Π) pressure, $W^\mu = \Delta^\mu_\alpha T^{\alpha\beta} u_\beta$ is the energy (or heat) current, $n_i = u_\mu j_i^\mu$ is the charge density, $V_i^\mu = \Delta^\mu{}_\nu j_i^\nu$ is the charge current, and $\pi^{\mu\nu} = \langle T^{\mu\nu}\rangle$ is the shear stress tensor. The angular brackets in the definition of the shear stress tensor $\pi^{\mu\nu}$ stand for the following operation:

$$\langle A^{\mu\nu}\rangle = \left[\frac{1}{2}(\Delta^\mu{}_\alpha \Delta^\nu{}_\beta + \Delta^\mu{}_\beta \Delta^\nu{}_\alpha) - \frac{1}{3}\Delta^{\mu\nu}\Delta_{\alpha\beta}\right] A^{\alpha\beta}. \tag{A4}$$

To further simplify our discussion, we restrict ourselves in the following to only the one conserved charge, the baryon number B, and denote the corresponding baryon current as $j^\mu \equiv j_B^\mu$. The various terms appearing in the decompositions (A2) and (A3) can then be grouped into ideal and dissipative parts as follows

$$T^{\mu\nu} = T_{id}^{\mu\nu} + T_{dis}^{\mu\nu} = [\epsilon u^\mu u^\nu - p_s \Delta^{\mu\nu}]_{id} + [-\Pi\Delta^{\mu\nu} + W^\mu u^\nu + W^\nu u^\mu + \pi^{\mu\nu}]_{dis} \tag{A5}$$

$$j^\mu = j_{id}^\mu + N_{dis}^\mu = [nu^\mu]_{id} + [V^\mu]_{dis}. \tag{A6}$$

Neglecting the dissipative parts, the energy-momentum conservation and the current conservation (A1) define ideal hydrodynamics. In this case (and for a single conserved charge), a solution of the hydrodynamical Equation (A1) for a given initial condition describes the spacetime evolution of the six variables—three state variables $\epsilon(x)$, $p(x)$, $n(x)$, and three space components of the flow velocity u^μ. However, since (A1) constitutes only five independent equations, the sixth equation relating p and ϵ, the EoS $p(\epsilon)$, has to be added by hand in order to solve them.

Two definitions of flow can be found in the literature, see e.g., Refs. [11,395]; one related to the flow of conserved charge (Eckart):

$$u_E^\mu = \frac{j^\mu}{\sqrt{j_\nu j^\nu}}, \tag{A7}$$

the other related to the flow of energy (Landau):

$$u_L^\mu = \frac{T^\mu{}_\nu u_L^\nu}{\sqrt{u_L^\alpha T_\alpha{}^\beta T_{\beta\gamma} u_L^\gamma}} = \frac{1}{e} T^\mu{}_\nu u_L^\nu. \tag{A8}$$

Let us note that $W^\mu = 0$ ($V^\mu = 0$) in the Landau (Eckart) frame. In the case of vanishing dissipative currents, both definitions represent a common flow. The Landau definition is more suitable when describing the evolution of matter at zero chemical potential, such as in the case of the mid-rapidity particle production in ultra-relativistic HIC at the LHC and at the top RHIC energy, or in the early universe. In this case, all momentum density is due to the flow of energy density, $u_\mu T_{id}^{\mu\nu} = \epsilon u^\nu$ and $u_\mu T_{dis}^{\mu\nu} = 0$, i.e., the heat conduction effects can be neglected.

Appendix B. Hydrodynamical Description of Dissipative Effects

In its modern formulation, relativistic fluid dynamics provides an effective description of a system that is in local thermal equilibrium, and it can be derived from the underlying kinetic description through Taylor expansion of the entropy four-current $S^\mu = su^\mu$ in gradients of the local thermodynamic variables [25]. In zeroth order in gradients, one obtains ideal fluid dynamics

$$\partial_\mu S^\mu = \partial_\mu (s u^\mu) = u^\mu \partial_\mu s + s \partial_\mu u^\mu = 0\,, \tag{A9}$$

and the evolution of the scale factor of the universe is driven solely by the entropy conservation $s(t)a^3(t) = \text{const}$. The higher orders describe effects due to irreversible thermodynamic processes such as the frictional energy dissipation between the fluid elements that are in relative motion or their heat exchange with its surroundings on its way to approach thermal equilibrium with the whole fluid.

When solving the hydrodynamic equations with the dissipative terms, it is customary to introduce the following two phenomenological definitions (so-called *constitutive equations*) for the shear stress tensor $\pi^{\mu\nu}$ and the bulk pressure Π appearing in Equation (A5) [395],

$$\pi^{\mu\nu} = 2\eta \langle \nabla^\mu u^\nu \rangle\,, \quad \Pi = -\zeta \partial_\mu u^\mu = -\zeta \nabla_\mu u^\mu\,, \tag{A10}$$

where the angular brackets $\langle \ldots \rangle$ are defined in Equation (A4) and $\nabla^\mu = \Delta^{\mu\nu}\partial_\nu$. Neglecting the charge current V_μ in Equation (A6), the first-order expansion of S^μ is completely determined by the *shear viscosity* η and *bulk viscosity* ζ coefficients [395]:

$$T\partial_\mu S^\mu = \pi_{\mu\nu}\langle \nabla^\mu u^\nu \rangle - \Pi \partial_\mu u^\mu = \frac{\pi_{\mu\nu}\pi^{\mu\nu}}{\eta} + \frac{\Pi^2}{\zeta} = 2\eta \langle \nabla^\mu u^\nu \rangle^2 + \zeta(-\partial_\mu u^\mu)^2 \geq 0\,. \tag{A11}$$

A well-known example of the flow involving both coefficients η and ζ is provided by boost-invariant one-dimensional expansion with the velocity in the z direction, v_z, proportional to z co-ordinate [397]

$$u^\mu_{\rm BJ} = \frac{x^\mu}{\tau} = \frac{t}{\tau}\left(1, 0, 0, \frac{z}{t}\right), \quad \tau = \sqrt{t^2 - z^2}\,. \tag{A12}$$

After inserting this solution into the constitutive Equation (A10), we arrive at the equation of motion [395]:

$$\frac{d\epsilon}{d\tau} = -\frac{\epsilon + p_s}{\tau}\left(1 - \frac{4}{3\tau T}\frac{\eta}{s} - \frac{1}{\tau T}\frac{\zeta}{s}\right). \tag{A13}$$

The last two terms on the right-hand side in Equation (A13) describe a compression of the energy density due to viscous corrections. Two dimensionless coefficients in the viscous correction, η/s and ζ/s, where s is the entropy density, reflect the intrinsic properties of the fluids. It is worth mentioning that neglecting η and ζ in Equation (A13), i.e., for the ideal fluid EoS with $p_s = \frac{1}{3}\epsilon$, one obtains the celebrated Bjorken solution of ideal hydrodynamics [397]

$$\frac{d\epsilon}{d\tau} = -\frac{\epsilon + p_s}{\tau} = -\frac{4}{3}\frac{\epsilon}{\tau} \qquad \Rightarrow \qquad \epsilon = \tau^{-4/3}\,, \tag{A14}$$

frequently used to discuss the salient features of the ultra-relativistic HICs.

The one-dimensional character of the Bjorken flow (A12) makes it possible to replace the z co-ordinate with the radial one r. Radial flow in the transverse direction, i.e., when $r_\perp = \sqrt{x^2 + y^2}$, was studied in Ref. [398]. For this case and the constant sound velocity c_s, analytic solutions of relativistic viscous hydrodynamics describing expanding fireballs were developed in Ref. [399]. In a three-dimensional case, a new class of exact fireball solutions of relativistic dissipative hydrodynamics for arbitrary shear and bulk viscosities, as well as for other dissipative coefficients, was studied in Ref. [400]. The common property of these solutions is the presence of the relativistic Hubble flow.

However, the analogy between the solutions describing HICs and expansion of the early universe must not be pushed too far, since in the latter case form of the energy-momentum tensor $T_{\mu\nu}$ and particle four-current j_μ of the matter, cf. Equations (A5) and (A6) is strongly constrained by the symmetries of the FLRW metric (38). In particular, due to the local momentum isotropy, the term $\langle\nabla^\mu u^\nu\rangle$ appearing in the viscous shear-stress tensor $\pi^{\mu\nu}$, cf. Equation (A10), vanishes [401]. Consequently, the term proportional to η in Equation (A13) disappears. The shear viscosity η also disappears in the theories with scalar perturbations of the metric tensor $g_{\mu\nu}$ [84,270]. There, the fluctuations of energy density destroy the homogeneity but not the isotropy of the early universe FLRW metrics. Hence, in the following discussion, we will consider mainly the bulk viscosity—the property of expanding matter arising typically in mixtures. They can be either of different species, as in a radiative fluid, or of the same species but with different energies, as in a Maxwell–Boltzmann gas. In each of these instances, the bulk viscosity provides the internal 'friction' that sets in due to the different cooling rates in the expanding mixture [401].

The relativistic Navier-Stokes description given by Equation (A10) accounts only for terms that are *linear* in velocity gradient. This leads, unfortunately, to severe problems. In particular, when the thermodynamic force $\langle\nabla^\mu u^\nu\rangle$ or $\nabla_\mu u^\mu$ is suddenly switched off/on, the corresponding thermodynamic flux $\pi^{\mu\nu}$ or Π which is a purely local function of the velocity gradient also instantaneously vanishes/appears [402]. The linear proportionality between dissipative fluxes and forces causes an instantaneous (acausal) influence on the dissipative currents, leading to numerical instabilities [403]. The solution of this problem requires the inclusion of terms that are second order in gradients [404]. The resulting equations for the dissipative fluxes $\pi^{\mu\nu}$ and Π then become the relaxation-type equations [395,405]. The latter encode the time delay between the appearance of thermodynamic gradients that drive the system out of local equilibrium and the associated build-up of dissipative flows in response to these gradients, thereby restoring causality [405]. Accounting for non-zero relaxation times at all stages of the evolution constrains departures from local equilibrium, thereby both stabilizing the theory and improving its quantitative precision.

Let us provide a few examples of this approach. The 2nd-order theory version of boost-invariant one-dimensional flow, cf. Equation (A13), can be found in Ref. [395]. Due to its length, we do not reproduce it here and refer the interested reader to the original publication. The second example can be found in Ref. [406], where the relaxation time τ_π proportional to the shear viscosity parameter η was used to study the evolution of the universe filled with QGP with nonzero shear viscosity. The authors argue that in general relativity, the following modification of the shear-stress tensor

$$\pi^{\mu\nu} \to \pi^{\mu\nu} + \tau_\pi\left[u^\alpha \pi^{\mu\nu}_{;\alpha} + \frac{4}{3}\pi^{\mu\nu}\nabla_\alpha u^\alpha\right] = 2\eta\langle\nabla^\mu u^\nu\rangle\,, \qquad \pi^{\mu\nu}_{;\alpha} \equiv \partial_\alpha \pi^{\mu\nu} + \Gamma^\mu_{\alpha\beta}\pi^{\beta\nu} + \Gamma^\nu_{\alpha\beta}\pi^{\mu\beta}\,, \tag{A15}$$

where $\pi^{\mu\nu}_{;\alpha}$ is a covariant derivative of $\pi^{\mu\nu}$ and $\Gamma^\mu_{\alpha\beta}$ are the Christoffel symbols, makes the resulting Navier-Stokes equations causal. Using the FLRW metric and taking into account that the compatibility with the isotropy and homogeneity of the universe demands $\pi^{\mu\nu}$ to be diagonal, the solution of Equation (A15) reads [406]

$$\pi^{00}(t) = \pi^{00}(t_0)\left[\frac{a(t_0)}{a(t)}\right]^4 e^{-\frac{t-t_0}{\tau_\pi}}\,, \qquad \pi^{ij}(t) = \pi^{ij}(t_0)\left[\frac{a(t_0)}{a(t)}\right]^6 e^{-\frac{t-t_0}{\tau_\pi}}\delta_{ij}\,. \tag{A16}$$

In the Friedmann equations, the effect of the traceless viscosity tensor shows up in the modification of the initial energy density $\epsilon(t_0)$ and in the behavior of the energy density at times $t \lesssim t_0 + \tau_\pi$, which at later times goes over to the standard expression [406]

$$\epsilon(t) = \left[\epsilon(t_0) + \pi^{00}(t_0)\right]\left[\frac{a(t_0)}{a(t)}\right]^4 - \pi^{00}(t_0)\left[\frac{a(t_0)}{a(t)}\right]^4 e^{-\frac{t-t_0}{\tau_\pi}}\,. \tag{A17}$$

The third example is provided in Ref. [407] where causal general relativistic viscous fluid theory with the inclusion of all dissipative contributions (shear viscosity η, bulk viscosity ζ, and heat flow W^μ) and the effects from nonzero baryon number are discussed. According to the authors, the applicability of this theory ranges from the modeling of viscous effects in neutron star mergers to low-energy HICs.

Appendix C. YM Equations of Motion in the Effective Action Approach

This section provides a short summary on the derivation of the YM equations of motion by means of variational methods with respect to the connections $\mathcal{A}^a_\mu$ when applied to the effective Lagrangian of Equation (84), as was performed in Ref. [66].

When varying the effective action with respect to $\mathcal{A}^a_\mu$ and $\partial_\nu \mathcal{A}^a_\mu$, one arrives at the Euler-Lagrange equations of motion:

$$\frac{\partial \mathcal{L}_{\rm eff}}{\partial \mathcal{A}^a_\mu} - \nabla_\nu \frac{\partial \mathcal{L}_{\rm eff}}{\partial(\partial_\nu \mathcal{A}^a_\mu)} = 0, \qquad \nabla_\nu \frac{\partial \mathcal{L}_{\rm eff}}{\partial(\partial_\nu \mathcal{A}^a_\mu)} = \frac{1}{\sqrt{-g}}\partial_\nu \left[\sqrt{-g}\frac{\partial \mathcal{L}_{\rm eff}}{\partial(\partial_\nu \mathcal{A}^a_\mu)}\right]. \tag{A18}$$

It is straightforward to compute the derivatives of the effective Lagrangian:

$$\frac{\partial \mathcal{L}_{\rm eff}}{\partial \mathcal{A}^a_\mu} = \frac{1}{4\bar{g}^2}\left[\frac{\partial \mathcal{J}}{\partial \mathcal{A}^a_\mu} - \frac{\mathcal{J}}{\bar{g}^2}\frac{\partial \bar{g}^2}{\partial \mathcal{A}^a_\mu}\right], \qquad \frac{\partial \mathcal{J}}{\partial \mathcal{A}^a_\mu} = \frac{4 f^{abc}\mathcal{F}^{b\,\mu\nu}\mathcal{A}^c_\nu}{\sqrt{-g}}, \qquad \frac{\mathcal{J}}{\bar{g}^2}\frac{\partial \bar{g}^2}{\partial \mathcal{A}^a_\mu} = \frac{\mathcal{J}}{\bar{g}^2}\frac{\partial \bar{g}^2}{\partial \mathcal{J}}\frac{\partial \mathcal{J}}{\partial \mathcal{A}^a_\mu}, \tag{A19}$$

$$\Rightarrow \quad \frac{\partial \mathcal{L}_{\rm eff}}{\partial \mathcal{A}^a_\mu} = \frac{1}{\bar{g}^2}\frac{f^{abc}\mathcal{F}^{b\,\mu\nu}\mathcal{A}^c_\nu}{\sqrt{-g}}\left[1-\frac{\beta}{2}\right], \tag{A20}$$

$$\frac{\partial \mathcal{L}_{\rm eff}}{\partial(\partial_\nu \mathcal{A}^a_\mu)} = \frac{4\mathcal{F}^{a\,\mu\nu}}{\sqrt{-g}}, \tag{A21}$$

$$\Rightarrow \quad \nabla_\nu \frac{\partial \mathcal{L}_{\rm eff}}{\partial(\partial_\nu \mathcal{A}^a_\mu)} = \frac{1}{\sqrt{-g}}\partial_\nu\left(\sqrt{-g}\frac{1}{\bar{g}^2}\frac{\mathcal{F}^{a\,\mu\nu}}{\sqrt{-g}}\left[1-\frac{\beta}{2}\right]\right), \tag{A22}$$

where the RG equation for the exact β-function that conveniently recasts Equation (86) as

$$\frac{\mathcal{J}}{\bar{g}^2}\frac{\partial \bar{g}^2}{\partial \mathcal{J}} \equiv \frac{\beta}{2} = \frac{\mathrm{d}\ln|\bar{g}^2|}{\mathrm{d}\ln|\mathcal{J}|/\mu_0^4}, \quad \beta = \beta(\bar{g}^2), \tag{A23}$$

has been inserted. If β is known to all-loop order, the running of the coupling and hence the solutions to the equations of motion may be found to all-loop order accuracy. In the expression above, an arbitrary dimensionful renormalisation parameter μ_0 has been explicitly introduced as a reference scale. The natural boundary condition is $\bar{g}(\mathcal{J}) \to g_{\rm YM}$ when $|\mathcal{J}| \to \mu_0^4$.

Inserting Equations (A20) and (A22) into Equation (A18), the resulting equations of motion is

$$\frac{1}{\bar{g}^2}\frac{f^{abc}\mathcal{F}^{b\,\mu\nu}\mathcal{A}^c_\nu}{\sqrt{-g}}\left[1-\frac{\beta}{2}\right] - \frac{1}{\sqrt{-g}}\partial_\nu\left(\sqrt{-g}\frac{1}{\bar{g}^2}\frac{\mathcal{F}^{a\,\mu\nu}}{\sqrt{-g}}\left[1-\frac{\beta}{2}\right]\right) = 0\,. \tag{A24}$$

This can be rewritten on the operator form

$$\hat{\mathcal{D}}^{ab}_\nu\left[\frac{\mathcal{F}^{b\,\mu\nu}}{\bar{g}^2\sqrt{-g}}\left(1-\frac{\beta}{2}\right)\right] = 0\,, \tag{A25}$$

where the differential operator $\hat{\mathcal{D}}$ is given by

$$\hat{\mathcal{D}}^{ab}_\nu \equiv \delta^{ab}\frac{\partial_\nu\sqrt{-g}}{\sqrt{-g}} - f^{abc}\mathcal{A}^c_\nu. \tag{A26}$$

The action of this differential operator on a function $h(x)$ is defined as follows:

$$\left[\hat{\mathcal{D}}h(x)\right]^{ab}_{\nu} \equiv \delta^{ab}\frac{\partial_\nu\left[\sqrt{-g}h(x)\right]}{\sqrt{-g}} - f^{abc}\mathcal{A}^c_\nu h(x)\,. \tag{A27}$$

Appendix D. One-Loop Effective YM Lagrangian

Let us briefly discuss the effective YM theory at the one-loop order. The usefulness of studying the one-loop case is further motivated by a comparison of the one-loop and all-loop order expansion in Section 4.5. The standard one-loop SU(N) β-function reads (see, e.g., Ref. [408])

$$\beta_1 \equiv -B_1\bar{g}_1^2\,, \quad B_1 = \frac{bN}{48\pi^2}\,, \quad b = 11\,, \tag{A28}$$

and the corresponding solution of the RG equation (Equation (A23)) is given by

$$\bar{g}^2(\mathcal{J}) = \frac{\bar{g}_1^2(\mu_0^4)}{1 + \frac{B_1}{2}\bar{g}_1^2(\mu_0^4)\ln\left(|\mathcal{J}|/\mu_0^4\right)}\,. \tag{A29}$$

Substituting this expression into the effective all-order Lagrangian in Equation (84), we obtain

$$\mathcal{L}^{(1)}_{\rm eff} = \frac{\mathcal{J}}{4\bar{g}_1^2(\mu_0^4)}\left[1 + \frac{B_1}{2}\bar{g}_1^2(\mu_0^4)\ln\left(\frac{|\mathcal{J}|}{\mu_0^4}\right)\right]. \tag{A30}$$

Making trivial substitutions,

$$\mathcal{J} \to -4\bar{g}_1^2(\mu_0^4)\mathcal{F}\,, \quad \mu_0^4 \to 2e\mu^4\,, \quad \bar{g}_1^2(\mu_0^4) \to g_{\rm YM}^2\,, \tag{A31}$$

one arrives at another form of the one-loop effective Lagrangian frequently used in the literature (e.g., Ref. [345] and references therein),

$$\mathcal{L}^{(1)}_{\rm eff} = -\mathcal{F} - \frac{bN}{96\pi^2}g_{\rm YM}^2\mathcal{F}\left[\ln\left(\frac{2|g_{\rm YM}^2\mathcal{F}|}{\mu^4}\right) - 1\right]. \tag{A32}$$

The compact form of the all-order effective Lagrangian used earlier in Equation (84) straightforwardly produces the standard representation of the one-loop effective Lagrangian upon the redefinitions of Equation (A31), which is reassuring. The usual covariant renormalization condition on the effective Lagrangian [345]

$$\left.\frac{\partial\mathcal{L}_{\rm eff}}{\partial\mathcal{F}}\right|_{t=0} = -1\,, \qquad t = \frac{1}{2}\ln\left(\frac{2|g_{\rm YM}^2\mathcal{F}|}{\mu^4}\right), \tag{A33}$$

is apparently satisfied for Equation (A32). Indeed,

$$\frac{\partial\mathcal{L}^{(1)}_{\rm eff}}{\partial\mathcal{F}} = -1 - \frac{bN}{96\pi^2}g_{\rm YM}^2\ln\left(\frac{2|g_{\rm YM}^2\mathcal{F}|}{\mu^4}\right) \to -1 \qquad \text{for} \qquad \ln\left(\frac{2|g_{\rm YM}^2\mathcal{F}|}{\mu^4}\right) \to 0\,. \tag{A34}$$

This condition has been originally employed in Refs. [326,344] to derive the generic form of the one-loop effective Lagrangian in Equation (A32) (see Ref. [345] and references therein, for a more elaborate review). In a compact notation of Equation (84), the latter condition reads

$$\left.\frac{\partial\mathcal{L}_{\rm eff}}{\partial\mathcal{J}}\right|_{t=0} = \frac{1}{4\bar{g}_1^2(\mu_0^4)}\,, \qquad t \equiv \frac{1}{2}\ln\left(\frac{e|\mathcal{J}|}{\mu_0^4}\right). \tag{A35}$$

Notes

1 The light cone four-vectors are related to Minkowski four-vectors in a standard way $k = (k^0, k_\perp, k_z) = [k^-, \boldsymbol{k}_\perp, k^+]$ where $k^\pm = k^0 \pm k^z$. The Minkowski dot product in light-cone coordinates is $k\cdot p = \frac{1}{2}(k^+p^- + k^-p^+) - \boldsymbol{k}_\perp\cdot\boldsymbol{p}_\perp$.

2 In the early universe with $\epsilon \sim T^4 \sim a^{-4}$, the saturation scale $Q_s^2(x) \sim \alpha_S(T) R_A(T) \sim [T \ln T]^{-1}$ was extremely small.

3 Let us recall that at the temperatures $T \gg T_c^{\rm QCD}$, most of the gluons are forming the condensate and are thus in the equilibrium but do not participate in two-particle scatterings.

4 It is worth mentioning that even though a fluid filling a FLRW universe (38) homogeneously is static in the comoving frame $u^\mu = (1,0,0,0)$, the expanding geometry induces a nonzero fluid expansion rate $\partial_\mu(\sqrt{-g})u^\mu/\sqrt{-g} = 3H(t)$, where $g = -a^6(t)$ is the determinant of the FLRW metric tensor $g_{\mu\nu}$ with $k = 0$.

5 One of the authors (M.Š.) would like to thank Petr Jizba for pointing out this analogy.

6 By means of the following ansatz: $a(t) = a^* \exp\left[f(t)\right]$, the equation for the scale factor can be rewritten as

$$\ddot{f} - \tilde{h}(t)(\dot{f})^2 + A\tilde{h}(t) = 0,$$

with $\tilde{h}(t) = \frac{1}{2}(g(t) - 3)$ and $A = \frac{\kappa\epsilon_{CC}}{3}$. The introduction of $m(t) \equiv \dot{f}$, results in a first-order equation that may be solved. It is explicitly

$$\dot{m} - \tilde{h}(t)m^2(t) + A\tilde{h}(t) = 0\,.$$

The scale factor is therefore found in terms of the integral of the solution for $m(t)$ as

$$a(t) = a^* \exp\left[\int_{t_0}^{t} \mathrm{d}t\, m(t)\right].$$

7 For the case of the simple background considered in Section 4.4, $\mathcal{A}_\mu = \bar{A}_\mu + a_\mu$, then $\Gamma_k^{(n,m)}\left[\bar{A}, a\right] = \frac{\delta^n}{(\delta\bar{A})^n}\frac{\delta^m}{(\delta a)^m}\Gamma_k\left[\bar{A}, a\right]$.

References

1. Shuryak, E.V. Quantum Chromodynamics and the Theory of Superdense Matter. *Phys. Rep.* **1980**, *61*, 71–158. [CrossRef]
2. Kapusta, J.; Muller, B.; Rafelski, J. *Quark-Gluon Plasma: Theoretical Foundations*; Elsevier: Amsterdam, The Netherlands, 2003.
3. Polyakov, A.M. Thermal Properties of Gauge Fields and Quark Liberation. *Phys. Lett. B* **1978**, *72*, 477–480. [CrossRef]
4. Olive, K.A. The Thermodynamics of the Quark—Hadron Phase Transition in the Early Universe. *Nucl. Phys. B* **1981**, *190*, 483–503. [CrossRef]
5. Witten, E. Cosmic Separation of Phases. *Phys. Rev. D* **1984**, *30*, 272–285. [CrossRef]
6. Ornik, U.; Weiner, R.M. Expansion of the Early Universe and the Equation of State. *Phys. Rev. D* **1987**, *36*, 1263. [CrossRef]
7. Busza, W.; Rajagopal, K.; van der Schee, W. Heavy Ion Collisions: The Big Picture, and the Big Questions. *Ann. Rev. Nucl. Part. Sci.* **2018**, *68*, 339–376. [CrossRef]
8. Pasechnik, R.; Šumbera, M. Phenomenological Review on Quark–Gluon Plasma: Concepts vs. Observations. *Universe* **2017**, *3*, 7. [CrossRef]
9. Lattimer, J.M.; Prakash, M. The Equation of State of Hot, Dense Matter and Neutron Stars. *Phys. Rep.* **2016**, *621*, 127–164. [CrossRef]
10. Shuryak, E. Strongly coupled quark-gluon plasma in heavy ion collisions. *Rev. Mod. Phys.* **2017**, *89*, 035001. [CrossRef]
11. Yagi, K.; Hatsuda, T.; Miake, Y. *Quark-Gluon Plasma: From Big Bang to Little Bang*; Cambridge University Press: Cambridge, UK, 2005; Volume 23.
12. Boyanovsky, D.; de Vega, H.J.; Schwarz, D.J. Phase transitions in the early and the present universe. *Ann. Rev. Nucl. Part. Sci.* **2006**, *56*, 441–500. [CrossRef]
13. Sanches, S.M.; Navarra, F.S.; Fogaça, D.A. The quark gluon plasma equation of state and the expansion of the early Universe. *Nucl. Phys. A* **2015**, *937*, 1–16. [CrossRef]
14. Zhitnitsky, A.R. Dynamical de Sitter phase and nontrivial holonomy in strongly coupled gauge theories in an expanding universe. *Phys. Rev. D* **2015**, *92*, 043512. [CrossRef]
15. Braun-Munzinger, P.; Wambach, J. The Phase Diagram of Strongly-Interacting Matter. *Rev. Mod. Phys.* **2009**, *81*, 1031–1050. [CrossRef]
16. McInnes, B. Trajectory of the cosmic plasma through the quark matter phase diagram. *Phys. Rev. D* **2016**, *93*, 043544. [CrossRef]
17. Campbell, J.; Huston, J.; Krauss, F. *The Black Book of Quantum Chromodynamics: A Primer for the LHC Era*; Oxford University Press: Oxford, UK, 2017.
18. Fodor, Z.; Hoelbling, C. Light Hadron Masses from Lattice QCD. *Rev. Mod. Phys.* **2012**, *84*, 449. [CrossRef]
19. Bazavov, A.; Ding, H.-T.; Hegde, P.; Kaczmarek, O.; Karsch, F.; Laermann, E.; Maezawa, Y.; Swagato Mukherjee, H.; Ohno, P.; Petreczky, H.; et al. The QCD Equation of State to $\mathcal{O}(\mu_B^6)$ from Lattice QCD. *Phys. Rev. D* **2017**, *95*, 054504. [CrossRef]
20. Philipsen, O. Constraining the phase diagram of QCD at finite temperature and density. *PoS* **2019**, *363*, 273. [CrossRef]
21. Shuryak, E. Lectures on nonperturbative QCD (Nonperturbative Topological Phenomena in QCD and Related Theories). *arXiv* **2018**, arxiv:hep-ph/1812.01509.
22. Pasechnik, R.; Šumbera, M. Different faces of confinement. *Universe* **2021**, *7*, 9. [CrossRef]
23. Poggio, E.C.; Quinn, H.R.; Weinberg, S. Smearing the Quark Model. *Phys. Rev. D* **1976**, *13*, 1958. [CrossRef]

24. Shifman, M.A. Quark hadron duality. In Proceedings of the 8th International Symposium on Heavy Flavor Physics, Southampton, UK, 25–29 July 1999; World Scientific: Singapore, 2000; Volume 3, pp. 1447–1494. [CrossRef]
25. Romatschke, P.; Romatschke, U. *Relativistic Fluid Dynamics In and Out of Equilibrium*; Cambridge Monographs on Mathematical Physics; Cambridge University Press: Cambridge, UK, 2019. [CrossRef]
26. Lundberg, T.; Pasechnik, R. Thermal Field Theory in real-time formalism: Concepts and applications for particle decays. *Eur. Phys. J. A* **2021**, *57*, 71. [CrossRef]
27. Hofmann, R. *The Thermodynamics of Quantum Yang-Mills Theory*; World Scientific: Singapore, 2011. [CrossRef]
28. Aarts, G. Introductory lectures on lattice QCD at nonzero baryon number. *J. Phys. Conf. Ser.* **2016**, *706*, 022004. [CrossRef]
29. Itoh, N. Hydrostatic Equilibrium of Hypothetical Quark Stars. *Prog. Theor. Phys.* **1970**, *44*, 291. [CrossRef]
30. Ivanenko, D.D.; Kurdgelaidze, D.F. Hypothesis concerning quark stars. *Astrophysics* **1965**, *1*, 251–252. [CrossRef]
31. Gross, D.J.; Wilczek, F. Ultraviolet Behavior of Nonabelian Gauge Theories. *Phys. Rev. Lett.* **1973**, *30*, 1343–1346. [CrossRef]
32. Politzer, H.D. Reliable Perturbative Results for Strong Interactions? *Phys. Rev. Lett.* **1973**, *30*, 1346–1349. [CrossRef]
33. Collins, J.C.; Perry, M.J. Superdense Matter: Neutrons Or Asymptotically Free Quarks? *Phys. Rev. Lett.* **1975**, *34*, 1353. [CrossRef]
34. Cabibbo, N.; Parisi, G. Exponential Hadronic Spectrum and Quark Liberation. *Phys. Lett. B* **1975**, *59*, 67–69. [CrossRef]
35. Shuryak, E.V. Theory of Hadronic Plasma. *Sov. Phys. JETP* **1978**, *47*, 212–219.
36. Shuryak, E.V. Quark-Gluon Plasma and Hadronic Production of Leptons, Photons and Psions. *Phys. Lett. B* **1978**, *78*, 150. [CrossRef]
37. Freedman, B.A.; McLerran, L.D. Fermions and Gauge Vector Mesons at Finite Temperature and Density. 3. The Ground State Energy of a Relativistic Quark Gas. *Phys. Rev. D* **1977**, *16*, 1169. [CrossRef]
38. Kapusta, J.I. Quantum Chromodynamics at High Temperature. *Nucl. Phys. B* **1979**, *148*, 461–498. [CrossRef]
39. Ichimaru, S. Strongly coupled plasmas: High-density classical plasmas and degenerate electron liquids. *Rev. Mod. Phys.* **1982**, *54*, 1017–1059. [CrossRef]
40. Braun-Munzinger, P.; Koch, V.; Schäfer, T.; Stachel, J. Properties of hot and dense matter from relativistic heavy ion collisions. *Phys. Rep.* **2016**, *621*, 76–126. [CrossRef]
41. Fukushima, K.; Hatsuda, T. The phase diagram of dense QCD. *Rep. Prog. Phys.* **2011**, *74*, 014001. [CrossRef]
42. Barrois, B.C. Superconducting Quark Matter. *Nucl. Phys. B* **1977**, *129*, 390–396. [CrossRef]
43. Bailin, D.; Love, A. Superfluidity and Superconductivity in Relativistic Fermion Systems. *Phys. Rep.* **1984**, *107*, 325. [CrossRef]
44. Ivanenko, D.; Kurdgelaidze, D.F. Remarks on quark stars. *Lett. Nuovo Cim.* **1969**, *2*, 13–16. [CrossRef]
45. Alford, M.G.; Rajagopal, K.; Wilczek, F. Color flavor locking and chiral symmetry breaking in high density QCD. *Nucl. Phys. B* **1999**, *537*, 443–458. [CrossRef]
46. Becker, W. *Neutron Stars and Pulsars*; Springer: Berlin/Heidelberg, Germany, 2009. [CrossRef]
47. McLerran, L.; Pisarski, R.D. Phases of cold, dense quarks at large N(c). *Nucl. Phys. A* **2007**, *796*, 83–100. [CrossRef]
48. McLerran, L.; Reddy, S. Quarkyonic Matter and Neutron Stars. *Phys. Rev. Lett.* **2019**, *122*, 122701. [CrossRef] [PubMed]
49. Migdal, A.B. Pion Fields in Nuclear Matter. *Rev. Mod. Phys.* **1978**, *50*, 107–172. [CrossRef]
50. Pisarski, R.D.; Rennecke, F. Signatures of Moat Regimes in Heavy-Ion Collisions. *Phys. Rev. Lett.* **2021**, *127*, 152302. [CrossRef] [PubMed]
51. Tejeda-Yeomans, M.E. Heavy-ion physics: Freedom to do hot, dense, exciting QCD. *CERN Yellow Rep. Sch. Proc.* **2021**, *2*, 137. [CrossRef]
52. Arsene, I.; Beardeng, I.G.; Beavisa, D.; Besliuj, C.; Budickf, B.; Bøggildg, H.; Chasmana, C.; Christenseng, C.H.; Christianseng, P.; Ciborc, J.; et al. Quark gluon plasma and color glass condensate at RHIC? The Perspective from the BRAHMS experiment. *Nucl. Phys. A* **2005**, *757*, 1–27. [CrossRef]
53. Back, B.B.; et al. [PHOBOS Collaboration] The PHOBOS perspective on discoveries at RHIC. *Nucl. Phys. A* **2005**, *757*, 28–101. [CrossRef]
54. Adams, J.; et al. [STAR Collaboration] Experimental and theoretical challenges in the search for the quark gluon plasma: The STAR Collaboration's critical assessment of the evidence from RHIC collisions. *Nucl. Phys. A* **2005**, *757*, 102–183. [CrossRef]
55. Adcox, K.; et al. [PHENIX Collaboration] Formation of dense partonic matter in relativistic nucleus-nucleus collisions at RHIC: Experimental evaluation by the PHENIX collaboration. *Nucl. Phys. A* **2005**, *757*, 184–283. [CrossRef]
56. Aprahamian, A.; Robert, A.; Caines, H.; Cates, G.; Cizewski, G.A.; Cirigliano, V.; Dean, D.J.; Deshpande, A.; Ent, R.; Fahey, F.; et al. Reaching for the Horizon: The 2015 Long Range Plan for Nuclear Science. 2015. Available online: https://inspirehep.net/literature/1398831 (accessed on 27 August 2022).
57. Bzdak, A.; Esumi, S.; Koch, V.; Liao, J.; Stephanov, M.; Xu, N. Mapping the Phases of Quantum Chromodynamics with Beam Energy Scan. *Phys. Rep.* **2020**, *853*, 1–87. [CrossRef]
58. Wilson, K.G. The renormalization group and critical phenomena. *Rev. Mod. Phys.* **1983**, *55*, 583–600. [CrossRef]
59. Stephanov, M.A.; Rajagopal, K.; Shuryak, E.V. Signatures of the tricritical point in QCD. *Phys. Rev. Lett.* **1998**, *81*, 4816–4819. [CrossRef]
60. Gupta, S.; Luo, X.; Mohanty, B.; Ritter, H.G.; Xu, N. Scale for the Phase Diagram of Quantum Chromodynamics. *Science* **2011**, *332*, 1525–1528. [CrossRef] [PubMed]
61. Sumbera, M. Results from STAR Beam Energy Scan Program. *Acta Phys. Polon. Supp.* **2013**, *6*, 429–436. [CrossRef]

62. Adamczyk, L.; et al. [STAR Collaboration] Bulk Properties of the Medium Produced in Relativistic Heavy-Ion Collisions from the Beam Energy Scan Program. *Phys. Rev. C* **2017**, *96*, 044904. [CrossRef]
63. Bellwied, R.; Borsanyi, S.; Fodor, Z.; Günther, J.; Katz, S.D.; Ratti, C.; Szabo, K.K. The QCD phase diagram from analytic continuation. *Phys. Lett. B* **2015**, *751*, 559–564. [CrossRef]
64. Ding, H.T.; Karsch, F.; Mukherjee, S. Thermodynamics of strong-interaction matter from Lattice QCD *Int. J. Mod. Phys. E* **2015**, *24*, 1530007. [CrossRef]
65. Pasechnik, R.; Beylin, V.; Vereshkov, G. Dark Energy from graviton-mediated interactions in the QCD vacuum. *J. Cosmol. Astropart. Phys.* **2013**, *6*, 11. [CrossRef]
66. Addazi, A.; Marcianò, A.; Pasechnik, R.; Prokhorov, G. Mirror Symmetry of quantum Yang-Mills vacua and cosmological implications. *Eur. Phys. J. C* **2019**, *79*, 251. [CrossRef]
67. Pasechnik, R. Quantum Yang–Mills Dark Energy. *Universe* **2016**, *2*, 4. [CrossRef]
68. Bailin, D.; Love, A. *Cosmology in Gauge Field Theory and String Theory*; Taylor & Francis: Milton Park, UK, 2004.
69. Gorbunov, D.S.; Rubakov, V.A. *Introduction to the Theory of the Early Universe: Cosmological Perturbations and Inflationary Theory*; World Scientific: Singapore, 2011. [CrossRef]
70. Oertel, M.; Hempel, M.; Klähn, T.; Typel, S. Equations of state for supernovae and compact stars. *Rev. Mod. Phys.* **2017**, *89*, 015007. [CrossRef]
71. Baym, G.; Hatsuda, T.; Kojo, T.; Powell, P.D.; Song, Y.; Takatsuka, T. From hadrons to quarks in neutron stars: A review. *Rep. Prog. Phys.* **2018**, *81*, 056902. [CrossRef] [PubMed]
72. Bertone, G.; Hooper, D. History of dark matter. *Rev. Mod. Phys.* **2018**, *90*, 045002. [CrossRef]
73. Green, A.M.; Kavanagh, B.J. Primordial Black Holes as a dark matter candidate. *J. Phys. G* **2021**, *48*, 043001. [CrossRef]
74. Addazi, A.; Marcianò, A.; Pasechnik, R. Time-crystal ground state and production of gravitational waves from QCD phase transition. *Chin. Phys. C* **2019**, *43*, 065101. [CrossRef]
75. Yang, C.N.; Mills, R.L. Conservation of Isotopic Spin and Isotopic Gauge Invariance. *Phys. Rev.* **1954**, *96*, 191–195. [CrossRef]
76. Nambu, Y.; Jona-Lasinio, G. Dynamical Model of Elementary Particles Based on an Analogy with Superconductivity. 1. *Phys. Rev.* **1961**, *122*, 345–358. [CrossRef]
77. Goldstone, J. Field Theories with Superconductor Solutions. *Nuovo Cim.* **1961**, *19*, 154–164. [CrossRef]
78. Goldstone, J.; Salam, A.; Weinberg, S. Broken Symmetries. *Phys. Rev.* **1962**, *127*, 965–970. [CrossRef]
79. Aad, G.; et al. [ATLAS Collaboration] Observation of a new particle in the search for the Standard Model Higgs boson with the ATLAS detector at the LHC. *Phys. Lett. B* **2012**, *716*, 1–29. [CrossRef]
80. Chatrchyan, S.; et al. [CMS Collaboration] Observation of a New Boson at a Mass of 125 GeV with the CMS Experiment at the LHC. *Phys. Lett. B* **2012**, *716*, 30–61. [CrossRef]
81. Giacosa, F.; Hofmann, R. Thermal ground state in deconfining Yang-Mills thermodynamics. *Prog. Theor. Phys.* **2007**, *118*, 759–767. [CrossRef]
82. Herbst, U.; Hofmann, R. Asymptotic freedom and compositeness. *ISRN High Energy Phys.* **2012**, *2012*, 373121. [CrossRef]
83. Linde, A.D. Phase Transitions in Gauge Theories and Cosmology. *Rep. Prog. Phys.* **1979**, *42*, 389. [CrossRef]
84. Mukhanov, V. *Physical Foundations of Cosmology*; Cambridge University Press: Oxford, UK, 2005.
85. Laine, M.; Vuorinen, A. *Basics of Thermal Field Theory*; Springer: Berlin/Heidelberg, Germany, 2016; Volume 925. [CrossRef]
86. Mazumdar, A.; White, G. Review of cosmic phase transitions: Their significance and experimental signatures. *Rep. Prog. Phys.* **2019**, *82*, 076901. [CrossRef] [PubMed]
87. D'Onofrio, M.; Rummukainen, K. Standard model cross-over on the lattice. *Phys. Rev. D* **2016**, *93*, 025003. [CrossRef]
88. Bazavov, A.; et al. [HotQCD Collaboration] Chiral crossover in QCD at zero and non-zero chemical potentials. *Phys. Lett. B* **2019**, *795*, 15–21. [CrossRef]
89. Borsanyi, S.; Fodor, Z.; Guenther, J.; Kampert, K.-H.; Katz, S.D.; Kawanai, T.; Kovacs, T.G. ; Mages, S. W.; Pasztor, A.; Pittler, F.; et al. Calculation of the axion mass based on high-temperature lattice quantum chromodynamics. *Nature* **2016**, *539*, 69–71. [CrossRef]
90. Laine, M.; Meyer, M. Standard Model thermodynamics across the electroweak crossover. *J. Cosmol. Astropart. Phys.* **2015**, *7*, 35. [CrossRef]
91. Kogut, J.B.; Wyld, H.W.; Karsch, F.; Sinclair, D.K. First Order Chiral Phase Transition in Lattice QCD. *Phys. Lett. B* **1987**, *188*, 353–358. [CrossRef]
92. Aoki, Y.; Endrodi, G.; Fodor, Z.; Katz, S.D.; Szabo, K.K. The Order of the quantum chromodynamics transition predicted by the standard model of particle physics. *Nature* **2006**, *443*, 675–678. [CrossRef]
93. Boeckel, T.; Schettler, S.; Schaffner-Bielich, J. The Cosmological QCD Phase Transition Revisited. *Prog. Part. Nucl. Phys.* **2011**, *66*, 266–270. [CrossRef]
94. Schettler, S.; Boeckel, T.; Schaffner-Bielich, J. Imprints of the QCD Phase Transition on the Spectrum of Gravitational Waves. *Phys. Rev. D* **2011**, *83*, 064030. [CrossRef]
95. Boeckel, T.; Schaffner-Bielich, J. A little inflation in the early universe at the QCD phase transition. *Phys. Rev. Lett.* **2010**, *105*, 041301; Erratum in: *Phys. Rev. Lett.* **2011**, *106*, 069901. [CrossRef] [PubMed]
96. Ayala, A.; Bashir, A.; Cobos-Martinez, J.J.; Hernandez-Ortiz, S.; Raya, A. The effective QCD phase diagram and the critical end point. *Nucl. Phys. B* **2015**, *897*, 77–86. [CrossRef]

97. Cui, Z.F.; Zhang, J.L.; Zong, H.S. Proper time regularization and the QCD chiral phase transition. *Sci. Rep.* **2017**, *7*, 45937. [CrossRef]
98. Burkert, V.D.; Elouadrhiri, L.; Girod, F.X. The pressure distribution inside the proton. *Nature* **2018**, *557*, 396–399. [CrossRef]
99. Diamantini, M.C.; Trugenberger, C.A.; Vinokur, V.M. Confinement and Asymptotic Freedom with Cooper pairs. *Commun. Phys.* **2018**, *1*, 77. [CrossRef]
100. Andronic, A.; Braun-Munzinger, P.; Redlich, K.; Stachel, J. Decoding the phase structure of QCD via particle production at high energy. *Nature* **2018**, *561*, 321–330. [CrossRef]
101. Pang, L.G.; Zhou, K.; Su, N.; Petersen, H.; Stöcker, H.; Wang, X.N. An equation-of-state-meter of quantum chromodynamics transition from deep learning. *Nat. Commun.* **2018**, *9*, 210. [CrossRef]
102. Du, Y.L.; Zhou, K.; Steinheimer, J.; Pang, L.G.; Motornenko, A.; Zong, H.S.; Wang, X.N.; Stöcker, H. Identifying the nature of the QCD transition in relativistic collision of heavy nuclei with deep learning. *Eur. Phys. J. C* **2020**, *80*, 516. [CrossRef]
103. Attems, M.; Bea, Y.; Casalderrey-Solana, J.; Mateos, D.; Triana, M.; Zilhão, M. Holographic Collisions across a Phase Transition. *Phys. Rev. Lett.* **2018**, *121*, 261601. [CrossRef] [PubMed]
104. Boiko, V.G.; Jenkovszky, L.L.; Sysoev, V.M. Thermodynamics of phase transitions in nuclear matter. *Fiz. Elem. Chast. Atom. Yadra* **1991**, *22*, 675–715. (In Russian)
105. Bonometto, S.A.; Pantano, O. Physics of the cosmological quark-hadron transition. *Phys. Rep.* **1993**, *228*, 175–252. [CrossRef]
106. Schwarz, D.J. The first second of the universe. *Ann. Phys.* **2003**, *12*, 220–270. [CrossRef]
107. Castorina, P.; Greco, V.; Plumari, S. QCD equation of state and cosmological parameters in the early universe. *Phys. Rev. D* **2015**, *92*, 063530. [CrossRef]
108. Tawfik, A. Cosmological Consequences of QCD Phase Transition(s) in Early Universe. *AIP Conf. Proc.* **2009**, *1115*, 239–247. [CrossRef]
109. Kaczmarek, O.; Karsch, F.; Lahiri, A.; Mazur, L.; Schmidt, C. QCD phase transition in the chiral limit. *arXiv* **2020**, arXiv:2003.07920.
110. Eidelman, S.; et al. [Particle Data Group] Review of particle physics. Particle Data Group. *Phys. Lett. B* **2004**, *592*, 1. [CrossRef]
111. Zhu, X.; Bleicher, M.; Huang, S.L.; Schweda, K.; Stoecker, H.; Xu, N.; Zhuang, P. D anti-D correlations as a sensitive probe for thermalization in high-energy nuclear collisions. *Phys. Lett. B* **2007**, *647*, 366–370. [CrossRef]
112. Hindmarsh, M.B.; Lüben, M.; Lumma, J.; Pauly, M. Phase transitions in the early universe. *SciPost Phys. Lect. Notes* **2021**, *24*, 1. [CrossRef]
113. Weinberg, S. Gauge and Global Symmetries at High Temperature. *Phys. Rev. D* **1974**, *9*, 3357–3378. [CrossRef]
114. Patel, H.H.; Ramsey-Musolf, M.J.; Wise, M.B. Color Breaking in the Early Universe. *Phys. Rev. D* **2013**, *88*, 015003. [CrossRef]
115. Ramsey-Musolf, M.J.; Winslow, P.; White, G. Color Breaking Baryogenesis. *Phys. Rev. D* **2018**, *97*, 123509. [CrossRef]
116. Byrnes, C.T.; Hindmarsh, M.; Young, S.; Hawkins, M.R.S. Primordial black holes with an accurate QCD equation of state. *J. Cosmol. Astropart. Phys.* **2018**, *8*, 41. [CrossRef]
117. Husdal, L. On Effective Degrees of Freedom in the Early Universe. *Galaxies* **2016**, *4*, 78. [CrossRef]
118. Florkowski, W. The realistic QCD equation of state in relativistic heavy-ion collisions and the early Universe. *Nucl. Phys. A* **2011**, *853*, 173–188. [CrossRef]
119. Guardo, G.L.; Greco, V.; Ruggieri, M. Energy density fluctuations in Early Universe. *AIP Conf. Proc.* **2014**, *1595*, 224–227. [CrossRef]
120. Weinberg, S. *Cosmology*; Oxford University Press: Oxford, UK, 2008.
121. Zel'dovich, Y.B.; Novikov, I.D. The Hypothesis of Cores Retarded during Expansion and the Hot Cosmological Model. *Sov. Astron.* **1967**, *10*, 602.
122. Hawking, S. Gravitationally collapsed objects of very low mass. *Mon. Not. Roy. Astron. Soc.* **1971**, *152*, 75. [CrossRef]
123. Abbott, B.P.; et al. [LIGO Scientific Collaboration and Virgo Collaboration] GW151226: Observation of Gravitational Waves from a 22-Solar-Mass Binary Black Hole Coalescence. *Phys. Rev. Lett.* **2016**, *116*, 241103. [CrossRef]
124. Carr, B.; Kuhnel, F.; Sandstad, M. Primordial Black Holes as Dark Matter. *Phys. Rev. D* **2016**, *94*, 083504. [CrossRef]
125. Carr, B.; Kuhnel, F. Primordial Black Holes as Dark Matter: Recent Developments. *Ann. Rev. Nucl. Part. Sci.* **2020**, *70*, 355–394. [CrossRef]
126. Biagetti, M.; De Luca, V.; Franciolini, G.; Kehagias, A.; Riotto, A. The formation probability of primordial black holes. *Phys. Lett. B* **2021**, *820*, 136602. [CrossRef]
127. Allahverdi, R.; Osiński, J.K. Early matter domination from long-lived particles in the visible sector. *Phys. Rev. D* **2022**, *105*, 023502. [CrossRef]
128. Carr, B.; Clesse, S.; García-Bellido, J.; Kühnel, F. Cosmic conundra explained by thermal history and primordial black holes. *Phys. Dark Univ.* **2021**, *31*, 100755. [CrossRef]
129. Hung, C.M.; Shuryak, E.V. Hydrodynamics near the QCD phase transition: Looking for the longest lived fireball. *Phys. Rev. Lett.* **1995**, *75*, 4003–4006. [CrossRef] [PubMed]
130. Jedamzik, K. Primordial black hole formation during the QCD epoch. *Phys. Rev. D* **1997**, *55*, 5871–5875. [CrossRef]
131. Jedamzik, K. Could MACHOS be primordial black holes formed during the QCD epoch? *Phys. Rep.* **1998**, *307*, 155–162. [CrossRef]
132. Borsanyi, S.; Endrodi, G.; Fodor, Z.; Jakovac, A.; Katz, S.D.; Krieg, S.; Ratti, C.; Szabo, K.K. The QCD equation of state with dynamical quarks. *J. High Energy Phys.* **2010**, *11*, 77. [CrossRef]
133. Bekenstein, J.D. Black holes and entropy. *Phys. Rev. D* **1973**, *7*, 2333–2346. [CrossRef]

134. Penrose, R. The Big Bang and its Dark-Matter Content: Whence, Whither, and Wherefore. *Found. Phys.* **2018**, *48*, 1177–1190. [CrossRef]
135. Wysocki, D.; Gerosa, D.; O'Shaughnessy, R.; Belczynski, K.; Gladysz, W.; Berti, E.; Kesden, M.; Holz, D.E. Explaining LIGO's observations via isolated binary evolution with natal kicks. *Phys. Rev. D* **2018**, *97*, 043014. [CrossRef]
136. Adler, R.J.; Bjorken, J.D.; Chen, P.; Liu, J.S. Simple analytic models of gravitational collapse. *Am. J. Phys.* **2005**, *73*, 1148–1159. [CrossRef]
137. García-Bellido, J.; Carr, B.; Clesse, S. Primordial Black Holes and a Common Origin of Baryons and Dark Matter. *Universe* **2021**, *8*, 12. [CrossRef]
138. Gross, D.J.; Wilczek, F. Asymptotically Free Gauge Theories-I. *Phys. Rev. D* **1973**, *8*, 3633–3652. [CrossRef]
139. CMS Collaboration. *Determination of the Strong Coupling Constant from the Measurement of Inclusive Multijet Event Cross Sections in pp Collisions at* $\sqrt{s}$ *= 8 TeV*; Tech. Rep. CMS-PAS-SMP-16-008; CERN: Geneva, Switzerland, 2017.
140. Ghiglieri, J.; Kurkela, A.; Strickland, M.; Vuorinen, A. Perturbative Thermal QCD: Formalism and Applications. *Phys. Rep.* **2020**, *880*, 1–73. [CrossRef]
141. Martin, A.D.; Stirling, W.J.; Thorne, R.S.; Watt, G. Parton distributions for the LHC. *Eur. Phys. J. C* **2009**, *63*, 189–285. [CrossRef]
142. Gelis, F.; Iancu, E.; Jalilian-Marian, J.; Venugopalan, R. The Color Glass Condensate. *Ann. Rev. Nucl. Part. Sci.* **2010**, *60*, 463–489. [CrossRef]
143. Fujii, H.; Kharzeev, D. Long range forces of QCD. *Phys. Rev. D* **1999**, *60*, 114039. [CrossRef]
144. Blaizot, J.P. Weakly and strongly coupled degrees of freedom in the quark-gluon plasma. *Acta Phys. Polon. Supp.* **2011**, *4*, 641–646. [CrossRef]
145. Lacey, R.A.; Ajitanand, N.N.; Alexander, J.M.; Chung, P.; Holzmann, W.G.; Issah, M.; Taranenko, A.; Danielewicz, P.; Stoecker, H. Has the QCD Critical Point been Signaled by Observations at RHIC? *Phys. Rev. Lett.* **2007**, *98*, 092301. [CrossRef] [PubMed]
146. Heinz, U.; Snellings, R. Collective flow and viscosity in relativistic heavy-ion collisions. *Ann. Rev. Nucl. Part. Sci.* **2013**, *63*, 123–151. [CrossRef]
147. Aamodt, K.; et al. [ALICE Collaboration] Elliptic flow of charged particles in Pb-Pb collisions at 2.76 TeV. *Phys. Rev. Lett.* **2010**, *105*, 252302. [CrossRef] [PubMed]
148. Aad, G.; et al. [ATLAS Collaboration] Measurement of the pseudorapidity and transverse momentum dependence of the elliptic flow of charged particles in lead-lead collisions at $\sqrt{s_{NN}} = 2.76$ TeV with the ATLAS detector. *Phys. Lett. B* **2012**, *707*, 330–348. [CrossRef]
149. Chatrchyan, S.; et al. [CMS Collaboration] Azimuthal anisotropy of charged particles at high transverse momenta in PbPb collisions at $\sqrt{s_{NN}} = 2.76$ TeV. *Phys. Rev. Lett.* **2012**, *109*, 022301. [CrossRef]
150. Gyulassy, M.; Plumer, M. Jet Quenching in Dense Matter. *Phys. Lett. B* **1990**, *243*, 432–438. [CrossRef]
151. Bielčíková, J. Jets and correlations in heavy-ion collisions. *PoS* **2015**, *234*, 22. [CrossRef]
152. Adare, A.; et al. [PHENIX Collaboration] Enhanced production of direct photons in Au+Au collisions at $\sqrt{s_{NN}} = 200$ GeV and implications for the initial temperature. *Phys. Rev. Lett.* **2010**, *104*, 132301. [CrossRef]
153. Shuryak, E. Physics of Strongly coupled Quark-Gluon Plasma. *Prog. Part. Nucl. Phys.* **2009**, *62*, 48–101. [CrossRef]
154. Thoma, M.H. Complex plasmas as a model for the quark-gluon-plasma liquid. *Nucl. Phys. A* **2006**, *774*, 307–314. [CrossRef]
155. Casalderrey-Solana, J.; Liu, H.; Mateos, D.; Rajagopal, K.; Wiedemann, U.A. *Gauge/String Duality, Hot QCD and Heavy Ion Collisions*; Cambridge University Press: Cambridge, UK, 2014. [CrossRef]
156. Maldacena, J.M. The Large N limit of superconformal field theories and supergravity. *Adv. Theor. Math. Phys.* **1998**, *2*, 231–252. [CrossRef]
157. Susskind, L.; Lindesay, J. *An Introduction to Black Holes, Information and the String Theory Revolution: The Holographic Universe*; World Scientific: Singapore, 2005.
158. McGreevy, J. Holographic duality with a view toward many-body physics. *Adv. High Energy Phys.* **2010**, *2010*, 723105. [CrossRef]
159. Altarelli, G.; Parisi, G. Asymptotic Freedom in Parton Language. *Nucl. Phys. B* **1977**, *126*, 298–318. [CrossRef]
160. Dokshitzer, Y.L. Calculation of the Structure Functions for Deep Inelastic Scattering and e+ e- Annihilation by Perturbation Theory in Quantum Chromodynamics. *Sov. Phys. JETP* **1977**, *46*, 641–653.
161. Gribov, V.N.; Lipatov, L.N. Deep inelastic e p scattering in perturbation theory. *Sov. J. Nucl. Phys.* **1972**, *15*, 438–450.
162. Ioffe, B.L.; Fadin, V.S.; Lipatov, L.N. *Quantum Chromodynamics: Perturbative and Nonperturbative Aspects*; Cambridge University Press: Cambridge, UK, 2010. [CrossRef]
163. Cea, P. The Higgs condensate as a quantum liquid. *Int. J. Theor. Phys.* **2020**, *59*, 3310–3323. [CrossRef]
164. McLerran, L. A Brief Introduction to the Color Glass Condensate and the Glasma. In Proceedings of the 38th International Symposium on Multiparticle Dynamics, Hamburg, Germany, 15–20 September 2008; pp. 3–18. [CrossRef]
165. Gribov, L.V.; Levin, E.M.; Ryskin, M.G. Semihard Processes in QCD. *Phys. Rep.* **1983**, *100*, 1–150. [CrossRef]
166. Kharzeev, D. Classical chromodynamics of relativistic heavy ion collisions. In *Cargese Summer School on QCD Perspectives on Hot and Dense Matter*; Springer: Berlin/Heidelberg, Germany, 2002; pp. 207–236.
167. Berges, J.; Heller, M.P.; Mazeliauskas, A.; Venugopalan, R. QCD thermalization: Ab initio approaches and interdisciplinary connections. *Rev. Mod. Phys.* **2021**, *93*, 035003. [CrossRef]
168. McLerran, L.D.; Venugopalan, R. Computing quark and gluon distribution functions for very large nuclei. *Phys. Rev. D* **1994**, *49*, 2233–2241. [CrossRef]

169. Kovner, A.; McLerran, L.D.; Weigert, H. Gluon production from nonAbelian Weizsacker-Williams fields in nucleus-nucleus collisions. *Phys. Rev. D* **1995**, *52*, 6231–6237. [CrossRef]
170. Tu, Z.; Kharzeev, D.E.; Ullrich, T. Einstein-Podolsky-Rosen Paradox and Quantum Entanglement at Subnucleonic Scales. *Phys. Rev. Lett.* **2020**, *124*, 062001. [CrossRef] [PubMed]
171. Ashtekar, A.; Corichi, A.; Kesavan, A. Emergence of classical behavior in the early universe. *Phys. Rev. D* **2020**, *102*, 023512. [CrossRef]
172. Vilenkin, A. Quantum Creation of Universes. *Phys. Rev. D* **1984**, *30*, 509–511. [CrossRef]
173. Martin, J.; Vennin, V. Quantum Discord of Cosmic Inflation: Can we Show that CMB Anisotropies are of Quantum-Mechanical Origin? *Phys. Rev. D* **2016**, *93*, 023505. [CrossRef]
174. Green, D.; Porto, R.A. Signals of a Quantum Universe. *Phys. Rev. Lett.* **2020**, *124*, 251302. [CrossRef] [PubMed]
175. Liddle, A.R. *An Introduction to Modern Cosmology*; Wiley, Hoboken, NJ, USA, 1998.
176. Linde, A. *Inflationary Cosmology after Planck 2013*; Oxford University Press: Oxford, UK, 2015; pp. 231–316. [CrossRef]
177. Sethna, J.P. *Statistical Mechanics: Entropy, Order Parameters, and Complexity*; Oxford University Press: Oxford, UK, 2006.
178. Dvali, G.; Gomez, C.; Zell, S. Quantum Break-Time of de Sitter. *J. Cosmol. Astropart. Phys.* **2017**, *6*, 28. [CrossRef]
179. Berezhiani, L.; Zantedeschi, M. Evolution of coherent states as quantum counterpart of classical dynamics. *Phys. Rev. D* **2021**, *104*, 085007. [CrossRef]
180. Langer, S.A.; Sethna, J.P. Entropy of Glasses. *Phys. Rev. Lett.* **1988**, *61*, 570–573. [CrossRef]
181. Carrington, M.E.; Czajka, A.; Mrowczynski, S. The energy-momentum tensor at the earliest stage of relativistic heavy-ion collisions. *Eur. Phys. J. A* **2022**, *58*, 5. [CrossRef]
182. Carrington, M.E.; Czajka, A.; Mrowczynski, S. Physical characteristics of glasma from the earliest stage of relativistic heavy ion collisions. *arXiv* **2021**, arxiv:2105.05327.
183. Weinberg, S. *The Quantum Theory of Fields. Volume 2: Modern Applications*; Cambridge University Press: Cambridge, UK, 2013.
184. Degrassi, G.; Di Vita, S.; Elias-Miro, J.; Espinosa, J.R.; Giudice, G.F.; Isidori, G.; Strumia, A. Higgs mass and vacuum stability in the Standard Model at NNLO. *J. High Energy Phys.* **2012**, *8*, 98. [CrossRef]
185. Markkanen, T.; Rajantie, A.; Stopyra, S. Cosmological Aspects of Higgs Vacuum Metastability. *Front. Astron. Space Sci.* **2018**, *5*, 40. [CrossRef]
186. Georgi, H.; Glashow, S.L. Unity of All Elementary Particle Forces. *Phys. Rev. Lett.* **1974**, *32*, 438–441. [CrossRef]
187. Nishino, H.; et al. [Super-Kamiokande Collaboration] Search for Proton Decay via p —> e+ pi0 and p —> mu+ pi0 in a Large Water Cherenkov Detector. *Phys. Rev. Lett.* **2009**, *102*, 141801. [CrossRef] [PubMed]
188. Langacker, P. Grand Unified Theories and Proton Decay. *Phys. Rep.* **1981**, *72*, 185. [CrossRef]
189. Nath, P.; Fileviez Perez, P. Proton stability in grand unified theories, in strings and in branes. *Phys. Rep.* **2007**, *441*, 191–317. [CrossRef]
190. Altarelli, G.; Meloni, D. A non supersymmetric SO(10) grand unified model for all the physics below M_{GUT}. *J. High Energy Phys.* **2013**, *8*, 21. [CrossRef]
191. Morais, A.P.; Pasechnik, R.; Porod, W. Grand Unified origin of gauge interactions and families replication in the Standard Model. *Universe* **2021**, *7*, 461. [CrossRef]
192. Croon, D.; Gonzalo, T.E.; Graf, L.; Košnik, N.; White, G. GUT Physics in the era of the LHC. *Front. Phys.* **2019**, *7*, 76. [CrossRef]
193. Dimopoulos, S.; Raby, S.; Wilczek, F. Supersymmetry and the Scale of Unification. *Phys. Rev. D* **1981**, *24*, 1681–1683. [CrossRef]
194. Linde, A.D. Particle physics and inflationary cosmology. *Contemp. Concepts Phys.* **1990**, *5*, 1–362.
195. Gangui, A. Topological Defects in Cosmology. arXiv **2001**, arXiv:astro-ph/0110285.
196. Caprini, C.; Durrer, R. Gravitational wave production: A Strong constraint on primordial magnetic fields. *Phys. Rev. D* **2001**, *65*, 023517. [CrossRef]
197. Caprini, C.; Durrer, R.; Servant, G. The stochastic gravitational wave background from turbulence and magnetic fields generated by a first-order phase transition. *J. Cosmol. Astropart. Phys.* **2009**, *12*, 24. [CrossRef]
198. Figueroa, D.G.; Hindmarsh, M.; Urrestilla, J. Exact Scale-Invariant Background of Gravitational Waves from Cosmic Defects. *Phys. Rev. Lett.* **2013**, *110*, 101302. [CrossRef]
199. Hindmarsh, M. Sound shell model for acoustic gravitational wave production at a first-order phase transition in the early Universe. *Phys. Rev. Lett.* **2018**, *120*, 071301. [CrossRef]
200. Kamionkowski, M.; Kosowsky, A.; Turner, M.S. Gravitational radiation from first order phase transitions. *Phys. Rev. D* **1994**, *49*, 2837–2851. [CrossRef]
201. Durrer, R.; Neronov, A. Cosmological Magnetic Fields: Their Generation, Evolution and Observation. *Astron. Astrophys. Rev.* **2013**, *21*, 62. [CrossRef]
202. Vilenkin, A.; Shellard, E.P.S. *Cosmic Strings and Other Topological Defects*; Cambridge University Press: Cambridge, UK, 2000.
203. Espinosa, J.R.; Quiros, M. The Electroweak phase transition with a singlet. *Phys. Lett. B* **1993**, *305*, 98–105. [CrossRef]
204. Iso, S.; Serpico, P.D.; Shimada, K. QCD-Electroweak First-Order Phase Transition in a Supercooled Universe. *Phys. Rev. Lett.* **2017**, *119*, 141301. [CrossRef]
205. Sikivie, P. Axion Cosmology. *Lect. Notes Phys.* **2008**, *741*, 19–50. [CrossRef]
206. Peccei, R.D. The Strong CP problem and axions. *Lect. Notes Phys.* **2008**, *741*, 3–17. [CrossRef]
207. Marsh, D.J.E. Axion Cosmology. *Phys. Rep.* **2016**, *643*, 1–79. [CrossRef]

208. Di Luzio, L.; Giannotti, M.; Nardi, E.; Visinelli, L. The landscape of QCD axion models. *Phys. Rep.* **2020**, *870*, 1–117. [CrossRef]
209. 't Hooft, G. Computation of the Quantum Effects Due to a Four-Dimensional Pseudoparticle. *Phys. Rev. D* **1976**, *14*, 3432–3450; Erratum in: *Phys. Rev. D* **1978**, *18*, 2199. [CrossRef]
210. Schäfer, T.; Shuryak, E.V. Instantons in QCD. *Rev. Mod. Phys.* **1998**, *70*, 323–426. [CrossRef]
211. Callan, C.G., Jr.; Dashen, R.F.; Gross, D.J. Toward a Theory of the Strong Interactions. *Phys. Rev. D* **1978**, *17*, 2717. [CrossRef]
212. Abel, C.; Afach, S.; Ayres, N.J.; Baker, C.A.; Ban, G.; Bison, G.; Zsigmond, G. Measurement of the permanent electric dipole moment of the neutron. *Phys. Rev. Lett.* **2020**, *124*, 081803. [CrossRef]
213. Peccei, R.D.; Quinn, H.R. CP Conservation in the Presence of Instantons. *Phys. Rev. Lett.* **1977**, *38*, 1440–1443. [CrossRef]
214. Dvali, G.; Zell, S. Classicality and Quantum Break-Time for Cosmic Axions. *J. Cosmol. Astropart. Phys.* **2018**, *7*, 64. [CrossRef]
215. Kawasaki, M.; Nakayama, K. Axions: Theory and Cosmological Role. *Ann. Rev. Nucl. Part. Sci.* **2013**, *63*, 69–95. [CrossRef]
216. Adler, S.L. Axial vector vertex in spinor electrodynamics. *Phys. Rev.* **1969**, *177*, 2426–2438. [CrossRef]
217. Bell, J.S.; Jackiw, R. A PCAC puzzle: $\pi^0 \to \gamma\gamma$ in the σ model. *Nuovo Cim. A* **1969**, *60*, 47–61. [CrossRef]
218. Di Vecchia, P.; Veneziano, G. Chiral Dynamics in the Large n Limit. *Nucl. Phys. B* **1980**, *171*, 253–272. [CrossRef]
219. Alles, B.; D'Elia, M.; Di Giacomo, A. Topological susceptibility in full QCD at zero and finite temperature. *Phys. Lett. B* **2000**, *483*, 139–143. [CrossRef]
220. Gattringer, C.; Hoffmann, R.; Schaefer, S. The Topological susceptibility of SU(3) gauge theory near T(c). *Phys. Lett. B* **2002**, *535*, 358–362. [CrossRef]
221. Bernard, V.; Descotes-Genon, S.; Toucas, G. Topological susceptibility on the lattice and the three-flavour quark condensate. *J. High Energy Phys.* **2012**, *6*, 051. [CrossRef]
222. Bonati, C.; D'Elia, M.; Panagopoulos, H.; Vicari, E. Change of θ Dependence in 4D SU(N) Gauge Theories Across the Deconfinement Transition. *Phys. Rev. Lett.* **2013**, *110*, 252003. [CrossRef]
223. Berkowitz, E.; Buchoff, M.I.; Rinaldi, E. Lattice QCD input for axion cosmology. *Phys. Rev. D* **2015**, *92*, 034507. [CrossRef]
224. Kitano, R.; Yamada, N. Topology in QCD and the axion abundance. *J. High Energy Phys.* **2015**, *10*, 136. [CrossRef]
225. Borsanyi, S.; Dierigl, M.; Fodor, Z.; Katz, S.D.; Mages, S.W.; Nogradi, D.; Redondo, J.; Ringwald, A.; Szabo, K.K. Axion cosmology, lattice QCD and the dilute instanton gas. *Phys. Lett. B* **2016**, *752*, 175–181. [CrossRef]
226. Bonati, C.; D'Elia, M.; Mariti, M.; Martinelli, G.; Mesiti, M.; Negro, F.; Sanfilippo, F.; Villadoro, G. Axion phenomenology and θ-dependence from $N_f = 2+1$ lattice QCD. *J. High Energy Phys.* **2016**, *3*, 155. [CrossRef]
227. Taniguchi, Y.; Kanaya, K.; Suzuki, H.; Umeda, T. Topological susceptibility in finite temperature (2+1)-flavor QCD using gradient flow. *Phys. Rev. D* **2017**, *95*, 054502. [CrossRef]
228. Petreczky, P.; Schadler, H.P.; Sharma, S. The topological susceptibility in finite temperature QCD and axion cosmology. *Phys. Lett. B* **2016**, *762*, 498–505. [CrossRef]
229. Grilli di Cortona, G.; Hardy, E.; Pardo Vega, J.; Villadoro, G. The QCD axion, precisely. *J. High Energy Phys.* **2016**, *1*, 34. [CrossRef]
230. Kibble, T.W.B. Topology of Cosmic Domains and Strings. *J. Phys. A* **1976**, *9*, 1387–1398. [CrossRef]
231. Saikawa, K. A review of gravitational waves from cosmic domain walls. *Universe* **2017**, *3*, 40. [CrossRef]
232. Penrose, R. Gravitational collapse: The role of general relativity. *Riv. Nuovo Cim.* **1969**, *1*, 252–276. [CrossRef]
233. Zel'dovich, Y.B. Generation of Waves by a Rotating Body. *Zh. Eksp. Teor. Fiz.* **1971**, *14*, 180.
234. Vinyoles, N.; Serenelli, A.; Villante, F.L.; Basu, S.; Redondo, J.; Isern, J. New axion and hidden photon constraints from a solar data global fit. *J. Cosmol. Astropart. Phys.* **2015**, *10*, 15. [CrossRef]
235. Dent, J.B.; Dutta, B.; Newstead, J.L.; Thompson, A. Inverse Primakoff Scattering as a Probe of Solar Axions at Liquid Xenon Direct Detection Experiments. *Phys. Rev. Lett.* **2020**, *125*, 131805. [CrossRef] [PubMed]
236. Ballesteros, G.; Redondo, J.; Ringwald, A.; Tamarit, C. Standard Model—Axion—Seesaw—Higgs portal inflation. Five problems of particle physics and cosmology solved in one stroke. *J. Cosmol. Astropart. Phys.* **2017**, *8*, 1. [CrossRef]
237. Arcadi, G.; Djouadi, A.; Raidal, M. Dark Matter through the Higgs portal. *Phys. Rep.* **2020**, *842*, 1–180. [CrossRef]
238. Ge, S.; Siddiqui, M.S.R.; Van Waerbeke, L.; Zhitnitsky, A. Impulsive radio events in quiet solar corona and axion quark nugget dark matter. *Phys. Rev. D* **2020**, *102*, 123021. [CrossRef]
239. Ipek, S.; Tait, T.M.P. Early Cosmological Period of QCD Confinement. *Phys. Rev. Lett.* **2019**, *122*, 112001. [CrossRef]
240. Berger, D.; Ipek, S.; Tait, T.M.P.; Waterbury, M. Dark Matter Freeze Out during an Early Cosmological Period of QCD Confinement. *J. High Energy Phys.* **2020**, *7*, 192. [CrossRef]
241. Addazi, A.; Marcianò, A.; Pasechnik, R.; Zeng, K.A. QCD surprises: Strong CP problem, neutrino mass, Dark Matter and Dark Energy. *Phys. Dark Univ.* **2022**, *36*, 101007. [CrossRef]
242. Zee, A. *Einstein Gravity in a Nutshell*; Princeton University Press: Princeton, NJ, USA, 2013.
243. Zel'dovich, Y.B. The equation of state at ultrahigh densities and its relativistic limitations. *Zh. Eksp. Teor. Fiz.* **1961**, *41*, 1609–1615.
244. Trojan, E.; Vlasov, G.V. Thermodynamics of exotic matter with constant $w = P/E$. *arXiv* **2011**, arXiv:1108.0824.
245. Kapusta, J.I. *Finite Temperature Field Theory*; Cambridge Monographs on Mathematical Physics; Cambridge University Press: Cambridge, UK, 1989.
246. Rhoades, C.E., Jr.; Ruffini, R. Maximum mass of a neutron star. *Phys. Rev. Lett.* **1974**, *32*, 324–327. [CrossRef]
247. Blaschke, D.; Cierniak, M. Studying the onset of deconfinement with multi-messenger astronomy of neutron stars. *Astron. Nachr.* **2021**, *342*, 227–233. [CrossRef]
248. Dutta, S.; Scherrer, R.J. Big Bang nucleosynthesis with a stiff fluid. *Phys. Rev. D* **2010**, *82*, 083501. [CrossRef]

249. Stiele, R.; Boeckel, T.; Schaffner-Bielich, J. Cosmological implications of a Dark Matter self-interaction energy density. *Phys. Rev. D* **2010**, *81*, 123513. [CrossRef]
250. Mathew, T.K.; Aswathy, M.B.; Manoj, M. Cosmology and thermodynamics of FLRW universe with bulk viscous stiff fluid. *Eur. Phys. J. C* **2014**, *74*, 3188. [CrossRef]
251. Banks, T.; Fischler, W. An Holographic cosmology. *arXiv* **2001**, arXiv:hep-th/0111142.
252. Miguelote, A.Y.; Tomimura, N.A.; Wang, A. Gravitational collapse of selfsimilar perfect fluid in 2+1 gravity. *Gen. Rel. Grav.* **2004**, *36*, 1883–1918. [CrossRef]
253. Dashen, R.; Ma, S.K.; Bernstein, H.J. S Matrix formulation of statistical mechanics. *Phys. Rev.* **1969**, *187*, 345–370. [CrossRef]
254. Welke, G.M.; Venugopalan, R.; Prakash, M. The Speed of sound in an interacting pion gas. *Phys. Lett. B* **1990**, *245*, 137–141. [CrossRef]
255. Lo, P.M. S-matrix formulation of thermodynamics with N-body scatterings. *Eur. Phys. J. C* **2017**, *77*, 533. [CrossRef]
256. Baacke, J. Thermodynamics of a Gas of MIT Bags. *Acta Phys. Polon. B* **1977**, *8*, 625.
257. Fogaca, D.A.; Ferreira Filho, L.G.; Navarra, F.S. Non-linear waves in a Quark Gluon Plasma. *Phys. Rev. C* **2010**, *81*, 055211. [CrossRef]
258. Chodos, A.; Jaffe, R.L.; Johnson, K.; Thorn, C.B.; Weisskopf, V.F. A New Extended Model of Hadrons. *Phys. Rev. D* **1974**, *9*, 3471–3495. [CrossRef]
259. DeTar, C.E.; Donoghue, J.F. BAG MODELS OF HADRONS. *Ann. Rev. Nucl. Part. Sci.* **1983**, *33*, 235–264. [CrossRef]
260. Pisarski, R.D. Effective Theory of Wilson Lines and Deconfinement. *Phys. Rev. D* **2006**, *74*, 121703. [CrossRef]
261. Megias, E.; Ruiz Arriola, E.; Salcedo, L.L. The Quark-antiquark potential at finite temperature and the dimension two gluon condensate. *Phys. Rev. D* **2007**, *75*, 105019. [CrossRef]
262. Zuo, F.; Gao, Y.H. Quadratic thermal terms in the deconfined phase from holography. *J. High Energy Phys.* **2014**, *7*, 147. [CrossRef]
263. Pisarski, R.D. Fuzzy Bags and Wilson Lines. *Prog. Theor. Phys. Suppl.* **2007**, *168*, 276–284. [CrossRef]
264. Schneider, R.A.; Weise, W. On the quasiparticle description of lattice QCD thermodynamics. *Phys. Rev. C* **2001**, *64*, 055201. [CrossRef]
265. Giacosa, F. Analytical study of a gas of gluonic quasiparticles at high temperature: Effective mass, pressure and trace anomaly. *Phys. Rev. D* **2011**, *83*, 114002. [CrossRef]
266. Shuryak, E.V. *The QCD Vacuum, Hadrons and the Superdense Matter*; World Scientific: Singapore, 2004; Volume 71. [CrossRef]
267. Castorina, P.; Miller, D.E.; Satz, H. Trace Anomaly and Quasi-Particles in Finite Temperature SU(N) Gauge Theory. *Eur. Phys. J. C* **2011**, *71*, 1673. [CrossRef]
268. Kou, F.F.; et al. [The FAST collaboration] Periodic and Phase-locked Modulation in PSR B1929+10 Observed with FAST. *Astrophys. J.* **2021**, *909*, 170. [CrossRef]
269. Bacon, D.J.; Battye, R.A.; Bull, P.; Camera, S.; Ferreira, P.G.; Harrison, I.; Parkinson , D.; Pourtsidou, A.; Santos, M.G.; Zuntz, J.; et al. Cosmology with Phase 1 of the Square Kilometre Array: Red Book 2018: Technical specifications and performance forecasts. *Publ. Astron. Soc. Austral.* **2020**, *37*, e007. [CrossRef]
270. Abe, K.T.; Tada, Y.; Ueda, I. Induced gravitational waves as a cosmological probe of the sound speed during the QCD phase transition. *J. Cosmol. Astropart. Phys.* **2021**, *6*, 48. [CrossRef]
271. Linde, A.D. Infrared Problem in Thermodynamics of the Yang-Mills Gas. *Phys. Lett. B* **1980**, *96*, 289–292. [CrossRef]
272. Gross, D.J.; Pisarski, R.D.; Yaffe, L.G. QCD and Instantons at Finite Temperature. *Rev. Mod. Phys.* **1981**, *53*, 43. [CrossRef]
273. Gynther, A.; Vepsalainen, M. Pressure of the standard model at high temperatures. *J. High Energy Phys.* **2006**, *1*, 60. [CrossRef]
274. Prokhorov, G.; Pasechnik, R.; Vereshkov, G. Wave fluctuations in the system with some Yang-Mills condensates. *Phys. Atom. Nucl.* **2016**, *79*, 1502–1504. [CrossRef]
275. Pasechnik, R.; Prokhorov, G.; Vereshkov, G. Conformal Evolution of Waves in the Yang-Mills Condensate: The Quasi-Classical Approach. *J. Mod. Phys.* **2014**, *5*, 209–229. [CrossRef]
276. Prokhorov, G.; Pasechnik, R.; Vereshkov, G. Dynamics of wave fluctuations in the homogeneous Yang-Mills condensate. *J. High Energy Phys.* **2014**, *7*, 3. [CrossRef]
277. Prokhorov, G.; Pasechnik, R. Light meson gas in the QCD vacuum and oscillating Universe. *J. Cosmol. Astropart. Phys.* **2018**, *1*, 17. [CrossRef]
278. Tawfik, A.N.; Mishustin, I. Equation of State for Cosmological Matter at and beyond QCD and Electroweak Eras. *J. Phys. G* **2019**, *46*, 125201. [CrossRef]
279. Laine, M.; Schroder, Y. Quark mass thresholds in QCD thermodynamics. *Phys. Rev. D* **2006**, *73*, 085009. [CrossRef]
280. Tawfik, A.N.; Greiner, C. Bulk viscosity in strong and electroweak matter. *Int. J. Mod. Phys. E* **2021**, *30*, 2150067. [CrossRef]
281. Tawfik, A.N.; Greiner, C. Early Universe Thermodynamics and Evolution in Nonviscous and Viscous Strong and Electroweak epochs: Possible Analytical Solutions. *Entropy* **2021**, *23*, 295. [CrossRef]
282. Bento, M.C.; Bertolami, O.; Sen, A.A. Generalized Chaplygin gas, accelerated expansion and dark energy matter unification. *Phys. Rev. D* **2002**, *66*, 043507. [CrossRef]
283. Chaplygin, S. On gas jets. *Sci. Mem. Mosc. Univ. Math. Phys.* **1904**, *21*, 1.
284. Kamenshchik, A.Y.; Moschella, U.; Pasquier, V. An Alternative to quintessence. *Phys. Lett. B* **2001**, *511*, 265–268. [CrossRef]
285. Shuryak, E.V.; Schäfer, T. The QCD vacuum as an instanton liquid. *Ann. Rev. Nucl. Part. Sci.* **1997**, *47*, 359–394. [CrossRef]
286. Kolb, E.W.; Turner, M.S. *The Early Universe*; Addison-Wesley: San Francisco, CA, USA, 1990; Volume 69.

287. Iorio, A.; Lambiase, G. Thermal relics in cosmology with bulk viscosity. *Eur. Phys. J. C* **2015**, *75*, 115. [CrossRef]
288. Myung, Y.S.; Cho, B.H. Entropy Production in a Hot Heterotic String. *Mod. Phys. Lett. A* **1986**, *1*, 37–41. [CrossRef]
289. Cheng, B. Bulk viscosity in the early universe. *Phys. Lett. A* **1991**, *160*, 329–338. [CrossRef]
290. Brevik, I.; Grøn, O.; de Haro, J.; Odintsov, S.D.; Saridakis, E.N. Viscous Cosmology for Early- and Late-Time Universe. *Int. J. Mod. Phys. D* **2017**, *26*, 1730024. [CrossRef]
291. Carter, G.W.; Ellis, P.J.; Rudaz, S. An Effective Lagrangian with broken scale and chiral symmetry. 3: Mesons at finite temperature. *Nucl. Phys. A* **1997**, *618*, 317–329. [CrossRef]
292. Carter, G.W.; Scavenius, O.; Mishustin, I.N.; Ellis, P.J. An Effective model for hot gluodynamics. *Phys. Rev. C* **2000**, *61*, 045206. [CrossRef]
293. Mocsy, A.; Mishustin, I.N.; Ellis, P.J. Role of fluctuations in the linear sigma model with quarks. *Phys. Rev. C* **2004**, *70*, 015204. [CrossRef]
294. Bowman, E.S.; Kapusta, J.I. Critical Points in the Linear Sigma Model with Quarks. *Phys. Rev. C* **2009**, *79*, 015202. [CrossRef]
295. Chen, H.X.; Imai, S.; Toki, H.; Geng, L.S. Study of hadrons using the Gaussian functional method in the O(4) linear σ model. *Chin. Phys. C* **2015**, *39*, 064103. [CrossRef]
296. Fukushima, K. Chiral effective model with the Polyakov loop. *Phys. Lett. B* **2004**, *591*, 277–284. [CrossRef]
297. Buballa, M. NJL model analysis of quark matter at large density. *Phys. Rep.* **2005**, *407*, 205–376. [CrossRef]
298. Blaschke, D.; Dubinin, A.; Buballa, M. Polyakov-loop suppression of colored states in a quark-meson-diquark plasma. *Phys. Rev. D* **2015**, *91*, 125040. [CrossRef]
299. Shifman, M.A.; Vainshtein, A.I.; Zakharov, V.I. QCD and Resonance Physics. Theoretical Foundations. *Nucl. Phys. B* **1979**, *147*, 385–447. [CrossRef]
300. Cornwall, J.M.; Jackiw, R.; Tomboulis, E. Effective Action for Composite Operators. *Phys. Rev. D* **1974**, *10*, 2428–2445. [CrossRef]
301. Norton, R.E.; Cornwall, J.M. On the Formalism of Relativistic Many Body Theory. *Ann. Phys.* **1975**, *91*, 106. [CrossRef]
302. Amelino-Camelia, G.; Pi, S.Y. Selfconsistent improvement of the finite temperature effective potential. *Phys. Rev. D* **1993**, *47*, 2356–2362. [CrossRef] [PubMed]
303. Gell-Mann, M.; Oakes, R.J.; Renner, B. Behavior of current divergences under SU(3) $\times$ SU(3). *Phys. Rev.* **1968**, *175*, 2195–2199. [CrossRef]
304. Ioffe, B.L. Calculation of Baryon Masses in Quantum Chromodynamics. *Nucl. Phys. B* **1981**, *188*, 317–341; Erratum in: *Nucl. Phys. B* **1981**, *191*, 591–592. [CrossRef]
305. Reinders, L.J.; Rubinstein, H.; Yazaki, S. Hadron Properties from QCD Sum Rules. *Phys. Rep.* **1985**, *127*, 1. [CrossRef]
306. Aghanim, N.; Akrami, Y.; Ashdown, M.; Aumont, J.; Baccigalupi, C.; Ballardini, M.; Banday, A.J.; Barreiro, R.B.; Bartolo, N.; Roudier, G.; et al. Planck 2018 results. VI. Cosmological parameters. *Astron. Astrophys.* **2020**, *641*, A6; Erratum in: *Astron. Astrophys.* **2021**, *652*, C4. [CrossRef]
307. Bull, P.; Akrami, Y.; Adamek, J.; Baker, T.; Bellini, E.; Jimenez, J.B.; Bentivegna, E.; Camera, S.; Clesse, S.; Winther, H.A. et al. Beyond ΛCDM: Problems, solutions, and the road ahead. *Phys. Dark Univ.* **2016**, *12*, 56–99. [CrossRef]
308. Sakharov, A.D. Early stage of Universe expansion and origin of matter inhomogenities. *Sov. Phys. JETP* **1966**, *22*, 241.
309. Sakharov, A.D. Vacuum quantum fluctuations in curved space and the theory of gravitation. *Dokl. Akad. Nauk Ser. Fiz.* **1967**, *177*, 70–71. [CrossRef]
310. Crewther, R.J. Nonperturbative evaluation of the anomalies in low-energy theorems. *Phys. Rev. Lett.* **1972**, *28*, 1421. [CrossRef]
311. Chanowitz, M.S.; Ellis, J.R. Canonical Trace Anomalies. *Phys. Rev. D* **1973**, *7*, 2490–2506. [CrossRef]
312. Collins, J.C.; Duncan, A.; Joglekar, S.D. Trace and Dilatation Anomalies in Gauge Theories. *Phys. Rev. D* **1977**, *16*, 438–449. [CrossRef]
313. Martin, J. Everything You Always Wanted To Know about the Cosmological Constant Problem (But Were Afraid to Ask). *Comptes Rendus Phys.* **2012**, *13*, 566–665. [CrossRef]
314. Sola, J. Cosmological constant and vacuum energy: Old and new ideas. *J. Phys. Conf. Ser.* **2013**, *453*, 012015. [CrossRef]
315. Pasechnik, R.; Beylin, V.; Vereshkov, G. Possible compensation of the QCD vacuum contribution to the dark energy. *Phys. Rev. D* **2013**, *88*, 023509. [CrossRef]
316. Pasechnik, R.; Prokhorov, G.; Teryaev, O. Mirror QCD and Cosmological Constant. *Universe* **2017**, *3*, 43. [CrossRef]
317. Polchinski, J. The Cosmological Constant and the String Landscape. In Proceedings of the 23rd Solvay Conference in Physics: The Quantum Structure of Space and Time, Brussels, Belgium, 1–3 December 2005; World Scientific: Singapore, 2006; pp. 216–236.
318. Copeland, E.J.; Sami, M.; Tsujikawa, S. Dynamics of dark energy. *Int. J. Mod. Phys. D* **2006**, *15*, 1753–1936. [CrossRef]
319. Weinberg, S. The Cosmological Constant Problem. *Rev. Mod. Phys.* **1989**, *61*, 1–23. [CrossRef]
320. Wilczek, F. Foundations and Working Pictures in Microphysical Cosmology. *Phys. Rep.* **1984**, *104*, 143. [CrossRef]
321. Ferroni, L.; Koch, V. Crossover transition in bag-like models. *Phys. Rev. C* **2009**, *79*, 034905. [CrossRef]
322. Novello, M.; Bergliaffa, S.E.P. Bouncing Cosmologies. *Phys. Rep.* **2008**, *463*, 127–213. [CrossRef]
323. Mukhanov, V.F.; Brandenberger, R.H. A Nonsingular universe. *Phys. Rev. Lett.* **1992**, *68*, 1969–1972. [CrossRef]
324. Szydlowski, M.; Godlowski, W.; Krawiec, A.; Golbiak, J. Can the initial singularity be detected by cosmological tests? *Phys. Rev. D* **2005**, *72*, 063504. [CrossRef]
325. Dabrowski, M.P. Oscillating Friedman cosmology. *Ann. Phys.* **1996**, *248*, 199–219. [CrossRef]

326. Savvidy, G.K. Infrared Instability of the Vacuum State of Gauge Theories and Asymptotic Freedom. *Phys. Lett. B* **1977**, *71*, 133–134. [CrossRef]
327. Batalin, I.A.; Matinyan, S.G.; Savvidy, G.K. Vacuum Polarization by a Source-Free Gauge Field. *Sov. J. Nucl. Phys.* **1977**, *26*, 214.
328. Nielsen, N.K.; Olesen, P. An Unstable Yang-Mills Field Mode. *Nucl. Phys. B* **1978**, *144*, 376–396. [CrossRef]
329. Olesen, P. On the QCD Vacuum. *Phys. Scr.* **1981**, *23*, 1000–1004. [CrossRef]
330. Shuryak, E.V. Theory and phenomenology of the QCD vacuum. *Phys. Rep.* **1984**, *115*, 151. [CrossRef]
331. Wilczek, F. Quantum Time Crystals. *Phys. Rev. Lett.* **2012**, *109*, 160401. [CrossRef] [PubMed]
332. Wilczek, F. Wilczek Reply. *Phys. Rev. Lett.* **2013**, *110*, 118902. [CrossRef] [PubMed]
333. Zhang, J.; Hess, P.W.; Kyprianidis, A.; Becker, P.; Lee, A.; Smith, J.; Pagano, G.; Potirniche, I.D.; Potter, A.C.; Monroe, C.; et al. Observation A Discret. Time Crystal. *Nature* **2017**, *543*, 217–220. [CrossRef] [PubMed]
334. Sacha, K.; Zakrzewski, J. Time crystals: A review. *Rep. Prog. Phys.* **2018**, *81*, 016401. [CrossRef]
335. Maiani, L.; Testa, M. Final state interactions from Euclidean correlation functions. *Phys. Lett. B* **1990**, *245*, 585–590. [CrossRef]
336. Glimm, J.; Jaffe, A.M. *Quantum Physics. A Functional Integral Point of View*; Springer: Berlin/Heidelberg, Germany, 1987.
337. Poland, D.; Rychkov, S.; Vichi, A. The Conformal Bootstrap: Theory, Numerical Techniques, and Applications. *Rev. Mod. Phys.* **2019**, *91*, 015002. [CrossRef]
338. Ferrara, S.; Grillo, A.F.; Gatto, R. Tensor representations of conformal algebra and conformally covariant operator product expansion. *Ann. Phys.* **1973**, *76*, 161–188. [CrossRef]
339. Polyakov, A.M. Nonhamiltonian approach to conformal quantum field theory. *Zh. Eksp. Teor. Fiz.* **1974**, *66*, 23–42.
340. Rychkov, S. *EPFL Lectures on Conformal Field Theory in D>= 3 Dimensions*; Springer Briefs in Physics; Springer: New York, NY, USA, 2016. [CrossRef]
341. Simmons-Duffin, D. The Conformal Bootstrap. In *Theoretical Advanced Study Institute in Elementary Particle Physics: New Frontiers in Fields and Strings*; World Scientific: Singapore, 2017; pp. 1–74. [CrossRef]
342. Rattazzi, R.; Rychkov, V.S.; Tonni, E.; Vichi, A. Bounding scalar operator dimensions in 4D CFT. *J. High Energy Phys.* **2008**, *12*, 31. [CrossRef]
343. Kos, F.; Poland, D.; Simmons-Duffin, D.; Vichi, A. Precision Islands in the Ising and $O(N)$ Models. *J. High Energy Phys.* **2016**, *8*, 36. [CrossRef]
344. Matinyan, S.G.; Savvidy, G.K. Vacuum Polarization Induced by the Intense Gauge Field. *Nucl. Phys. B* **1978**, *134*, 539–545. [CrossRef]
345. Savvidy, G. From Heisenberg–Euler Lagrangian to the discovery of Chromomagnetic Gluon Condensation. *Eur. Phys. J. C* **2020**, *80*, 165. [CrossRef]
346. Callan, C.G., Jr. Broken scale invariance in scalar field theory. *Phys. Rev. D* **1970**, *2*, 1541–1547. [CrossRef]
347. Symanzik, K. Small distance behavior in field theory and power counting. *Commun. Math. Phys.* **1970**, *18*, 227–246. [CrossRef]
348. Agasian, N.O. Low-energy relation for the trace of the energy-momentum tensor in QCD and the gluon condensate in a magnetic field. *JETP Lett.* **2016**, *104*, 71–74. [CrossRef]
349. Aharonov, Y.; Casher, A.; Yankielowicz, S. Instantons and Confinement. *Nucl. Phys. B* **1978**, *146*, 256–272. [CrossRef]
350. Vinciarelli, P. Fluxon Solutions in Nonabelian Gauge Models. *Phys. Lett. B* **1978**, *78*, 485–488. [CrossRef]
351. Ambjorn, J.; Olesen, P. A Color Magnetic Vortex Condensate in QCD. *Nucl. Phys. B* **1980**, *170*, 265–282. [CrossRef]
352. Del Debbio, L.; Faber, M.; Greensite, J.; Olejnik, S. Casimir scaling versus Abelian dominance in QCD string formation. *Phys. Rev. D* **1996**, *53*, 5891–5897. [CrossRef]
353. Del Debbio, L.; Faber, M.; Greensite, J.; Olejnik, S. Center dominance and Z(2) vortices in SU(2) lattice gauge theory. *Phys. Rev. D* **1997**, *55*, 2298–2306. [CrossRef]
354. Faber, M.; Greensite, J.; Olejnik, S. Casimir scaling from center vortices: Towards an understanding of the adjoint string tension. *Phys. Rev. D* **1998**, *57*, 2603–2609. [CrossRef]
355. Engelhardt, M.; Langfeld, K.; Reinhardt, H.; Tennert, O. Interaction of confining vortices in SU(2) lattice gauge theory. *Phys. Lett. B* **1998**, *431*, 141–146. [CrossRef]
356. Greensite, J. *An Introduction to the Confinement Problem*; Springer: Berlin/Heidelberg, Germany, 2020; Volume 972. [CrossRef]
357. Wetterich, C. Exact evolution equation for the effective potential. *Phys. Lett. B* **1993**, *301*, 90–94. [CrossRef]
358. Reuter, M.; Wetterich, C. Gluon condensation in nonperturbative flow equations. *Phys. Rev. D* **1997**, *56*, 7893–7916. [CrossRef]
359. Gies, H. Running coupling in Yang-Mills theory: A flow equation study. *Phys. Rev. D* **2002**, *66*, 025006. [CrossRef]
360. Eichhorn, A.; Gies, H.; Pawlowski, J.M. Gluon condensation and scaling exponents for the propagators in Yang-Mills theory. *Phys. Rev. D* **2011**, *83*, 045014; Erratum in: *Phys. Rev. D* **2011**, *83*, 069903. [CrossRef]
361. Khvedelidze, A.M.; Pavel, H.P. Unconstrained Hamiltonian formulation of SU(2) gluodynamics. *Phys. Rev. D* **1999**, *59*, 105017. [CrossRef]
362. Khvedelidze, A.M.; Mladenov, D.M.; Pavel, H.P.; Ropke, G. Unconstrained SU(2) Yang-Mills theory with topological term in the long wavelength approximation. *Phys. Rev. D* **2003**, *67*, 105013. [CrossRef]
363. Cervero, J.; Jacobs, L. Classical Yang-Mills Fields in a Robertson-walker Universe. *Phys. Lett. B* **1978**, *78*, 427–429. [CrossRef]
364. Henneaux, M.; Shepley, L.C. Lagrangians for spherically symmetric potentials. *J. Math. Phys.* **1982**, *23*, 2101–2107. [CrossRef]
365. Hosotani, Y. Exact Solution to the Einstein Yang-Mills Equation. *Phys. Lett. B* **1984**, *147*, 44–46. [CrossRef]
366. Morris, T.R. The Exact renormalization group and approximate solutions. *Int. J. Mod. Phys. A* **1994**, *9*, 2411–2450. [CrossRef]

367. Papenbrock, T.; Wetterich, C. Two loop results from one loop computations and nonperturbative solutions of exact evolution equations. *Z. Phys. C* **1995**, *65*, 519–535. [CrossRef]
368. Fischer, C.S.; Alkofer, R. Infrared exponents and running coupling of SU(N) Yang-Mills theories. *Phys. Lett. B* **2002**, *536*, 177–184. [CrossRef]
369. Fischer, C.S.; Pawlowski, J.M. Uniqueness of infrared asymptotics in Landau gauge Yang-Mills theory. *Phys. Rev. D* **2007**, *75*, 025012. [CrossRef]
370. Fischer, C.S.; Pawlowski, J.M. Uniqueness of infrared asymptotics in Landau gauge Yang-Mills theory II. *Phys. Rev. D* **2009**, *80*, 025023. [CrossRef]
371. Fischer, C.S.; Maas, A.; Pawlowski, J.M. On the infrared behavior of Landau gauge Yang-Mills theory. *Ann. Phys.* **2009**, *324*, 2408–2437. [CrossRef]
372. Ellwanger, U.; Hirsch, M.; Weber, A. Flow equations for the relevant part of the pure Yang-Mills action. *Z. Phys. C* **1996**, *69*, 687–698. [CrossRef]
373. Ellwanger, U.; Hirsch, M.; Weber, A. The Heavy quark potential from Wilson's exact renormalization group. *Eur. Phys. J. C* **1998**, *1*, 563–578. [CrossRef]
374. Bergerhoff, B.; Wetterich, C. Effective quark interactions and QCD propagators. *Phys. Rev. D* **1998**, *57*, 1591–1604. [CrossRef]
375. Pawlowski, J.M.; Litim, D.F.; Nedelko, S.; von Smekal, L. Infrared behavior and fixed points in Landau gauge QCD. *Phys. Rev. Lett.* **2004**, *93*, 152002. [CrossRef] [PubMed]
376. Donà, P.; Marcianò, A.; Zhang, Y.; Antolini, C. Yang-Mills condensate as dark energy: A nonperturbative approach. *Phys. Rev. D* **2016**, *93*, 043012. [CrossRef]
377. Savvidy, G. Gauge field theory vacuum and cosmological inflation without scalar field. *Ann. Phys.* **2022**, *436*, 168681. [CrossRef]
378. Ford, L.H. Inflation Driven by a Vector Field. *Phys. Rev. D* **1989**, *40*, 967. [CrossRef]
379. Golovnev, A.; Mukhanov, V.; Vanchurin, V. Vector Inflation. *J. Cosmol. Astropart. Phys.* **2008**, *6*, 9. [CrossRef]
380. Maleknejad, A.; Sheikh-Jabbari, M.M.; Soda, J. Gauge Fields and Inflation. *Phys. Rep.* **2013**, *528*, 161–261. [CrossRef]
381. Maleknejad, A.; Sheikh-Jabbari, M.M. Gauge-flation: Inflation From Non-Abelian Gauge Fields. *Phys. Lett. B* **2013**, *723*, 224–228. [CrossRef]
382. Maleknejad, A.; Sheikh-Jabbari, M.M. Non-Abelian Gauge Field Inflation. *Phys. Rev. D* **2011**, *84*, 043515. [CrossRef]
383. Alexander, S.; Marciano, A.; Spergel, D. Chern-Simons Inflation and Baryogenesis. *J. Cosmol. Astropart. Phys.* **2013**, *4*, 46. [CrossRef]
384. Koksma, J.F.; Prokopec, T. Fermion Propagator in Cosmological Spaces with Constant Deceleration. *Class. Quant. Grav.* **2009**, *26*, 125003. [CrossRef]
385. Domcke, V.; von Harling, B.; Morgante, E.; Mukaida, K. Baryogenesis from axion inflation. *J. Cosmol. Astropart. Phys.* **2019**, *10*, 32. [CrossRef]
386. Alexander, S.; Jyoti, D.; Kosowsky, A.; Marciano, A. Dynamics of Gauge Field Inflation. *J. Cosmol. Astropart. Phys.* **2015**, *5*, 5. [CrossRef]
387. Stueckelberg, E.C.G. Interaction forces in electrodynamics and in the field theory of nuclear forces. *Helv. Phys. Acta* **1938**, *11*, 299–328.
388. Adshead, P.; Wyman, M. Chromo-Natural Inflation: Natural inflation on a steep potential with classical non-Abelian gauge fields. *Phys. Rev. Lett.* **2012**, *108*, 261302. [CrossRef]
389. Barvinsky, A.O.; Kamenshchik, A.Y. Cosmological landscape from nothing: Some like it hot. *J. Cosmol. Astropart. Phys.* **2006**, *9*, 14. [CrossRef]
390. Barvinsky, A.O.; Kamenshchik, A.Y. Thermodynamics via Creation from Nothing: Limiting the Cosmological Constant Landscape. *Phys. Rev. D* **2006**, *74*, 121502. [CrossRef]
391. Barvinsky, A.O. Why there is something rather than nothing (out of everything)? *Phys. Rev. Lett.* **2007**, *99*, 071301. [CrossRef]
392. Barvinsky, A.O.; Zhitnitsky, A.R. Inflation and gauge field holonomy. *Phys. Rev. D* **2018**, *98*, 045008. [CrossRef]
393. Zhitnitsky, A.R. Inflaton as an auxiliary topological field in a QCD-like system. *Phys. Rev. D* **2014**, *89*, 063529. [CrossRef]
394. Zhitnitsky, A.R. Cosmological perturbations in \barQCD- inflation. Estimates confronting the observations, including BICEP2. *Phys. Rev. D* **2014**, *90*, 043504. [CrossRef]
395. Hirano, T.; van der Kolk, N.; Bilandzic, A. Hydrodynamics and Flow. *Lect. Notes Phys.* **2010**, *785*, 139–178. [CrossRef]
396. Kovtun, P. Lectures on hydrodynamic fluctuations in relativistic theories. *J. Phys. A* **2012**, *45*, 473001. [CrossRef]
397. Bjorken, J.D. Highly Relativistic Nucleus-Nucleus Collisions: The Central Rapidity Region. *Phys. Rev. D* **1983**, *27*, 140–151. [CrossRef]
398. Chojnacki, M.; Florkowski, W.; Csorgo, T. On the formation of Hubble flow in little bangs. *Phys. Rev. C* **2005**, *71*, 044902. [CrossRef]
399. Csanád, M.; Nagy, M.I.; Jiang, Z.F.; Csörgő, T. A simple family of solutions of relativistic viscous hydrodynamics for fireballs with Hubble flow and ellipsoidal symmetry. In *Gribov-90 Memorial Volume*; World Scientific: Singapore, 2019; pp. 275–296. [CrossRef]
400. Csorgo, T.; Kasza, G. New, multipole solutions of relativistic, viscous hydrodynamics. In *Gribov-90 Memorial Volume*; World Scientific: Singapore, 2020; pp. 297–318. [CrossRef]
401. Maartens, R. Causal thermodynamics in relativity. *arXiv* **1996**, arXiv:astro-ph/9609119.

402. Muronga, A. Causal theories of dissipative relativistic fluid dynamics for nuclear collisions. *Phys. Rev. C* **2004**, *69*, 034903. [CrossRef]
403. Hiscock, W.A.; Lindblom, L. Stability and causality in dissipative relativistic fluids. *Ann. Phys.* **1983**, *151*, 466–496. [CrossRef]
404. Gale, C.; Jeon, S.; Schenke, B. Hydrodynamic Modeling of Heavy-Ion Collisions. *Int. J. Mod. Phys. A* **2013**, *28*, 1340011. [CrossRef]
405. Jaiswal, A.; Roy, V. Relativistic hydrodynamics in heavy-ion collisions: General aspects and recent developments. *Adv. High Energy Phys.* **2016**, *2016*, 9623034. [CrossRef]
406. Bravo Medina, S.; Nowakowski, M.; Batic, D. Viscous Cosmologies. *Class. Quant. Grav.* **2019**, *36*, 215002. [CrossRef]
407. Bemfica, F.S.; Disconzi, M.M.; Noronha, J. First-Order General-Relativistic Viscous Fluid Dynamics. *Phys. Rev. X* **2022**, *12*, 021044. [CrossRef]
408. Deur, A.; Brodsky, S.J.; de Teramond, G.F. The QCD Running Coupling. *Nucl. Phys.* **2016**, *90*, 1. [CrossRef]

Review

New Physics of Strong Interaction and Dark Universe

Vitaly Beylin [1], Maxim Khlopov [1,2,3,*], Vladimir Kuksa [1] and Nikolay Volchanskiy [1,4]

1 Institute of Physics, Southern Federal University, Stachki 194, 344090 Rostov on Don, Russia; beylinv@sfedu.ru (V.B.); vkuksa47@mail.ru (V.K.); nvolchanskiy@sfedu.ru (N.V.)
2 CNRS, Astroparticule et Cosmologie, Université de Paris, F-75013 Paris, France
3 National Research Nuclear University "MEPHI" (Moscow State Engineering Physics Institute), 31 Kashirskoe Chaussee, 115409 Moscow, Russia
4 Bogoliubov Laboratory of Theoretical Physics, Joint Institute for Nuclear Research, Joliot-Curie 6, 141980 Dubna, Russia
* Correspondence: khlopov@apc.univ-paris7.fr; Tel.:+33-676380567

Received: 18 September 2020; Accepted: 21 October 2020; Published: 26 October 2020

Abstract: The history of dark universe physics can be traced from processes in the very early universe to the modern dominance of dark matter and energy. Here, we review the possible nontrivial role of strong interactions in cosmological effects of new physics. In the case of ordinary QCD interaction, the existence of new stable colored particles such as new stable quarks leads to new exotic forms of matter, some of which can be candidates for dark matter. New QCD-like strong interactions lead to new stable composite candidates bound by QCD-like confinement. We put special emphasis on the effects of interaction between new stable hadrons and ordinary matter, formation of anomalous forms of cosmic rays and exotic forms of matter, like stable fractionally charged particles. The possible correlation of these effects with high energy neutrino and cosmic ray signatures opens the way to study new physics of strong interactions by its indirect multi-messenger astrophysical probes.

Keywords: cosmology; particle physics; physics beyond the standard model; particle symmetry; stable particles; dark matter; cosmic rays

1. Introduction

The modern standard model of cosmology, involving inflation, baryosynthesis and dark matter/energy finds its basis beyond the standard model (BSM) of electroweak (EW) and strong interactions, thus moving the physics of the universe to the dark side of the fundamental physics. These phenomena determine the history of the cosmological evolution that resulted in the modern structure of the universe [1–8].

BSM physics not only provides a physical basis for the standard elements of cosmological construction, but also inevitably involves new nonstandard cosmological and astrophysical features (see [9,10] for review and references). Here, we study such features, which may appear as signatures for the new physics of strong interactions.

In the strong interaction of the standard model (QCD), new physics comes from colored states with new quantum numbers. If these new charges are conserved, the lightest particle which possesses this property is stable and can have important cosmological impact. An interesting feature of new stable heavy quarks is their binding by chromo-Coulomb forces in heavy quark clusters with strongly suppressed hadronic interaction, making their properties more close to the features of leptons.

BSM models can involve additional non-abelian symmetry, giving rise to new composite particles, whose constituents are bound by QCD-like confinement. Such states may have exotic features of multiple or fractional charge leptons.

In the present review we discuss predictions of QCD and QCD-like models for possible new forms of stable matter and dark matter candidates (Section 2) as well as their effects and multi-messenger probes in the galaxy (Section 3).

2. New Physics from QCD and QCD-Like Models

2.1. General Features of New Physics of Strong Interactions in Dark Cosmology

Models adding new symmetry to the symmetry of the standard model (SM) predict new conserved quantum numbers, which provide stability of the lightest particle that possesses them. Such a particle can also possess QCD color and constitute new stable hadrons.

The addition of new non-abelian symmetry involves QCD-like interactions binding new QCD-like constituents in new composite states. Such composite states do not have ordinary hadronic interaction and look like leptons in their interaction with baryonic matter.

These new stable particles can be dark matter (DM) candidates, since even hadronic interaction with cosmological plasma is not sufficiently strong to hinder the decoupling of gas of stable hadrons.

2.1.1. New Stable Quarks

Hadronic dark matter (HaDM) is one of the natural variants of strongly interacting dark matter scenario. The most popular candidates, weakly interacting massive particles (WIMP), have not yet been discovered. Moreover, strong restrictions on WIMP-nucleon scattering cross section [11] exclude some variants of WIMP scenarios. Some scenarios of DM with strongly interacting massive particles (SIMP) are presented in Refs. [10,12–15]. As was noted in the introduction, usually strongly self-interacting dark matter scenarios are realized by introducing extra groups of gauge symmetry and additional sets of fields. Here, we consider the scenario with hadronic DM, which contains a minimal set of new fields. Namely, heavy singlet quark. This scenario is realized in the framework of grand unification scheme (for example, E_6-theory or $SU(5)$ supersymmetric extension contain $SU(2)$-singlet quark). We should note that hadronic DM is not only self-interacting; hadron-type DM particles strongly interact with ordinary matter. Here, we give a brief description of the origin and main properties of new stable quarks which enter the new heavy hadrons as DM particles.

In the scenario with heavy HaDM, new hadrons consist of new heavy stable quarks Q and light standard ones, q. New quarks possess standard strong (QCD-type) interactions and, together with standard quarks, form mesonic $M = (qQ)$ and fermionic $F_1 = (qqQ), F_2 = (qQQ), F_3 = (QQQ)$ composite states. New heavy quarks arise in the extension with singlet quarks [14], chiral-symmetric models [15,16] and 4th generation standard model extensions [17–19]. Principal properties of new heavy hadrons and their phenomenology were presented in Refs. [10,14,15]. It was underlined in these works that the repulsive character of DM-nucleon interactions makes it possible to escape rigid cosmochemical restrictions on the relative concentration of anomalous hydrogen and helium [10,16]. In this subsection, we briefly present a theoretical base of the HaDM scenario, which is constructed in the framework of the extension with a singlet quark [14] and chiral-symmetric model [15].

The minimal Lagrangian of the extension of SM with new stable quarks is as follows:

$$L = L^{SM} + L^{Q}, \tag{1}$$

where L^{Q} describes interaction of new quarks Q with gauge bosons. In the case of singlet quark Q_s Lagrangian is defined in standard way:

$$L_s^Q = i\bar{Q}_s\gamma^{\mu}(\partial_\mu - ig_1 q_Q V_\mu - ig_{st} t_a G^a_\mu)Q_s - M_Q \bar{Q}_s Q_s. \tag{2}$$

In (2), the matrix $t_a = \lambda_a/2$ are generators of $SU_C(3)$-group and M_Q is the mass parameter of Q.

The chiral-symmetric extension of SM has an additional set of up and down quarks with anti-symmetric (with respect to standard one) chiral structure:

$$Q = \{Q_R = (U_R, D_R);\ \ U_L,\ \ D_L\} \tag{3}$$

The structure of covariant derivatives follows from this definition:

$$\begin{aligned} D_\mu Q_R =&(\partial_\mu - ig_1 Y_Q V_\mu - \frac{ig_2}{2}\tau_a V_\mu^a - ig_3 t_i G_\mu^i)Q_R; \\ D_\mu U_L =&(\partial_\mu - ig_1 Y_U V_\mu - ig_3 t_i G_\mu^i)U_L, \\ D_\mu D_L =&(\partial_\mu - ig_1 Y_D V_\mu - ig_3 t_i G_\mu^i)D_L. \end{aligned} \tag{4}$$

In the equations in (4), the values Y_A $(A = Q, U, D)$ are hypercharges of quarks' doublets and singlets and t_i are generators of the $SU_C(3)$-group. The gauge fields V_μ^a are superheavy chiral partners of standard gauge fields. Further, we consider some specific variant of the chiral-symmetric model with vector-like interaction of new quarks with gauge bosons [10].

The structure of interactions in the extension with singlet quarks and in the chiral-symmetric model has a vector nature in both cases. So, we use universal expression for the Lagrangian model of EW interactions:

$$L(Q_a, A, Z) = g_a(c_w A_\mu - s_w Z_\mu)\bar{Q}_a \gamma^\mu Q_a,\ \ Q_a = Q_s, U, D, \tag{5}$$

where $g_a = g_1 q_a$ and $q_a = 2/3$ or $q_a = -1/3$ for the case of up or down new quark. The definitions of charges for quarks U and D in chiral-symmetrical scenario are standard also. These assumptions make it possible to form neutrally coupled states with standard quarks. In the models under consideration, vector interaction of new heavy quarks with gauge vector fields gives small contributions to polarizations. So, the contributions of new quarks into Peskin–Takeuchi (PT) parameters S, T, U, which describe their effects in electroweak physics, are small too. Using standard definitions of the PT parameters and vertices in (5), by a straightforward calculation, we get for the parameters S and U $(T = 0)$ the following expressions:

$$S = -U = \frac{ks_w^4}{9\pi}[-\frac{1}{3} + 2(1 + 2\frac{M_Q^2}{M_Z^2})(1 - \sqrt{\beta}\arctan\frac{1}{\sqrt{\beta}})]. \tag{6}$$

Here, $\beta = 4M_Q^2/M_Z^2 - 1$, $k = 16(4)$ for the case of singlet quark model with the charge $q = 2/3(-1/3)$, and $k = 20$ for the case of chiral-symmetric model. From Equation (6), it follows that for heavy new quarks with mass $M_Q > 500$ GeV the value of parameters $S = -U < 10^{-2}$ is significantly less than the experimental limits [20]:

$$S = 0.00 + 0.11(-0.10),\ \ U = 0.08 \pm 0.11,\ \ T = 0.02 + 0.11(-0.12). \tag{7}$$

Thus, the scenarios with vector-like new heavy quarks are not excluded by EW experimental restrictions. There are, also, EW restrictions which follow from the presence of flavor-changing neutral currents (FCNC). In the scenario under consideration, new quarks do not mix with standard quarks and FCNC at the tree level are absent. So, there are no additional restrictions which follow from the rare processes, such as rare decays of mesons and the mixing in the systems of neutral mesons (oscillations).

The potential of interaction of new heavy mesons and nucleons, as was shown in Ref. [10], has repulsive character. So, new hadrons do not form coupled states with ordinary matter and this effect excludes the formation of anomalous hydrogen and helium. Thus the scenario with hadronic DM does not contradict rigid cosmochemical restrictions on the relative concentration of these elements [10].

2.1.2. QCD-Like Models

For almost forty years, many efforts have been invested in developing SM extensions conjecturing the existence of new nonperturbative BSM physics. In particular, such models can postulate that the Higgs boson is a composite object consisting of some new fundamental constituents held together by an analog of strong force. There are many brilliant reviews surveying different aspects of this huge field in detail, e.g., [21–29].

In this section, we consider some particular variants of models that extend SM by introducing an additional strong sector with heavy vector-like fermions, hyperquarks (H-quarks), charged under an H-color gauge group [30–41]. Depending on H-quark quantum numbers, such models can encompass scenarios with composite Higgs doublets (see, e.g., [42]) or a small mixing between fundamental Higgs fields of SM and composite hadron-like states of new strong sector making the Higgs boson partially composite. Models of this class leave room for the existence of DM candidates whose decays are forbidden by accidental symmetries. Besides, H-color models comply well with electroweak precision constraints, since H-quarks are assumed to be vector-like.

Among the simplest realizations of the scenario described are models with two or three vector-like H-flavors confined by strong H-color force $Sp(2\chi_{\tilde{c}})$, $\chi_{\tilde{c}} \geqslant 1$. The models with H-color group SU(2) [37,43] are included as particular cases in this consideration due to isomorphism $SU(2) = Sp(2)$ [37,43]. The global symmetry group of the strong sector with symplectic H-color group is larger than for the special unitary case—it is the group $SU(2n_F)$ broken spontaneously to $Sp(2n_F)$, with n_F being a number of H-flavors. Going beyond the simplest (two-flavor) model is of interest because the phenomenology of such models is richer involving new fractionally charged states that can be stable. We posit that the extensions of SM under consideration preserve the elementary Higgs doublet in the set of Lagrangian field operators. This doublet mixes with H-hadrons, which makes the physical Higgs partially composite. The same coset $SU(2n_F)/Sp(2n_F)$ can be used to construct composite two Higgs doublet model [42] or little Higgs models [44–49].

It should be also noted that there are multiple options of assigning electroweak quantum numbers to new H-quarks charged under symplectic color gauge group. In the two-flavor case, for example, there are two possibilities. Except for the model with vector-like H-quarks considered in this paper, one can also build a model with one left-handed doublet and two right-handed quark singlets (e.g., see [21]).

Let us consider a model with the symmetry $G = G_{SM} \times Sp(2\chi_{\tilde{c}})$, $\chi_{\tilde{c}} \geqslant 1$, with G_{SM} and $Sp(2\chi_{\tilde{c}})$ being the SM gauge group and a symplectic hypercolor group respectively. In its field content, the model introduces a doublet and a singlet of heavy vector-like H-quarks charged under H-color group. Then, in the renormalizable case, the most general Lagrangian invariant under G reads

$$\mathscr{L} = \mathscr{L}_{SM} - \frac{1}{4} H^{\mu\nu}_{\underline{a}} H^{\underline{a}}_{\mu\nu} + i\bar{Q}\not{D}Q - m_Q\bar{Q}Q + i\bar{S}\not{D}S - m_S\bar{S}S + \delta\mathscr{L}_Y, \tag{8}$$

$$D^\mu Q = \left[\partial^\mu + \frac{i}{2}g_1 Y_Q B^\mu - \frac{i}{2}g_2 W^\mu_a \tau_a - \frac{i}{2}g_{\tilde{c}} H^\mu_{\underline{a}} \lambda_{\underline{a}}\right] Q, \qquad D^\mu S = \left[\partial^\mu + i g_1 Y_S B^\mu - \frac{i}{2}g_{\tilde{c}} H^\mu_{\underline{a}} \lambda_{\underline{a}}\right] S, \tag{9}$$

where $H^\mu_{\underline{a}}$, $\underline{a} = 1 \ldots \chi_{\tilde{c}}(2\chi_{\tilde{c}}+1)$ are hypergluon fields and $H^{\mu\nu}_{\underline{a}}$ are their strength tensors; τ_a are the Pauli matrices; $\lambda_{\underline{a}}$, $\underline{a} = 1 \ldots \chi_{\tilde{c}}(2\chi_{\tilde{c}}+1)$ are $Sp(2\chi_{\tilde{c}})$ generators satisfying the relation

$$\lambda^{T}_{\underline{a}}\omega + \omega\lambda_{\underline{a}} = 0, \tag{10}$$

where T stands for "transpose", ω is an antisymmetric $2\chi_{\tilde{c}} \times 2\chi_{\tilde{c}}$ matrix, $\omega^T\omega = 1$. Hereafter, all underscored indices correspond to representations of the H-color group $Sp(2\chi_{\tilde{c}})$. In the

Lagrangian (8), the contact Yukawa couplings $\delta\mathscr{L}_{Y}$ of the H-quarks and the SM Higgs doublet $\mathscr{H}$ are permitted by the symmetry G if the hypercharges Y_Q and Y_S satisfy an additional linear relation:

$$\delta\mathscr{L}_{Y} = y_{L}\,(\bar{Q}_{L}\mathscr{H})\,S_{R} + y_{R}\,(\bar{Q}_{R}\varepsilon\bar{\mathscr{H}})\,S_{L} + \text{h.c.} \quad \text{for } \frac{Y_Q}{2} - Y_S = +\frac{1}{2}; \tag{11}$$

$$\delta\mathscr{L}_{Y} = y_{L}\,(\bar{Q}_{L}\varepsilon\bar{\mathscr{H}})\,S_{R} + y_{R}\,(\bar{Q}_{R}\mathscr{H})\,S_{L} + \text{h.c.} \quad \text{for } \frac{Y_Q}{2} - Y_S = -\frac{1}{2}. \tag{12}$$

It is a simple exercise to prove that the hypercolor part of the H-quark Lagrangian (8) can be rewritten in terms of a left-handed sextet as follows:

$$\delta\mathscr{L}_{\text{H-quarks, kin}} = i\bar{P}_{L}\not{D}P_{L}, \qquad P_{L} = \begin{pmatrix} Q_{L} \\ \epsilon\omega Q_{R}{}^{C} \\ S_{L} \\ -\omega S_{R}{}^{C} \end{pmatrix}, \qquad D^{\mu}P_{L} = \left[\partial^{\mu} - \frac{i}{2}g_{\tilde{c}}H^{\mu}_{\underline{a}}\lambda_{\underline{a}}\right]P_{L}, \tag{13}$$

where $\epsilon = i\tau_2$, the operation C denotes the charge conjugation. Equation (13) makes it obvious that, in the absence of the electroweak interactions, the H-quark Lagrangian is invariant under an extension of the chiral symmetry—a global SU(6) symmetry, dubbed the Pauli–Gürsey symmetry [50,51]. The subgroups of the SU(6) symmetry include:

- The chiral symmetry $SU(3)_L \times SU(3)_R$;
- SU(4) subgroup corresponding to the two-flavor model without singlet H-quark S;
- Two-flavor chiral group $SU(2)_L \times SU(2)_R$, which is a subgroup of both former subgroups.

The global symmetry is broken both explicitly and dynamically:

- Explicitly—by the electroweak and Yukawa interactions, (9) and (11), and the H-quark masses;
- Dynamically—by H-quark condensate [52,53]:

$$\langle\bar{Q}Q + \bar{S}S\rangle = \frac{1}{2}\langle\bar{P}_{L}M_{0}P_{R} + \bar{P}_{R}M_{0}^{\dagger}P_{L}\rangle, \qquad P_{R} = \omega P_{L}{}^{C}, \qquad M_{0} = \begin{pmatrix} 0 & \varepsilon & 0 \\ \varepsilon & 0 & 0 \\ 0 & 0 & \varepsilon \end{pmatrix}. \tag{14}$$

The condensate (14) is invariant under $Sp(6) \subset SU(6)$ transformations U that satisfy a condition

$$U^{T}M_{0} + M_{0}U = 0, \tag{15}$$

i.e., the global SU(6) symmetry is broken dynamically to its Sp(6) subgroup. The mass terms of H-quarks in (8) could break the symmetry further to $Sp(4) \times Sp(2)$:

$$\delta\mathscr{L}_{\text{H-quarks, masses}} = -\frac{1}{2}\bar{P}_{L}M_{0}'P_{R} + \text{h.c.}, \qquad M_{0}' = -M_{0}'^{T} = \begin{pmatrix} 0 & m_{Q}\varepsilon & 0 \\ m_{Q}\varepsilon & 0 & 0 \\ 0 & 0 & m_{S}\varepsilon \end{pmatrix}. \tag{16}$$

It should be noted that the model under consideration is free of the gauge anomalies and can be easily reconciled with the electroweak precision constraints, since the H-quarks are vector-like, i.e., their electroweak interactions are chirally symmetric.

The effective interactions of H-hadrons can be described in a linear σ-model involving the fundamental (not composite) Higgs doublet $\mathscr{H}$ and constituent H-quarks as independent degrees of freedom. The Lagrangian of the model can be broken down into four sectors—(1) a sector of the constituent H-quarks (containing interactions of the quarks with gauge bosons), (2) Yukawa interactions of the (pseudo)scalars with the H-quarks, (3) a sector of (pseudo)scalar fields (which produces their self-interactions and interactions with the Higgs boson), and (4) terms communicating the explicit breaking of the SU($2n_F$) global symmetry to the effective fields:[1]

$$\mathscr{L}_{\mathrm{L}\sigma} = \mathscr{L}_{\text{H-quarks}} + \mathscr{L}_{\mathrm{Y}} + \mathscr{L}_{\text{scalars}} + \mathscr{L}_{\mathrm{SB}}, \tag{17}$$

$$\mathscr{L}_{\text{H-quarks}} = i\bar{P}_{\mathrm{L}} \not{D} P_{\mathrm{L}}, \qquad \mathscr{L}_{\mathrm{Y}} = -\sqrt{2}\varkappa \left(\bar{P}_{\mathrm{L}} M P_{\mathrm{R}} + \bar{P}_{\mathrm{R}} M^{\dagger} P_{\mathrm{L}} \right), \tag{18}$$

$$\mathscr{L}_{\text{scalars}} = D_{\mu}\mathscr{H}^{\dagger} \cdot D^{\mu}\mathscr{H} + \operatorname{Tr} D_{\mu}M^{\dagger} \cdot D^{\mu}M - U_{\text{scalars}}, \qquad \mathscr{L}_{\mathrm{SB}} = -\zeta\langle \bar{Q}Q + \bar{S}S \rangle (u + \sigma'). \tag{19}$$

Here, $\varkappa$ is a coupling constant; the parameter ζ is proportional to the current mass m_Q of the H-quarks (see [57,58], for example); M is a complex antisymmetric $2n_F \times 2n_F$ matrix of (pseudo)scalar fields whose singlet component develops a v.e.v. $u \sim -\langle \mathrm{Tr}\,(MM_0)\rangle$; the multiplets $P_{\mathrm{L,\,R}}$ correspond now to the constituent H-quarks that are postulated not to interact with H-gluons but interact with the intermediate gauge bosons in the same way as the fundamental H-quarks. The fields transform under the global symmetry SU($2n_F$) as follows:

$$M \to UMU^{\mathrm{T}}, \qquad P_{\mathrm{L}} \to UP_{\mathrm{L}}, \qquad P_{\mathrm{R}} \to \bar{U}P_{\mathrm{R}}, \qquad U \in \mathrm{SU}(2n_F), \tag{20}$$

where $\bar{U}$ is the complex conjugate of U. These transformation laws allow one to define the covariant derivatives of the fields $P_{\mathrm{L,\,R}}$ and M easily (see explicit expressions in [10]).

The physical (pseudo)scalar components of the field M are listed in Table 1. They include heavier analogs of all the light mesons of 3-flavor QCD and a set of H-baryons (H-diquarks)—singlets A and B, and doublets $\mathscr{A}$ and $\mathscr{B}$. The Lagrangian of H-quark–H-hadron interactions reads

$$\begin{aligned}
&\mathscr{L}_{\text{H-quarks}} + \mathscr{L}_{\mathrm{Y}} = i\bar{Q}\not{D}Q + i\bar{S}\not{D}S - \varkappa u\,(\bar{Q}Q + \bar{S}S) \\
&-\varkappa\bar{Q}\left[\sigma' + \frac{1}{\sqrt{3}}f + i\left(\eta + \frac{1}{\sqrt{3}}\eta'\right)\gamma_5 + (a_a + i\pi_a\gamma_5)\,\tau_a\right]Q - \varkappa\bar{S}\left[\sigma' - \frac{2}{\sqrt{3}}f + i\left(\eta - \frac{2}{\sqrt{3}}\eta'\right)\gamma_5\right]S \\
&-\sqrt{2}\varkappa\left[(\bar{Q}\mathscr{K}^{\star})\,S + i\,(\bar{Q}\mathscr{K})\,\gamma_5 S + \text{h.c.}\right] - \sqrt{2}\varkappa\left[(\bar{Q}\mathscr{A})\,\omega S^{\mathrm{C}} + i\,(\bar{Q}\mathscr{B})\,\gamma_5\omega S^{\mathrm{C}} + \text{h.c.}\right] \\
&-\frac{\varkappa}{\sqrt{2}}\left(A\bar{Q}\varepsilon\omega Q^{\mathrm{C}} + iB\bar{Q}\gamma_5\varepsilon\omega Q^{\mathrm{C}} + \text{h.c.}\right),
\end{aligned} \tag{21}$$

$$D_{\mu}Q = \partial_{\mu}Q + \frac{i}{2}g_1 Y_Q B_{\mu}Q - \frac{i}{2}g_2 W^a_{\mu}\tau_a Q, \qquad D_{\mu}S = \partial_{\mu}S + ig_1 Y_S B_{\mu}S, \tag{22}$$

where $\mathscr{K}^{\star}$, $\mathscr{K}$ and $\mathscr{A}$, $\mathscr{B}$ are SU$(2)_{\mathrm{L}}$ doublets of H-mesons and H-baryons respectively. The Lagrangian for the case of two-flavor model, $n_F = 2$, is obtained by simply neglecting all terms with the singlet H-quark S in Equation (21).

The kinetic terms of the (pseudo)scalars in the Lagrangian (19) produce interactions of the H-hadrons with the gauge bosons:

$$\mathscr{T}_{\text{scalars}} = \frac{1}{2}\sum_{\varphi} D_{\mu}\varphi \cdot D^{\mu}\varphi + \sum_{\Phi}\left(D_{\mu}\Phi\right)^{\dagger} D^{\mu}\Phi + D_{\mu}\bar{A} \cdot D^{\mu}A + D_{\mu}\bar{B} \cdot D^{\mu}B, \tag{23}$$

where $\varphi = h,\, h_a,\, \pi_a,\, a_a,\, \sigma,\, f,\, \eta,\, \eta'$ are singlet and triplet fields, $\Phi = \mathscr{K},\, \mathscr{K}^{\star},\, \mathscr{A},\, \mathscr{B}$ are doublets. The fields h and h_a, $a = 1, 2, 3$ are components of the fundamental Higgs doublet $\mathscr{H} = \frac{1}{\sqrt{2}}(h + ih_a\tau_a)\,\binom{0}{1}$. The covariant derivatives in the Lagrangian (23) are defined as follows:

[1] Although the non-invariant terms responsible for the explicit symmetry breaking can be chosen in a variety of ways [54–56], we constrain ourselves to the most obvious tadpole-like one (as in [57,58], for example).

$$D_\mu h = \partial_\mu h + \frac{1}{2}(g_1\delta_3^a B_\mu + g_2 W_\mu^a)h_a, \qquad D_\mu\phi = \partial_\mu\phi, \quad \phi = \sigma, f, \eta, \eta', \tag{24}$$

$$D_\mu h_a = \partial_\mu h_a - \frac{1}{2}(g_1\delta_3^a B_\mu + g_2 W_\mu^a)h - \frac{1}{2}e_{abc}(g_1\delta_3^b B_\mu - g_2 W_\mu^b)h_c, \tag{25}$$

$$D_\mu M_a = \partial_\mu M_a + g_2 e_{abc} W_\mu^b M_c, \quad M = \pi, a, \qquad D_\mu Z = \partial_\mu Z + ig_1 Y_Q B_\mu Z, \quad Z = A, B, \tag{26}$$

$$D_\mu \mathscr{K} = \left[\partial_\mu + ig_1\left(\frac{Y_Q}{2} - Y_S\right)B_\mu - \frac{i}{2}g_2 W_\mu^a \tau^a\right]\mathscr{K}, \tag{27}$$

$$D_\mu \mathscr{K}^\star = D_\mu \mathscr{K}\big|_{\mathscr{K}\to\mathscr{K}^\star}, \quad D_\mu \mathscr{A} = D_\mu \mathscr{K}\Big|_{\substack{\mathscr{K}\to\mathscr{A} \\ Y_S\to -Y_S}}, \quad D_\mu \mathscr{B} = D_\mu \mathscr{K}\Big|_{\substack{\mathscr{K}\to\mathscr{B} \\ Y_S\to -Y_S}}. \tag{28}$$

Table 1. The lightest (pseudo)scalar H-hadrons in Sp($2\chi_c$) model with two and three flavors of H-quarks (in the limit of vanishing mixings). The lower half of the table lists the states present only in the three-flavor version of the model containing the singlet H-quark S. T is the weak isospin. $\tilde{G}$ denotes hyper-G-parity of a state. $\tilde{B}$ is the H-baryon number. $Q_{\rm em}$ is the electric charge (in units of the positron charge $e = |e|$). The H-quark charges are $Q^U_{\rm em} = (Y_Q+1)/2$, $Q^D_{\rm em} = (Y_Q-1)/2$, and $Q^S_{\rm em} = Y_S$, which is seen from (22).

State	H-Quark Current	$T^{\tilde{G}}(J^{PC})$	$\tilde{B}$	$Q_{\rm em}$
σ	$\bar{Q}Q + \bar{S}S$	$0^+(0^{++})$	0	0
η	$i(\bar{Q}\gamma_5 Q + \bar{S}\gamma_5 S)$	$0^+(0^{-+})$	0	0
a_k	$\bar{Q}\tau_k Q$	$1^-(0^{++})$	0	$\pm1, 0$
π_k	$i\bar{Q}\gamma_5\tau_k Q$	$1^-(0^{-+})$	0	$\pm1, 0$
A	$\bar{Q}^C\varepsilon\omega Q$	$0\ (0^-\ )$	1	Y_Q
B	$i\bar{Q}^C\varepsilon\omega\gamma_5 Q$	$0\ (0^+\ )$	1	Y_Q
f	$\bar{Q}Q - 2\bar{S}S$	$0^+(0^{++})$	0	0
η'	$i(\bar{Q}\gamma_5 Q - 2\bar{S}\gamma_5 S)$	$0^+(0^{-+})$	0	0
$\mathscr{K}^\star$	$\bar{S}Q$	$\frac{1}{2}\ (0^+\ )$	0	$Y_Q/2 - Y_S \pm 1/2$
$\mathscr{K}$	$i\bar{S}\gamma_5 Q$	$\frac{1}{2}\ (0^-\ )$	0	$Y_Q/2 - Y_S \pm 1/2$
$\mathscr{A}$	$\bar{S}^C\omega Q$	$\frac{1}{2}\ (0^-\ )$	1	$Y_Q/2 + Y_S \pm 1/2$
$\mathscr{B}$	$i\bar{S}^C\omega\gamma_5 Q$	$\frac{1}{2}\ (0^+\ )$	1	$Y_Q/2 + Y_S \pm 1/2$

For simplicity, we consider only renormalizable interactions of the scalar fields—the Higgs boson and (pseudo)scalar H-hadrons. Therefore, the corresponding potential can be written as follows:

$$U_{\rm scalars} = \sum_{i=0}^{4}\lambda_i I_i + \sum_{0=i\leqslant k=0}^{3}\lambda_{ik} I_i I_k. \tag{29}$$

Here, I_i, $i = 0, 1, 2, 3, 4$ are the lowest dimension invariants

$$I_0 = \mathscr{H}^\dagger\mathscr{H}, \qquad I_1 = \mathrm{Tr}\left(M^\dagger M\right), \qquad I_2 = \mathrm{Re\,Pf}\,M, \qquad I_3 = \mathrm{Im\,Pf}\,M, \qquad I_4 = \mathrm{Tr}\left[\left(M^\dagger M\right)^2\right], \tag{30}$$

where Pf M is the Pfaffian of M defined as

$$\mathrm{Pf}\,M = \frac{1}{2^2 2!}\varepsilon_{abcd}M_{ab}M_{cd} \quad \text{for } n_F = 2, \qquad \mathrm{Pf}\,M = \frac{1}{2^3 3!}\varepsilon_{abcdef}M_{ab}M_{cd}M_{ef} \quad \text{for } n_F = 3, \tag{31}$$

with ε being the $2n_F$-dimensional Levi-Civita symbol ($\varepsilon_{12\ldots(2n_F)} = +1$). In the potential (29), $\lambda_{i2} = \lambda_{i3} = 0$ for all i if $n_F = 3$; the invariant I_3 is CP odd, i.e., $\lambda_3 = 0$ as well as $\lambda_{i3} = 0$ for $i = 0, 1, 2$. Besides we can always set $\lambda_{22} = 0$ because of the identity $I_1^2 - 4I_2^2 - 4I_3^2 - 2I_4 = 0$ that holds for $n_F = 2$. (For $n_F = 3$, the corresponding term is nonrenormalizable and, thus, not taken into account.) One can derive and solve tadpole equations and diagonalize the quadratic forms in the scalar potential to obtain the mass spectrum of the (pseudo)scalar H-hadrons (see [10]). Note that the term $I_0 I_1$ leads to a small mixing of the Higgs field and the singlet H-meson σ' making the Higgs boson partially composite in this model.

If the hypercharges of H-quarks are set to zero, the Lagrangian (8) is invariant under an additional symmetry—hyper G-parity [59,60]:

$$Q^{\tilde{G}} = \varepsilon\omega Q^{C}, \qquad S^{\tilde{G}} = \omega S^{C}. \tag{32}$$

Since H-gluons and all SM fields are left intact by (32), the lightest $\tilde{G}$-odd H-hadron becomes stable. It happens to be the neutral H-pion π^0.

Besides, the numbers of doublet and singlet quarks are conserved in the model (8), because of two global U(1) symmetry groups of the Lagrangian. This makes two H-baryon states stable—the neutral singlet H-baryon B and the lightest state in doublet $\mathscr{B}$, which carries a charge of $\pm 1/2$.

2.2. Exotic States of New Colored Objects

2.2.1. Fractons

Mixed states with nontrivial electroweak and dark QCD charges (or vice versa, dark electroweak and ordinary QCD charges) can appear as fractionally charged colorless states (fractons), as first proposed in [61]. It should be noted that the term "fracton" has appeared later in condensed matter physics [62] defining the density of states on fractals. Here we use the original notion of fracton as fractionally charged colorless state.

One can distinguish leptonic and hadronic X particles, forming correspondingly leptonic and hadronic fractons.

Leptonic fractons are originated from a new lepton like state X-lepton with fractional electromagnetic charge and dark QCD color. Dark QCD confinement binds it with dark QCD quarks in a dark colorless state, which possess fractional electromagnetic charge. Created in early universe X-leptons and their antiparticles are bound with corresponding dark QCD quarks and antiquarks to form leptonic fractons. Similar to the case of free quarks, studied in [63], negatively charged fractons can be bound with ordinary positively charged nuclei in stars thus protecting positively charged fractons from their annihilation. As a result, the amount of primordial fractons cannot decrease and as it is the case that free quarks [63] exceed by several orders of magnitude experimental constraints in the search for fractionally charged particles in the terrestrial matter.

Hadronic fractons appear when X-quarks, having dark electroweak and ordinary QCD charges bind with ordinary quarks in fractional charged colorless states. In baryon asymmetric universe $\bar{X}$ antiquarks are bound with ordinary quarks u in fractionally charged stable meson $\bar{X}u$, while X-quark forms fractionally charged Xud baryon. If dark electromagnetic attraction can overcome ordinary electromagnetic repulsion, in the dense baryonic matter objects X-meson and X-baryon can recombine in charmonium-like $\bar{X}X$, decaying to ordinary particles, and reduce the abundance of fractons below the experimental upper limit.

2.2.2. Fractionally Charged States in QCD-Like Models

As it follows from the above, in the hypercolor SM extension with three doublets of additional H-quarks and in the case of zero hypercharges, the Lagrangian contains interacting field of hypermeson $\mathscr{B}$, the lightest state in doublet, carrying fractional charges $\pm 1/2$. These new H-mesons contain singlet H-quark, s. The problem of such hyperparticle interpretation in $SU(6)$ extension is aggravated by the fact that the model invariance with respect to two $U(1)$ groups ensures of these objects stability (simultaneously with the one for neutral singlet H-baryon B). At the same time, very strict restrictions are imposed on the concentration of fractionally charged particles in the modern universe. It can be said that within the framework of this $SU(6)$ scenario, such fractionally charged particles should either be created in some bound states (possibly, these are an analog of QCD tetraquarks), or effectively annihilated.

To describe their arising as some bound states from the very beginning, it is reasonable to try to rewrite the Lagrangian model in such a manner that compound H-quark objects could interact with other fields. In other words, we should construct effective vertices for the H-tetraquark interactions with gauge bosons and other H-hadrons. Obviously, only electromagnetic interaction of these fractionally charged components cannot explain an appearance of such multi-H-quark states at the early stage of universe evolution. It means that, for this procedure, we need to analyze how hyperstrong interactions of H-quarks and H-gluons can work at very high temperatures and densities producing strongly connected systems instead of using effective σ-model construction. In other words, we need to consider at high energy scale some hyper-QCD with analogous problems of the bound-state description in the framework of effective Lagrangian approach, with an integration over some degrees of freedom or introducing of hyper-vacuum v.e.v.'s for the analysis of the bound states of H-quarks with sum rules method, for example. Then, we would come to consideration of hyper-tetraquarks as H-quark bags, repeating procedures and approaches of orthodox QCD at other scale. It means an investigation of the dynamics and magnitudes of H-quark and H-gluon vacuum condensates, providing an existence of H-quark "bags" with some mass, structure and specific interactions with fields of matter in a hot and dense universe. So, this mechanism of H-strong interaction as the basis for the explaining of neutral H-quark states formation should be carefully considered in detail.

It can be, however, assumed that the symmetry breaking occurs and, correspondingly, fractionally charged objects appear at an early stage of evolution, apparently at the initial stage of inflation when their creation should be accompanied by a rapid transformation (annihilation) into "ordinary" stable neutral carriers of DM and fields of matter. Due to the presence of vertices describing their connection with W and Z-bosons in the σ-model framework, there is both the creation of (still massless) leptons and standard quarks, as well as the transition of fractionally charged particles into neutral components of the DM. Possible quick disappearing of charged stable particles in annihilation reactions can be suggested as an alternative way to save the scenario with $U(1)$ accident symmetries.

It also should be suggested, when the annihilation can be—before or after the hyper-QCD symmetry breaking? In other words, when all masses are zero or when already massive fractionally charged particles transform into the set of H-neutral mesons and ordinary quarks and leptons? Because these variants of high-energy sector of H-fields remains unstudied yet, the question whether some traces of the initial process exist in the modern observable universe, namely, creation and following transformation of fractionally charged stable particles does not have an exact answer.

If we suppose that the process of intensive annihilation into neutral particles and/or states with integer charges takes place mostly through H-strong channels where H-quarks interact via H-gluons, it should be before these exotic H-mesons will scatter across the universe due to inflation. Early possible initial inhomogeneities in distribution of this type of matter should effectively transform into neutral stable objects of the orthodox DM, and in these regions of increased density (prototypes of clumps) also should be produced photons with energies $\sim M_{DM}$ and streams of neutrinos and ordinary leptons and mesons. Perhaps inflation will keep some fingerprints of active processes of creation of photons and neutrino radiation from initial high density domains. However, the long evolution of the universe after inflation will inevitably try to hide the total and exact history of DM clumps' origin accompanied with high energy photons and neutrinos.

An important channel of $\mathscr{B}$-mesons connection with the SM world follows from (23) and (24)—these fractionally charged exotic stable states have EW interactions with W-bosons and also interact with Z-bosons and photons. It means that they participate in the standard set of reactions with ordinary matter, i.e., they annihilate or produced with the EW cross sections $\sim(10^2$–$10^4)$ pb.

There is, however, some specific feature of the scenario—two $\mathscr{B}$-states with the same fractional charges can annihilate into one charged W-boson, which decays into quarks or leptons conserving the charge. Assuming an existence of some initial asymmetry between these exotic states, it can be supposed that this is a source of the following asymmetry between standard quarks and, consequently, between baryons. Certainly, then we shift the baryon asymmetry origin to an early stage of evolution relating the symmetry breaking possibly with initial stage of inflation and processes of the hidden mass generation.

Besides, these stable objects (as the other stable particles in H-color model), if they were to exist, can be produced at collider by decaying virtual vector bosons. These reactions should have small (electroweak) cross sections manifesting itself as events with large missed energy and/or in generation of hadronic jets with corresponding fractional charge associated with H-quarks.

It can be concluded, in the hypercolor extension, as, probably, in any extension with additional (heavy) fermions, new types of stable particles may appear. They can arise as the QCD-like bound states of additional quarks (mesons, baryons or diquarks) or in the framework of the extended σ-model. Dynamics of these new objects is defined by the model gauge symmetry, but spectra of their masses and the scale of their manifestations are unclear in advance. An example is the appearance of fractionally charged stable objects when we extend the hypercolor symmetry from $SU(4)$ to $SU(6)$. The peculiarities of such scenarios need to be considered carefully by studying the burn out kinetics of these stable particles, scales of symmetry breaking induced by new types of vacuum condensates. The H-color symmetry allows to analyze specific features of these extended scenarios predicting the specific and measurable signals of new objects.

2.2.3. Multiple Charged States in QCD and QCD-Like Models

Δ-like states of new stable heavy quarks Q are stable and bound much stronger than ordinary hadrons since their chromo-Coulomb binding energy $\alpha_c^2 m_Q$ exceeds the energy of confinement $\Lambda \sim$ 300 MeV at $m_Q > 7.5$ GeV for QCD running constant $\alpha_c = 0.2$ [64]. If such stable states are charged, they should avoid geochemical constraints on anomalous isotopes [10]. It was first noticed in [65] that this problem can be solved in the case of stable quark of the 4th family, if the U quark is the lightest and thus most stable quark in this family and the generation of baryon asymmetry simultaneously provides generation of excess of $\bar{U}$ quarks. Δ^{--}-like ($\bar{U}\bar{U}\bar{U}$) can be effectively hidden in nuclear interacting dark atom bound with primordial helium, as soon as it is formed in Big Bang Nucleosynthesis.

Models with new nonabelian symmetry can predict much wider class of multiple charged stable particles. Such models can provide non-supersymmetric composite Higgs solution for the SM problems of divergent mass of Higgs boson and of the origin of the scale of electroweak symmetry breaking. This approach acquires special interest in the lack of positive results of the searches for supersymmetric particles at the LHC (see, e.g., [66] for review and references).

The Walking TechniColor (WTC) model involves two techniquarks, U and D transforming under the adjoint representation of a SU(2) technicolor gauge group [67–71]. A neutral techniquark–antiquark state is associated with the Higgs boson. The accompanying prediction is the existence of bosonic technibaryons UU, UD, DD, and their antiparticles. Conservation of the technibaryon number TB leads to stability of the lightest technibaryon.

Electric charges of UU, UD and DD are given in terms of an arbitrary real number q as $q+1$, q, and $q-1$, respectively [7,10,72]. Compensation of anomalies requires existence of technileptons ν' and ζ with charges $(1-3q)/2$ and $(-1-3q)/2$, respectively. If technilepton number L' is conserved the lightest technilepton is stable.

In the early universe sphaleron transitions provide equilibrium relationship between TB, baryon number B, of lepton number L, and L'. Freezing out of these transitions results in a balance between the stable techniparticle excess and the observed baryon asymmetry of the universe [10,72,73]. Stable negatively charged techniparticles are bound in nuclear interacting dark atoms with primordial helium.

In the case of $q = 1$, stable double charged techniparticles are possible [7,10,72,73].

Another choice of parameter q results in a possibility of multiple $-2n$ charged stable techniparticles for $n > 1$. Possible stable multiple charged techniparticles are marked bold in Table 2 [10].

Table 2. List of possible integer charged techniparticles. Candidates for even charged constituents of dark atoms are marked bold [10].

q	$UU(q+1)$	$UD(q)$	$DD(q-1)$	$\nu'(\frac{1-3q}{2})$	$\zeta(\frac{-1-3q}{2})$
1	**2**	1	0	−1	**−2**
3	**4**	3	**2**	**−4**	−5
5	**6**	5	**4**	−7	**−8**
7	**8**	7	**6**	**−10**	−11

2.3. *Strongly Interacting Dark Matter Candidates*

2.3.1. Stable Heavy Quark Hadrons

New heavy quarks possess strong QCD-type interaction, so they can form at hadronization phase of evolution the coupled states—new heavy mesons, $(q\bar{Q})$, and fermions, (qqQ), (qQQ), (QQQ). Classification and the main properties of these new hadrons were considered in [10] for the case of up, U, and down, D, type of new quark Q. As was noted earlier, the lightest neutral meson appears in the scenario with new quark of up type. In Table 3, we represent the main quantum characteristics and quark content of new heavy mesons and fermions which contain new quarks of up type, U.

Table 3. Characteristics and quark content of new hadrons.

J^P	T	Isotopic Content	Quark Content
0^-	$\frac{1}{2}$	$M = (M^0\, M^-)$	$M^0 = \bar{U}u,\ M^- = \bar{U}d$
$\frac{1}{2}$	1	$B_1 = (B_1^{++}\, B_1^{+}\, B_1^{0})$	$B_1^{++} = Uuu, B_1^{+} = Uud, B_1^{0} = Udd$
$\frac{1}{2}$	$\frac{1}{2}$	$B_2 = (B_2^{++}\, B_2^{+})$	$B_2^{++} = UUu, B_2^{+} = UUd$
$\frac{1}{2}$	0	(B_3^{++})	$B_3^{++} = UUU$

In Table 3, the mesons M^0 and baryons B_1^+, B_2^{++}, B_3^{++} are stable and the lightest of them, neutral meson M^0, can be proposed as the DM candidate. The evolution of these new heavy hadrons was briefly considered in Ref. [10], where the process of burning out of heavy baryons was analyzed. Here, we represent the main properties of new heavy mesons, $M^0 = (u\bar{U})$ and $M^- = (d\bar{U})$, which can lead to the characteristic signals of the hadronic dark matter.

We have determined the mass of new mesons from the data on DM relic density with the help of the following equality:

$$(\sigma(M)v_r)^{Mod} = (\sigma v_r)^{Exp}. \tag{33}$$

In Equation (33), the left part of equality is the model value of annihilation cross section and the right part follows from the data on the modern relic concentration of DM, $(\sigma v_r)^{Exp} = 2\times 10^{-9}$ GeV^{-2}. To calculate the model cross section $\sigma(M)$, which is a function of the mass M of meson, we take into account the fact that the freezing-out temperature $T_{freez} \approx M/30$ for the case of heavy DM particles is much greater than the temperature of QCD phase transition, $T_{QCD} \approx 0.15$ GeV. So, there are no coupled hadron states at freezing-out stage and the dynamics of this stage is defined by the process of annihilation of new quark-antiquark pairs, $Q\bar{Q} \to gg, q\bar{q}$, where g is a gluon and q is a standard quark. The total cross section of these strong channels of annihilation was calculated in the limit of massless quarks in the final states and presented in Ref. [10]:

$$(\sigma(M))^{Mod} = \sigma(Q\bar{Q} \to gg, q\bar{q}) \approx \frac{44\pi}{9}\frac{\alpha_s^2}{M^2}. \tag{34}$$

From the expression (34) and equality (33) the estimation of the new quarks mass follows, $M = 10$ TeV, which defines mass scale of new hadrons. Electroweak channels of annihilation, $Q\bar{Q} \to \gamma\gamma,\ ZZ,\ W^+W^-$, give small contribution into the total value of the annihilation cross section. It is known that the Sommerfeld–Gamov–Sakharov (SGS) enhancement effect can significantly modify the value of a cross section. Such enhancement takes place at hadronization stage of evolution, when SGS effect is caused by the light meson exchange [10]. At quark-gluon stage this effect can be caused by γ and Z-boson exchange only. As was shown by numerical calculations [74], in this case the coefficient of enhancement is of the order of unity, i.e., the effect is small.

The value of mass-splitting in the doublet of neutral, $M^0 = (u\bar{U})$, and charged, $M^- = (d\bar{U})$, new heavy mesons plays an important role in HaDM description. We define the value of mass-splitting as follows:

$$\Delta m = m(M^-) - m(M^0). \tag{35}$$

In the case of standard heavy-light mesons the value Δm is of the order of MeV, besides this value is positive for the case of D-meson (up type heavy quark) and negative for the case of K- and B-mesons (down type of heavy quark). New heavy mesons M^0 and M^- are just the case of heavy-light meson, $m_Q \gg m_q$. From the heavy quark symmetry, a direct analogy with standard heavy-light mesons follows, so, in the case under consideration, we can assume that Δm is positive and $\Delta m \sim$ MeV. Moreover, the condition of instability of the charged meson M^- leads to inequality $\Delta m > m_e$, where m_e is the mass of electron. So, the charged partner of neutral DM particle has a unique decay channel with very small phase space in a final state, $M^- \to M^0 e^- \bar{\nu}_e$. The expression for the width of the charged meson is as follows [10]:

$$\Gamma(M^-) = \frac{G_F^2}{60\pi^3}|U_{ud}|^2(\Delta m^5 - m_e^5), \tag{36}$$

where U_{ud} is the element of CKM matrix, which defines the charged transition $d \to uW$. From the expression (36) one can see that at $\Delta m \to m_e$, the value of width $\Gamma(M^-) \to 0$, i.e., the lifetime can be arbitrary large. For instance, at $\Delta m \sim 1$ MeV the lifetime $\tau \sim 10^5$ s. Thus, in the scenario with hadronic DM, new neutral meson M^0, as a DM candidate, has charged metastable partner with the same mass, which should be taken into account in the process of co-annihilation. New heavy charged meson appears in the process of collision of DM with ordinary matter, namely with leptons or nucleons. The cross sections of these processes were calculated in Ref. [75], where the main signals of hadronic DM manifestation are considered.

One more principal feature of hadronic DM scenario is the effect of hyperfine splitting of excited states of new heavy hadrons. First of all, we should note that, in contrast to fine splitting, which is caused by change of quark content ($d \to u$) and has the value of the order of MeV, hyperfine splitting takes place for the states with the same quark content and has much smaller value (of the order of KeV).

Further, we describe the effect of hyperfine splitting of ground and excited states, $\delta M_q = m(M_q^*) - m(M_q)$, where M_q^* is an excited state of heavy meson M. Here, we consider the simplest case of the lowest excited states of the new mesons $M_q = (q\bar{U})$. In a direct analogy with the standard heavy-light (HL) mesons, $D_q = (cq)$ and $B_q = (\bar{b}q)$, we define the ground and excited states in the terms S_0^1 and S_1^1 (the classification with quantum numbers L_J^{2s+1}), or $\frac{1}{2}(0^-)$ and $\frac{1}{2}(1^+)$ (the classification $I(J^P)$). Here, L, s, J, I and $P = (-1)^{1+L}$ are orbital momentum, spin, total momentum of the system, isospin and parity. We designate the ground states $\frac{1}{2}(0^-)$ of the HL mesons as D_q, B_q and M_q, while the excited states—as D_q^*, B_q^* and M_q^*. We evaluate the mass-splitting between the excited and ground states, M_q^* and M_q, in analogy with standard splitting mechanism. This possibility is provided by the heavy quark symmetry which is the basic assumption of heavy quark effective theory (HQET). Heavy quark symmetry leads to the relations between the masses of excited states of B and D mesons [76]:

$$m(B_2) - m(B_1) \approx \frac{m_c}{m_b}(m(D_2) - m(D_1)), \tag{37}$$

where $m(B_k)$ and $m(D_k)$ are masses of B_k and D_k, m_c and m_b are masses of constituent quarks. This expression successfully describes the relation of splitting between the lowest excited $\frac{1}{2}(1^-)$ and ground states $\frac{1}{2}(0^-)$ of B and D mesons:

$$\frac{m(B^*) - m(B)}{m(D^*) - m(D)} \approx \frac{m_c}{m_b} \longrightarrow 0.32 \approx 0.32\,(0.28). \tag{38}$$

In (38), we used $m(B^*) - m(B) = 45$ MeV and $m(D^*) - m(D) = 142$ MeV (see [20]), $m_c = 1.55$ GeV and $m_b = 4.88$ GeV [76]. The value of the relation in parentheses, (0.28), follows from the data $m_c = 1.32$ GeV and $M_b = 4.74$ GeV [77]. In order to evaluate the mass-splitting in the doublet of new mesons $M_q = (q\bar{U})$, we used the relation (37) and took into consideration the equality $m(U) \approx m(M_q) = M$. Using the value of mass $M = 10$ TeV, we get:

$$\frac{\delta m(M)}{\delta m(B)} = \frac{m(M^*) - m(M)}{m(B^*) - m(B)} \approx \frac{m_b}{M} \longrightarrow \delta m(M) \approx \delta m(B)\frac{m_b}{M} \approx 2\,\text{KeV}. \tag{39}$$

Thus, we get very small mass-splitting (hyperfine splitting) δm, which is much less than the fine splitting, $\delta m \ll \Delta m$. This effect follows from the HQFT prediction and is caused by very large mass of new hadrons, i.e., the hyperfine splitting is a specific property of hadronic DM.

The excited state of hadronic DM particles can manifest itself in the processes of interaction of neutral meson M^0 with radiation. Transition to the first excited state of the meson $M^0 = (u\bar{U})$ can be realized through the absorption of photons in KeV range which corresponds to the wavelength $\lambda \sim 10^{-9}$ cm. If we assume that the meson $M^0 = (u\bar{U})$ has the size of the order of nucleon radius, $R_M \sim 10^{-13}$ cm, then $R_M \ll \lambda_{trans}$ and interaction of M^0 with photons is described by the higher terms of multipole expansion of the charge distribution in the composite system $(u\bar{U})$. So, the cross section of γM^0 scattering is small and these mesons can be interpreted as dark matter particles. At $\lambda_{trans} \ll R_M$, i.e., $E_\gamma \gg 10$ MeV, the cross section of interaction γM^0 becomes large and dark matter becomes not absolutely "dark".

Low-energy interaction of new heavy hadrons with the standard leptons is described in spectator approach by the effective Lagrangian of WMM interaction in the differential form [75]:

$$L^{eff}(WMM) = iG_{WM}U_{ik}W^{+\mu}(\bar{M}_{ui}\partial_\mu M_{dk} - \partial_\mu \bar{M}_{ui}M_{dk}) + h.c., \tag{40}$$

where $ui = u, c, t$; $dk = d, s, b$; U_{ik} is corresponding element of CKM matrix, $M_{ui} = (ui\bar{U})$, $M_{dk} = (dk\bar{U})$, and $G_{WM} = gU_{ud}/2\sqrt{2}$. The value of effective coupling constant G_{WM} is equal to the fundamental constant in W-boson interaction with quark. Thus, the spectator approach, which directly follows from the structure of process at the fundamental (quark) level, is valid for the case of low-energy interactions.

The structure of low-energy Lagrangian of Z-boson interaction $L^{eff}(ZMM)$ can be represented in analogy with $L^{eff}(WMM)$ by the simplest differential expression with regard to the preservation of flavor $(qi \to qi)$. In contrast to (40), effective coupling G_{ZM} is caused by interactions of Z with quarks Q and q. So, in this case, we meet the problem of an effective coupling definition.

Inelastic scattering of the low-energy leptons on the new heavy M particles is defined by the t-channel diagram with W-boson in the intermediate state. In the limit of zero lepton mass and mass-splitting ΔM we get the cross section in the form:

$$\sigma(l^- M^0 \to \nu_l M^-) = \frac{3g^4|U_{ud}|^2}{2^{10}\pi M_W^4}s(1 - \frac{M^2}{s})^2, \tag{41}$$

where $\sqrt{s}$ is full energy in the CMS. In the non-relativistic case, expression (41) can be represented in the form:

$$\sigma(l^- M^0 \to \nu_l M^-) = \frac{3G_F^2|U_{ud}|^2}{8\pi}(E_l + W)^2, \tag{42}$$

where E_l is the energy of lepton and $W = Mv^2/2$ is kinetic energy of the non-relativistic M-particle. The process of lepton scattering on M^0 taking into account final states is as follows: $l^- M^0 \to \nu_l M^- \to \nu_l M^0 e^- \bar{\nu}_e$. So, in this process, the neutrino with energy $E_\nu \sim E_l$ appears together with $e^-\bar{\nu}_e$-pair (with total energy $E \sim \delta M$). For the cross section of the process $\nu_l M^0 \to l^- M^+$ we get the same expression due to neglecting lepton mass in Equation (41).

Heavy hadronic DM particles at the modern stage of evolution are non-relativistic, they have an average velocity $\sim 10^{-3}$ with respect to the galaxy. From the kinematics of the heavy DM particles and nucleon collisions it follows (see Ref. [10]) that low-energy interaction can be described by the effective meson-exchange approach. The nucleon-meson interaction was considered in [78] on the basis of the gauge scheme realization of symmetry $U(1) \times SU(3)$. This scheme was developed and applied to the interaction of new heavy mesons with ordinary vector mesons [16,79]. The part of physical Lagrangian which describes the interaction of nucleons and M-mesons with ordinary vector mesons consists of two terms:

$$L_{NMV} = L_{NV} + L_{MV}. \tag{43}$$

In Equation (43) the first term describes interaction of nucleon with usual mesons:

$$\begin{aligned} L_{NV} = {} & g_\omega \omega_\mu (\bar{p}\gamma^\mu p + \bar{n}\gamma^\mu n) + \frac{1}{2} g \rho^0_\mu (\bar{p}\gamma^\mu p - \bar{n}\gamma^\mu n) \\ & + \frac{1}{\sqrt{2}} g \rho^+_\mu \bar{p}\gamma^\mu n + \frac{1}{\sqrt{2}} g \rho^-_\mu \bar{n}\gamma^\mu p, \end{aligned} \tag{44}$$

where $g_\omega = \sqrt{3}g/2\sin\theta$, $g^2/4\pi \approx 3.4$ and $\sin\theta \approx 0.78$.

The second term in Equation (43) describes the interaction of M particles with ordinary vector mesons:

$$\begin{aligned} L_{MV} = {} & iG_{\omega M}\omega^\mu (\bar{M}^0 M^0_{,\mu} - \bar{M}^0_{,\mu} M^0 + M^+_{,\mu} M^- - M^+ M^-_{,\mu}) \\ & + \frac{ig}{2}\rho^0_\mu (\bar{M}^0 M^0_{,\mu} - \bar{M}^0_{,\mu} M^0 + M^+_{,\mu} M^- - M^+ M^-_{,\mu}) \\ & + \frac{ig}{\sqrt{2}}\rho^{+\mu} (\bar{M}^0 M^-_{,\mu} - \bar{M}^0_{,\mu} M^-) + \frac{ig}{\sqrt{2}}\rho^{-\mu} (M^+ M^0_{,\mu} - M^+_{,\mu} M^0). \end{aligned} \tag{45}$$

In Equation (45), the coupling constant $G_{\omega M} = g_\omega/3$. In Ref. [10], it was shown that scalar mesons give very small contribution into NM interaction and we omit it here. Note, the interactions of new mesons with ordinary pseudoscalar mesons (for instance, π-mesons) are absent due to parity conservation. This is an important property which differs new heavy hadrons from the nucleons.

Low-energy scattering of nucleons on new mesons is described by t-channel diagrams with ordinary vector and scalar mesons in the intermediate states. The diagrams with pseudoscalar mesons in the intermediate states are absent at the tree level, while the contribution of scalar mesons is negligible. So, the dominant contribution is given by the vector-meson exchange (the vector mesons being ω and ρ mesons).

Now, we consider the kinematics of elastic scattering process $MN \to MN$, where $M = (M^0, M^-)$ and $N = (p, n)$. For the case of non-relativistic particles, the maximal value of momentum transfer $Q^2 = -q^2$ is $Q^2_{max} = (pk)^2 \approx 4m_N^2 v_r^2$. So, $Q_{max} \approx m_N v_r \sim 10^{-3} m_N$, the value Q_{max} is much less than the mass of the vector mesons m_v ($m_v \sim m_N$) and the meson-exchange model is relevant.

Using the expressions for the vertices from the expressions (44) and (45) we calculated the cross section of the process $N_a M_b \to N_a M_b$:

$$\sigma(N_a M_b \to N_a M_b) = \frac{g^4 m_p^2}{16\pi m_v^4}(1 + \frac{k_{ab}}{\sin^2\theta})^2, \tag{46}$$

where $N_a = (p, n)$, $M_b = (M^0, M^-)$, $g^2/4\pi \approx 3.4$, $\sin\theta = 1/\sqrt{3}$ and $k_{ab} = \pm 1$ for the case of proton, p, and neutron, n. Equation (46) implies a rather large cross section, for example $\sigma(pM^0 \to pM^0) \approx 0.9$ barn. We should note that large cross section of NM-scattering can stipulate large interaction of dark matter halo and galaxy at some stage of their evolution. The problem of interaction between galaxies and dark matter halo was considered in details in Ref. [80]. Analysis of the low-energy elastic scattering $N_a M_b \to N_a M_b$ reveals an important peculiarity of the NM-interaction. Using the connection of potential and amplitude in Born approximation we show that the potential of M-nucleon interaction at large distances ($d \sim m_\rho^{-1}$) has repulsive character [10,16]. So, new heavy hadrons as DM particles do not form coupled states with nucleon at low energy, i.e., at the modern stage of evolution. This effect makes it possible to escape the problem of anomalous hydrogen and helium [10].

The processes of non-elastic scattering of type $N_a M_b \to N_c M_d$, where $N_a = (p, n)$ and $M_b = (M^0, M^-)$, have kinematics which is an analog of elastic scattering kinematics. In this case, the dominant contribution is caused by t-channel diagram with charged $\rho^\pm$-meson in the intermediate state. The structure of expression for the cross section explicitly describes the presence of threshold:

$$\sigma(N_a M_b \to N_c M_d) = \frac{g^4 m}{8\pi v_r m_v^4}\sqrt{2m}[E_a - \Delta_{ab}]^{1/2}, \tag{47}$$

where $E_a \approx m_a v_r^2/2$, $m(N_a) = m_a \approx m_b \approx m$, Δ_{ab} is some combination of mass-splitting $\Delta M = m(M^+) - m(M^0)$ and $\Delta m = m_n - m_p \approx 1.4$ MeV, which depends on the structure of the initial and final states (see Table 2). Expression (47) can be represented in another form:

$$\sigma(N_a M_b \to N_c M_d) = \frac{g^4 m^2}{8\pi m_v^4}[1 - \frac{\Delta_{ab}}{E_p}]^{1/2}. \tag{48}$$

From (47) it can be seen that the process of scattering has a threshold $E_p^{thr} = \Delta_{ab}$ when $\Delta_{ab} > 0$.

In Table 4, we present the expressions for the threshold in the case of basic reactions, namely $pM^0 \to nM^+$, $nM^+ \to pM^0$, $nM^0 \to pM^-$ and $pM^- \to nM^0$.

Table 4. The threshold parameters Δ_{ab}.

$N_a M_b \to N_c M_d$	$\Delta_{ab} = f(\Delta M, \Delta m)$	Signum Δ_{ab}
$pM^0 \to nM^+$	$\Delta_{p0} = \Delta M + \Delta m$	$\Delta_{p0} > 0$ (threshold)
$nM^+ \to pM^0$	$\Delta_{n+} = -\Delta M - \Delta m$	$\Delta_{n+} < 0$ (non-threshold)
$nM^0 \to pM^-$	$\Delta_{n0} = \Delta M - \Delta m$	$\Delta_{n0} > 0$ $(\Delta M > \Delta m)$
$pM^- \to nM^0$	$\Delta_{p-} = -\Delta M + \Delta m$	$\Delta_{p-} > 0$ $(\Delta M < \Delta m)$

Consider, for example, the first reaction, $pM^0 \to nM^+$, where $E_p \approx m_p v_r^2/2$. The expression for threshold $E_p^{thr} = \Delta M + \Delta m \equiv \Delta_{p0}$ gives the value of corresponding relative velocity $v_r^{thr} = \sqrt{2\Delta_{p0}/m_p}$. For the case $\Delta_{p0} = 10$ MeV we get the value of velocity $v_r^{thr} = 0.1$ which is significantly greater than the DM velocity now, $v_r \sim 10^{-3}$. So, this reaction is kinematically forbidden at the modern stage of evolution. The third and fourth processes can be both threshold and non-threshold depending on the value $\Delta M/\Delta m$. The first, third and fourth processes lead to the intermediate (final) states with unstable particles. These reactions go through two stages, for example, $pM^0 \to nM^+ \to pe^-\bar{\nu}_e M^0 e^-\bar{\nu}_e$ and $nM^0 \to pM^- \to pM^0 e^-\bar{\nu}_e$. We should note that the reaction $nM^0 \to pM^- \to pM^0 e^-\bar{\nu}_e$ is the most interesting due to presence of long-lived charge particle M^-. Note, the indirect evidences of these particles were reported in Ref. [16] (and references therein). Thus, we get an interesting phenomenology of low-energy nucleon-DM scattering which has a specific signature.

The process of annihilation $M^0\bar{M}^0 \to X$, where a standard light particle appears in the final state X, has some peculiarities in HDM scenario. DM particles M^0, in this scenario, are composite and annihilation proceeds through both strong and EW channels. Note, the theory of high-energy interaction of M particles is unknown, however this reaction at the sub-process level, $Q\bar{Q} \to q\bar{q}$, $gg \to$ jets, can be approximately described. With the help of this approach, we estimated the value of strong part of annihilation cross section, which is described by the formula (34). Here, we note that the dominant products of annihilation are the pairs of stable particles, $p\bar{p}$ and small fraction of e^+e^-, $\nu\bar{\nu}$, 2γ with total energy $E_{tot} \approx 2M$.

In this subsection, we considered the principal phenomenological consequences of the hadronic composite DM—low-energy strong interaction, fine and hyperfine splitting of excited states. Strong interaction of DM with ordinary matter raises the problem of the connection between galaxies and their DM halos, which can contribute additional information to the modern understanding of galaxy formation. Fine and hyperfine splitting in the set of new heavy mesons leads to the presence of metastable charged hadron and luminosity of hadronic DM. The generation and possibility of registration of heavy charged hadron in cosmic rays was briefly described. We also noted that the effect of hyperfine splitting generates the processes of electromagnetic transition and recombination. These processes can be launched by the interactions of new hadrons with the ordinary matter and cosmic rays and lead to the effect of hadronic DM luminosity. Note, the problems of hyperfine splitting and luminous DM become pressing now in view of the results of underground experiment XENON1T.

2.3.2. Dark Atoms with Primordial Helium

Natural choice of parameters of sphaleron transitions between baryons, leptons and stable techniparticles leads in WTC model to balance between baryon symmetry and the excess of stable $-2n$ charged techniparticles, corresponding to the observed dark matter density at the mass of these particles in the TeV range.

For the sequential 4th family with electroweak SU(2) charges, similar balance can be established between the excess of stable $\bar{U}$ quarks and baryon asymmetry.

Stable techniparticles behave like charged multiple charged leptons. $(\bar{U}\bar{U}\bar{U})$ states are also lepton-like, since their hadronic interaction is strongly suppressed [10,64].

Excessive negatively even charged particles bind with primordial helium as soon as it is formed in the Big Bang Nucleosynthesis.

Double charged particles form OHe dark atoms—a very nontrivial Bohr like atomic system with heavy lepton-like core and nuclear interacting helium shell with Bohr radius nearly equal to the size of helium nucleus.

$-2n$ charged techniparticles bind with n primordial helium nuclei in Thomson like XHe atoms with heavy lepton-like particle inside a nuclear droplet of n helium nuclei.

The Dark atom model has an advantage to explain the puzzles of direct dark matter searches by annual modulations of their low energy binding with Na nuclei. Strongly interacting shell of dark atoms provides their slowing down in terrestrial matter making this form of dark matter elusive for strategy of WIMP searches, involving significant nuclear recoil. However, dark atom interaction with nuclei can provide a low energy binding and the corresponding effect experiences annual modulations.

This explanation [81] is based on the following picture of OHe interaction with nuclei. OHe is a neutral atom in the ground state, perturbed by the Coulomb and nuclear forces of the approaching nucleus. The sign of OHe polarization changes with the distance. At larger distances, Stark-like effect takes place—the nuclear Coulomb force polarizes OHe so that the nucleus is attracted by the induced dipole moment of OHe, while as soon as the perturbation by the nuclear force starts to dominate, the nucleus polarizes OHe in the opposite way so that He is situated more closely to the nucleus, resulting in the repulsive effect of the helium shell of OHe. Qualitatively, it leads to a shallow potential well with a low energy bound state in OHe-Na system, while such a state does not exist for heavy nuclei like xenon. A quantitative description of OHe-nucleus interaction with self-consistent account for the effects of nuclear and Coulomb forces is crucial to prove this explanation and such a description can lead to nuclear physics of OHe (or XHe), which determines their physical and astrophysical effects.

3. New Physics of Strong Interaction in the Galaxy

3.1. New Components of Cosmic Rays

3.1.1. UHECR Interaction with Dark Matter

Mutual transformations of SM particles and their bound states in numerous reactions governing by conservation laws and known dynamics are necessarily supplemented by interactions with DM objects in the universe. In addition to the gravitational interaction, the main role is played by electroweak physics, ensuring the annihilation of dark matter (WIMPs) into standard particles. So, processes in which DM candidates disappear generating fluxes of (unstable) mesons and baryons, nuclei, leptons and photons can also be induced by strong interaction. In any case, macroscopic cross sections of such annihilation processes are proportional to the DM density squared and provide the main set of possible signals carrying information about the spatial distribution and dynamics of the hidden mass. Photons, charged leptons and neutrinos can be considered as the most important carriers of such information; their spectra are measured by space detectors and telescopes with ever increasing accuracy and completeness. Obviously, due to lack of hopeful results from collider experiments, our constant efforts to study the DM characteristics in the recent time inevitably reduce to different ideas and suggestions on indirect searches of DM manifestations in astrophysical data [82–88].

Actually, all final SM particles arise either directly in the processes of annihilation (or decay) of the DM particles or at the stage of secondary mesons (π^0, $\pi^{\pm}, K$ etc.) decays. The search for such signals which unambiguously produced by the hidden mass objects has been going on for a long time, but there are no reliably confirmed signals yet. It should be noted that the DM annihilation with the maximum cross section occurs in regions of increased DM density, i.e., in the central regions of the halo near active galaxy nuclei (AGN), the process can be also amplified in the possible DM clumps [89–93]. In other regions of galaxies, the efficiency of annihilation signals decreases due to the low density of DM.

Does not considering the decaying super-heavy DM, besides annihilation reactions, the presence of hidden mass can manifest itself in reactions of high-energy particles scattering on the DM particles. Such high-energy fluxes of cosmic rays permeate the entire universe, their composition, in addition to the main components—protons and nuclei of various elements—includes electrons, photons, neutrinos. Furthermore, aside from the question of how particles of such high energies are generated and distributed across the universe [94–99], the processes of quasi-elastic and inelastic scattering of ultra-high energy cosmic rays (UHECR) by DM particles can give important information on the hidden mass dynamics and its spatial structures. Despite the fact that in this case the macroscopic cross sections are proportional only to the first power of the DM density, the low probability of such processes is partially compensated by the specificity of the signature of the final states, i.e., the unique nature of the signals [100–106].

Obviously, intergalactic magnetic fields strongly affect the propagation of charged leptons; therefore, information on the sources, conditions, and principles of UHECR generation is mainly contained in neutrino and photon differential fluxes. Thus, the cosmic rays interactions with hidden mass particles should be considered as a useful additional tool for studying DM in the universe. Analysis of such processes should contain not only the cross sections of CR interactions with hidden mass objects calculated in various DM scenarios, but also an assessment of how the energy distributions and composition of UHECR can be changed due to quasi-elastic or inelastic interactions with the DM particles.

As a testing ground for evaluating and discussing the processes of UHECR interaction with the DM particles, we consider the above described scenario of the SM extension due to additional fermions—hyperquarks. As we have seen, in the minimal version of the extension ($SU(4) \to Sp(4)$ H-color model), the DM candidates are stable neutral hyperpion (its charged unstable partners in the triplet of hyperpions have a masses greater by $\approx$160 MeV and the neutral stable diquark, B^0 (see above,

Section 2.1.2). Here we will assume the masses of these two different DM candidates are practically equal; the case of an asymmetric DM, when one of the components is heavier than the other, can also be analyzed [35]. The reason is the calculated mass difference for these components clearly depends on the scale of renormalization and can be made nonzero. The dependence itself on the renormalization follows from different H-quark currents corresponding to these stable objects in the scenario. As it results from numerical analysis, the mass splitting between two components of the DM cannot be more than (10–15) GeV for reasonable values of renormalization scale, μ = (100–1000) GeV. It, however, means that one type of the DM components can effectively transform into another with simultaneous production of soft leptons, neutrino and diffuse photons (see also [107] where the DM components masses are rather different).

The reason of this soft radiation is that the dominant decay channel of secondary charged hyperpions is the decay $\tilde{\pi}^{\pm} \to \tilde{\pi}^0 \pi^{\pm}$ with subsequent leptonic decay of ordinary $\pi^{\pm}$. Therefore, the signature of such a process—creation and subsequent decay of a charged hyperpion—is the appearance of a (decaying) muon and a muonic (anti) neutrino. Appearance and decay of charged H-pions is a characteristic feature of the scenario where additional heavy fermions form new H-hadron states with possibilities of their transformations into each other. Higher unstable H-quark bound states are not considered here under the assumption that their masses are sufficiently large (some interesting results about spectra of masses in H-color extensions basing on the lattice calculations can be found in [108,109]). Specifics of H-color dynamics in this type of models, with two DM components, is that one of the DM candidates interacts via standard gauge bosons with the SM particles, but the other one uses for such interactions only scalar exchanges by partly composite Higgs boson and its more heavier partner, $\tilde{\sigma}$-meson.

Thus, the scenario under consideration represents of possible types of hidden mass components in the SM extensions with additional fermions. Due to the presence of DM components that are different in their nature and origin makes it possible to analyze various channels of interaction of these DM components with cosmic rays.

Note that in the pioneering work on studying the channels of leptons scattering on the dark matter [100,102], supersymmetry (SUSY) scenario was used as the basis, and, accordingly, neutralinos were considered as obvious candidates for the DM particles. In this case, the main subject of analysis was the secondary photons emitted during the scattering process. It seems to us that reactions in which not only leptons, as a (small) part of cosmic ray flux, but also neutrinos play an active role are very informative. Besides, we mean not only neutrino-initiated processes of quasi-elastic interaction, but also analysis of secondary neutrinos and photons produced in such scattering events.

Masses of the DM components can be extracted from the system of five kinetic equations for the DM density assuming the existence of freeze-out point for the annihilating DM candidates [105]. Using cross sections in all possible annihilation channels for the DM components, the system of equations has been solved numerically. Then, we get some regions depending on the model parameters where the DM relic density is correct (in Figures 1–3 these areas are crosshatched by vertical and horizontal lines) with H-pions fraction that is less than 25 percents: ($0.1047 \leq \Omega h^2_{HP} + \Omega h^2_{HB} \leq 0.1228$ and $\Omega h^2_{HP}/(\Omega h^2_{HP} + \Omega h^2_{HB}) \leq 0.25$). The crosshatching with oblique lines correspond to areas where all parameters are exactly the same, but here H-pions fraction make up just over a quarter of the DM ($0.1047 \leq \Omega h^2_{HP} + \Omega h^2_{HB} \leq 0.1228$ and $0.25 \leq \Omega h^2_{HP}/(\Omega h^2_{HP} + \Omega h^2_{HB}) \leq 0.4$). It is understood why B-component of DM dominates in all valid regions—this is because B^0-baryons interact with the world of ordinary particles only via H-quark and H-pion loops or through exchanges with scalar states, but H-pions have tree level interactions with weak vector bosons, so they burn out much faster.

Further, hatching with horizontal lines denotes regions where recent DM relic abundance is not explained by H-color candidates. Regions with vertical hatching are forbidden by underground-experiment data from XENON collaboration.

Thus, results of kinetic analysis of two-component dark matter can be presented in the $M_{\tilde{\sigma}} - M_{\tilde{\pi}}$ plane in Figure 1 as three allowed areas for the DM masses and other model parameters (v.e.v. u, mixing angle θ), namely:

Region 1: $M_{\tilde{\sigma}} > 2m_{\tilde{\pi}^0}$ and $u \geq M_{\tilde{\sigma}}$. At small angles of mixing, s_θ, and large masses of H-pions it is possible to obtain a significant fraction of H-pions.

Region 2: the same relation between $M_{\tilde{\sigma}}$, $m_{\tilde{\pi}^0}$, u but the H-pion mass is smaller, $m_{\tilde{\pi}} \approx$ 300–600 GeV, H-pion fraction is small here.

Region 3: $M_{\tilde{\sigma}} < 2m_{\tilde{\pi}}$. This domain is always possible and it is presented in all figures. Note, decay $\tilde{\sigma} \to \tilde{\pi}\tilde{\pi}$ is prohibited. H-pion fraction in the DM relic can be large if the mass $m_{\tilde{\pi}^0}$ is large and the mixing angle is small. In Figures 2 and 3, we illustrate the regions changing for lower values of the v.e.v. u and $\sin\theta$.

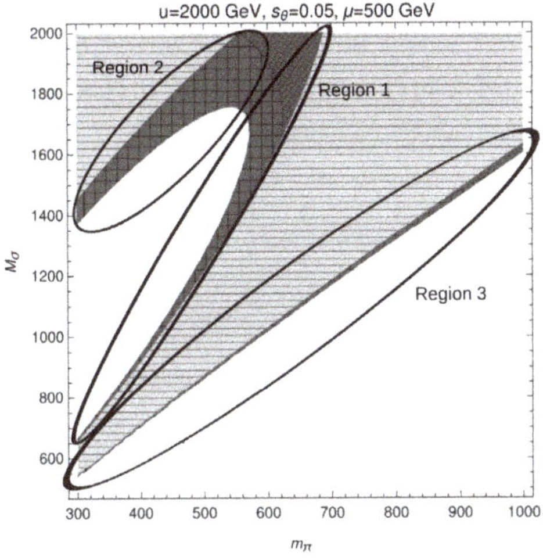

Figure 1. Numerical solution of kinetic equations system in a phase diagram in terms of $M_{\tilde{\sigma}}$ and $m_{\tilde{\pi}}$, other parameters are also indicated.

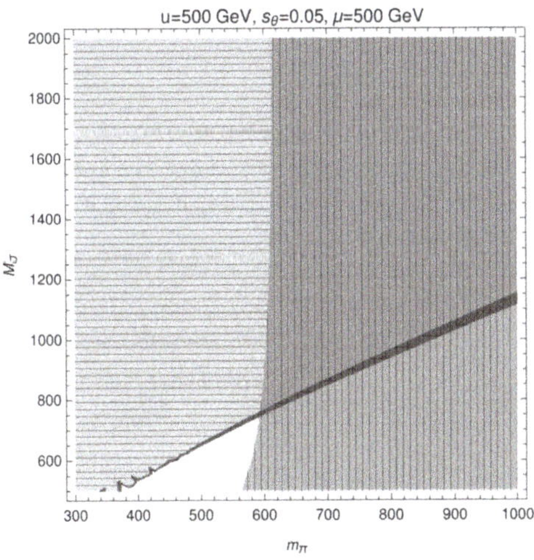

Figure 2. Analogous phase diagram in terms of $M_{\tilde{\sigma}}$ and $m_{\tilde{\pi}}$, but for much smaller u.

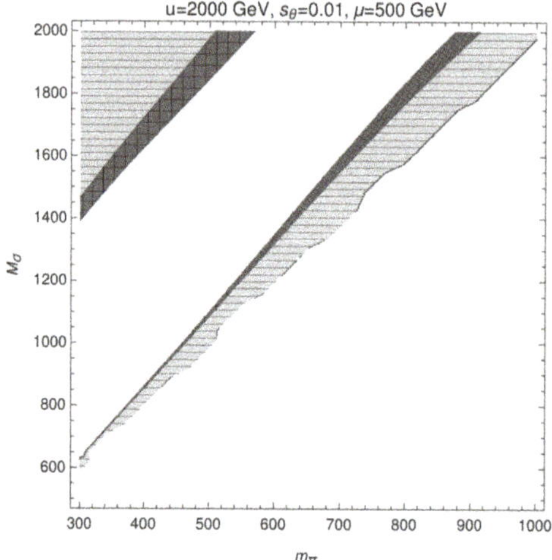

Figure 3. Phase diagram in terms of $M_{\tilde{\sigma}}$ and $m_{\tilde{\pi}}$, the same u but the mixing is smaller.

Thus, knowing tree level cross sections of DM components annihilation, from kinetics we estimate masses of stable hypercolor particles, $\tilde{\pi}^0$ and B^0, $\bar{B}^0$ in the range 0.6–1.2 TeV. Here we consider the case $\Delta M_{B\tilde{\pi}} \approx 0$. Note, due to connection between masses of H-pions and $\tilde{\sigma}$-meson for zero $h - \tilde{\sigma}$ mixing, mass of this scalar partner of "nearly standard" Higgs boson is also constrained in some range. Now, with the estimation of all masses in hand, we can analyze most simple process of quasi-elastic scattering $e^- \tilde{\pi}^0 \to \nu_e \tilde{\pi}^-$ [105] with a following decay of the charged H-pion.

Cross section was calculated supposing the target (DM particle) gets a small portion of projectile energy, so W-boson in t-channel is close to its mass shell, the momentum transfer is small and the final $\tilde{\pi}^0$ is not accelerated in this process (we return to discussion of this important possibility—acceleration of the DM objects up to TeV energies [110–116]—somewhat later). Then the secondary neutrino and leptons from the W decay (in the channel $\tilde{\pi}^- \to \tilde{\pi}^0 W^- \to \tilde{\pi}^0 e^- \bar{\nu}_e$) or from dominant channel of decay $\tilde{\pi}^- \to \tilde{\pi}^0 \pi^- \to \tilde{\pi}^0 \mu^- \bar{\nu}_\mu$ should have energies $\sim 10^2$ GeV or even less (see Ref. [43] for details of the charged hyperpion decays). Then, the cross section of the process and also distributions on energy and angle of emission for secondary neutrino were found as and the number of possible neutrino events at IceCube produced by this reaction [105]. Indeed, the electron scattering on the DM objects can be interesting as a source of high-energy neutrino from $We\nu$ vertex or accelerated DM particles with masses ~ 1 TeV, when the momentum transferred is sufficiently large.

However, the electron flux is only a small part of the cosmic ray total flux especially at energies $\geq 10^2$ TeV. As a result, we predict a very small fluxes of secondary neutrinos and, consequently, small probability to detect such events at IceCube [105,106].

At this moment, an important feature of H-color model emerges—we have two DM components and the process of UHECR scattering should be considered for both types of neutral stable objects. Moreover, H-baryons B^0 are the largest part of the total DM amount, as it follows from kinetic equations analysis. It follows from the absence of direct interaction of B^0-baryons with the standard gauge bosons and, consequently, with the SM matter. So, the burning out of this component is much slower than for $\tilde{\pi}^0$ particles.

It seems that there is a chance to introduce the B^0 interaction through H-pion and/or H-quark loops, however for the scattering channels these loops are exactly zero [105]. Thus, we need to consider more complex tree diagrams, in particular, tree diagrams with the exchange of Higgs boson and its partner, $\tilde{\sigma}$-meson, in t-channel give dominant non-zero contribution to the process $e^- B \to \nu_e W^- B$. Virtual W^--bosons eventually decay to $l\bar{\nu}_l$ or into light ordinary mesons. Of course, there is similar scattering reaction with the scalar-state exchange, $e^- \tilde{\pi}^0 \to \nu_e W^- \tilde{\pi}^0$, whose amplitude is smaller by half as it is seen from the model Lagrangian. Here, we do not take into account small contributions from diagrams with H-quark loops, hhZ and other multi-scalar vertices [106].

Note, diagrams of this type were recently considered and suggested for the analysis of neutrino scattering off nucleons [117], their significant contributions were confirmed by direct calculations. We, however, found that these diagrams present dominant tree level part of cosmic particles scattering cross section off the DM. To calculate total width of the process with the final state $B^0 e^- \nu\bar{\nu}$ or $\tilde{\pi}^0 e^- \nu\bar{\nu}$, we have used factorization method [118,119] considering independently amplitudes squared of subprocesses with intermediate W and Z-bosons and then estimating the (negative) interference of these contributions. The approach allows us to estimate with reasonable accuracy (no worse than $\sim$10% due to approximate estimation of the interference) the cross section of an "averaged" process where the final electron and neutrinos are produced by different vertices, $W \to l\nu_l$ and $Z \to \nu_l\bar{\nu}_l$, which practically coincide for the massless leptons.

So, without using complex computer programs we get the value of total cross section and can estimate also the possibility to detect at IceCube the neutrino signal produced by the process of electron scattering off the DM. Again, for these reactions we do not consider those phase space regions which correspond to acceleration of the initial DM particle (so called up-scattered dark matter). In other words, the final H-baryon (or neutral H-pion) is slow; nearly all energy of the incident electron is distributed between three final massless particles (electron and pair of neutrinos). Approximately, energies of secondary neutrinos are in the interval $\sim(E_e/3 - E_e)$.

In calculations, we use two values of masses of H-baryon and $\tilde{\sigma}$-meson. Note that in the model there is a correlation between masses of $m_{\tilde{\sigma}}$ and $m_{\tilde{\pi}}$: $m_{\tilde{\sigma}}^2 \approx 3 \cdot m_{\tilde{\pi}}^2$. This is an exact equality for zero mixing of Higgs boson and $\tilde{\sigma}$. Here, we suppose the splitting between masses of H-baryon and $\tilde{\pi}$ is very small. It was found that the cross section strongly depends on the mass of DM particle and grows with the mass increasing.

Now, we should note an interesting type of the scattering processes produced by high-energy neutrino off the DM components. Namely, there is a small probability to find in the UHECR content electrons of very high energies, which will initiate creation of high-energy neutrino in the scattering. Nevertheless, we calculate cross section values for high energy region despite the fact that the cosmic electrons flux noticeably decreases for these energies; the hope is based on estimation of effective areas for IceCube—high-energy neutrinos can be detected with a larger probability. Unfortunately, cosmic electron motion is strongly affected by galactic magnetic fields, so their sources are hardly identified both in direction and in intensity. Certainly, if we could separate, in all experimental data, signals from high-energy electron scattering with specific set of final states, it would manifest itself on the target with defined properties; and, perhaps, it would have been the DM object. Alas, such signals are practically unobserved for the ground neutrino telescopes due to large and constant background from the Sun neutrino and neutrinos resulting from decays of mesons and baryons produced by cosmic rays interactions with nuclei in the Earth atmosphere. We also note that in processes with the production of secondary neutrinos, acceleration of DM particles can occur simultaneously; such processes can themselves be initiated by incident high-energy neutrinos [120–122]. In other words, reactions with participation of high-energy neutrinos which can be accompanied with the DM accelerated seem as informative and important, especially because both of these particles are, in fact, messengers from regions of high DM density—regions near AGN or from possible DM inhomogeneities of some other nature—and early epoch of the universe [123].

In more detail, the secondary neutrino fluxes calculated are very small in comparison with expected neutrino fluxes from AGN which can be $\sim$ 10^5 cm^{-2}s^{-1}sr^{-1}. Atmospheric neutrino fluxes with neutrino energies $\leq$2 TeV are also much larger [124–127]: $\sim 10^{-10}$–10^{-9} cm^{-2}sr^{-1}s^{-1}. Namely, we get that the secondary neutrino flux resulted from the cosmic electrons scattering is $\sim 10^{-19}$–10^{-22} cm^{-2}sr^{-1}s^{-1}[106]. If, however, intergalactic neutrinos with very high energy come to the Earth (their sources can be jets from blazars), their scattering on the DM from halo can be marked by the specific reaction: $\nu_l + \mathrm{DM} \to \bar{\nu}_l + Z^* + \mathrm{DM} \to \bar{\nu}_l + \nu_k\bar{\nu}_k + \mathrm{DM}$. Of course, a virtual Z-boson can decay into light mesons producing a number of soft leptons, neutrino and photons, but some correlation in energies and directions between high-energy secondary neutrinos can be detected. Obviously, cross section of creation of secondary quarks or leptons by virtual Z-boson is resonantly amplified when the Z-boson is near its mass shell. In any case, for this process we evaluate cross section as $\sim$(20–300) pb for initial neutrino energies $\sim$(10–100) TeV.

Let us note some points which are important for study of the cosmic rays scattering off the DM. We suppose that, independently of the type of the SM extension, possibility of scalar exchange in the scattering channel results in a strong dependence of the cross section on the DM particle mass. It was found in the H-color extension, namely, the changing (increasing) of the DM component mass of 10% leads to the cross section growth by up to 50%. The opening of channels with scalar exchanges allows us to consider new ways to produce secondary high-energy leptons, neutrino and photons by UHECR scattering off the DM. Besides, these reactions can result in acceleration of the DM particles despite that this process takes place in t-channel. Now, resonant amplifying [120] is possible only due to that part of total amplitude which describes producing of leptons or quark pairs in virtual gauge boson decays.

3.1.2. Creation of New Components in the UHECR Sources

Cosmic rays of ultrahigh energies have a low intensity which rapidly decreases with increasing energy; for example, at the border of atmosphere, a detector with an area of 1 m^2 can detect about 100 cosmic particles per year with $E \geq 10^3$ TeV. With such small fluxes of UHECR, the analysis of extended air showers (EAS) generated by high-energy particles in the Earth's atmosphere becomes an effective method of studying them. Composition of the initial radiation changes due to creation and decay of new particles that generate a nuclear-electromagnetic cascade. The path traversed by particles in the atmosphere is much larger than the average range of inelastic interactions between protons and nuclei. Secondary particles forming an EAS, can be detected at large distances, up to 10^3 m or even more from the shower axis.

Standard picture of EAS production is based mostly on physics of high-energy protons (or light nuclei from cosmic rays) interaction with nuclei in atmosphere; then, secondary nucleons, pions, kaons, hyperons, leptons and photons are born. In the initial act of EAS generation leading secondary particle keeps about $\sim$50% of its initial energy, so is able to interact several times in the atmosphere. At high energies, $\geq$100 TeV, a significant part of unstable secondary particles (π- and K-mesons) do not decay on the path of the order of one path of the inelastic nuclear interaction, they again interact with the nuclei, forming new decaying charged and neutral mesons which produce (decaying with lifetime $\sim 2 \times 10^{-6}$ s) muons, neutrinos and photons.

Of course, along with a shower of nuclear-active particles, it develops an electron–photon cascade in the atmosphere due to fast decays of neutral pions into two gammas. So, an electron–photon component of the shower arises and evolves. Then, photons produce electron–positron pairs interacting with the medium, and these charged leptons again give high-energy photons due to bremsstrahlung on the nuclei. Obviously, the multiplication of particles in these showers is defined by energy dissipation processes for every type of interaction between mesons, nucleons and nuclei or mesons decays. These interactions have been studied with all necessary detail. The passage of an EAS through the atmosphere is also accompanied by optical radiation: Cherenkov and ionization radiation. So, this standard picture of the emergence and development of EAS generated by high-energy protons or nuclei must be supplemented since EAS can also be generated by high-energy neutrinos and, possibly, by high-energy accelerated neutral DM particles [128–130].

Neutrinos generate electromagnetic and hadronic showers due to deep inelastic (or quasi-elastic) scattering off nucleons in case of charged or neutral current interaction. Note that in the last case there arise some special contributions providing interactions of B^0 component with standard fermions through scalar exchanges. These diagrams, which are called as "trident" [105,106,117]), are especially important for the cosmic rays (electrons, neutrino, protons) interaction not only with nucleons but with the DM objects, in particular, for various multi-component DM models with scalar interaction of the DM with gauge bosons currents.

It is important to note that the showers generated by neutrinos and nucleons can be effectively discriminated due to small cross section of neutrino interaction with nuclei in atmosphere (this argument can also be used for separation of possible showers producing by high-energy neutral heavy DM particle [131,132]). As a result, EAS induced by neutral particles begins its development deeply in atmosphere (where the density increases significantly) in comparison with EAS generated by cosmic protons. Produced by galactic neutrinos EAS are highly inclined, despite this, there is a set of parameters (Cherenkov light, depth of shower maximum in the atmosphere, duration of the shower in dependence on the altitude and, correspondingly, density) allowing to separate neutrino showers from proton ones [128,131,133]. These characteristics of EAS can also be used to mark EAS which are produced by boosted DM particles.

Remind, the possibility to accelerate light DM particles in the scattering of high-energy cosmic rays off the DM was supposed and numerically analyzed in [110–116,129,130,134]. This possibility to boost the DM was also confirmed for heavy DM objects with mass $\sim$1 TeV, more exactly, we have considered and approximately calculated scattering of protons of high energy, up to 200 TeV on the DM particles from halo, in the region of most large DM density—near AGN [135]. In other words, we consider interaction of protons from blazar's jets with heavy DM particles. In the framework of the H-color scenario, in this interaction of protons with two DM components a significant part of protons energy can be transferred due to charged current to heavy H-pion and to both DM component in the trident type diagrams involving scalar exchange.

For initial protons with energy 200 TeV cross section of the scattering process where final charged H-pion is produced with energies (40–50) TeV is $\sim$(10–15) pb. This charged H-pion decays predominantly as $\tilde{\pi}^{\pm} \to \tilde{\pi}^{0}\pi^{\pm}$ with the width $\Gamma \to 3 \times 10^{-15}$ GeV. So, we get also secondary muon (which again decays) and muonic antineutrino.

In fact, in this deep inelastic reaction the main charged component of UHECR (protons) originated from blazar jet disappear transforming finally into flux of high energy neutrino and leptons. More exactly, our estimations demonstrates that $\sim$(10–25)% of the proton energy is transferred to heavy neutral component of the DM with cross section $\approx$(10–100) pb. In the scattering channel described by "trident" diagrams total cross section is of the same order but there appear additional neutrinos, for example, generated in the resonant decay of intermediate Z-boson.

Despite the cross section not being large, it is an example of accelerating even heavy DM particles up to significant energies. Most importantly, this boosted neutral particle, as light neutrino, will pass away from the DM halo moving in the constant direction because it interacts with the matter very slowly. So, this rare process when the charge component of cosmic rays can be ruined in the deep inelastic reaction and as a result neutral DM particle moves like a neutrino towards the Earth. Remind that above considered high-energy electron scattering off the DM can also accelerate the DM but in quasi-elastic process high energy neutrino are generated with more probability.

Thus, from this brief description of some processes of scattering of high-energy cosmic ray particles off the DM we can conclude that these reactions can enrich the cosmic rays composition with boosted heavy neutral DM particles [136]. At energies of these projectiles $\sim$(10–100) TeV cross sections of their interactions with nucleons and nuclei, $\sim(10^{-34}$–$10^{-37})$ cm^2, are compared with cross sections of neutrino-nucleons scattering. In this deep-inelastic process nucleons or nuclei are transformed a multiparticle final states consisting of charged leptons, photons and neutrino. Additional neutrinos are generated by the charged H-pion decay and in the processes with resonant decay of Z-boson. So, the accelerated neutral DM components can produce rare events - specific types of EAS. Certainly, these EAS in the atmosphere should be separated in some manner from other types of showers.

It is known that as usually cosmic rays generate a shower of secondary particles which are mainly muons, electrons and photons. They go to ground detectors and can be fixed as measured signals registering also due to fluorescence and Cherenkov light, and radio emission generated by charged component, electrons, in atmosphere of the Earth. It seems, such type of shower is similar to neutrino induced shower and its initial point also should be deeply in atmosphere, however, the neutral DM particle can not disappear from the EAS composition and will interact with the ground detector producing some radiation from secondary electrons or from excited nuclei in the detector. The DM showers, as they generated by intergalactic DM objects which were accelerated by UHECR or AGN jets from halo of other galaxies, or DM particles boosted from halo of our galaxy by intergalactic UHECR do not have to be mostly inclined or nearly horizontal. It is supposed, these accelerated DM components and EAS produced by them should be distributed more or less isotropic. May be, the EAS axis can be connected with direction to some blazar, as it was found for some very high-energy neutrino events at IceCube.

So, we can conclude that EAS from heavy DM particles are distinguished from EAS generated by protons or neutrino because in the former event the shower contains in his composition neutral stable object up to the final moment when this fast DM particle scattered on nucleon in the detector (see also [137] and references therein). To the contrary, in composition of EAS which was induced by neutrino or protons (or light nuclei), there is no any heavy stable particles, only leptons, photons and neutrino are detected as final states. Note also that interaction of DM components with nucleons in detector should have specific signature: the scattering in charged current channel is accompanied with creation and following decay of charged H-pion, so, the event can be seen due to charged lepton bremsstrahlung. We hope that observing and measuring the characteristics new types of EAS containing heavy neutral stable particles will be possible at modern complex LHAASO [138], in other words, the DM candidates can manifest itself in a specific types of EAS.

Ionization of dark atoms by high energy cosmic rays or in supernova explosions can lead to formation of an anomalous cosmic ray flux of stable multiple charged leptons. The search for such component and the possibility to discriminate the corresponding air-showers are challenges for LHAASO experiments.

3.2. *Multimessenger Probes for New Physics Effects*

As was noted above, UHECR can result in specific interactions with DM particles, which are concentrated in halo with the largest density near AGN, giving some special signatures. First, in the scattering processes the target (slow heavy stable DM component) can receive a significant portion of energy of initial projectile. So, the DM particle can be effectively accelerated mostly in the direction close to the projectile direction. However, in the two-component DM scenario considered, if the interaction is provided by charged current with the W-exchange neutral H-pion component transforms into charged H-pion decaying with generation of lepton and neutrinos from $\pi^{\pm}$ and muon decays. This charged H-pion, in fact, decays in flight, so it is possible to estimate energies of final particles and their angles of emission. Indeed, these reactions are rare because of small density of DM, especially if we consider the intergalactic UHECR scattering off the DM halo of our galaxy.

Although secondary neutrinos (prompt neutrino from $Z\nu\bar{\nu}$ vertex, neutrino from $\tilde{\pi}^{\pm}$ decay and neutrinos from decaying mesons generated by initial proton) are emitted and scattered at different angles, there is non-zero probability to detect neutrino from this reaction at IceCube together with observation of specific shower associated with the DM component at LHAASO, for example. Certainly, the probability of such events which are separated by definite interval of time (its magnitude is defined by energies and velocities of the neutrino and DM particle) is evaluated as very small. However, if such twin event would be detected, it could be an important manifestation of the UHECR interaction with the DM. More precisely, measuring the characteristics of such an event—the delay time between signals (registration of a high-energy neutrino and a shower of particles), the energy release of the EAS, its composition, as well as the establishment of the fact of interaction of neutral object (heavy DM particle) with the substance of the detector (for example, by low-energy radiation)—could improve our understanding of the DM nature and the features of its interaction with neutrinos, leptons and nucleons. These probes can shed light on the possible hadronic and hadron-like particles of dark matter and their spectroscopy.

4. Conclusions

Over recent decades, the mainstream of BSM physics has been concentrated on the search for direct and indirect effects of supersymmetry (SUSY). SUSY partners of SM particles with masses in the range of several hundred GeV–1 TeV were expected to be found at the LHC. The lightest stable neutral SUSY particle could nicely implement the WIMP miracle and was considered as the preferable candidate the cosmological dark matter.

However, SUSY particles were not found up to now at the LHC. The results of direct underground WIMP searches are controversial and the positive result of DM searches by DAMA/NaI and DAMA/LIBRA experiments can hardly be interpreted in the terms of WIMPs.

It is reasonable to extend the field of studies of the new physics and, in particular, to consider non-SUSY BSM models. New stable color particles can provide candidates for dark matter with hadronic interaction, while new BSM nonabelian symmetry can increase the list of WIMP-like dark matter candidates.

We have presented in this review various nontrivial features of new physics of strong interaction and their possible physical, astrophysical and cosmological signatures. Experimental probes for these signature will shed new light on possible role of various forms of strong interaction in the dark universe.

Author Contributions: Review by V.B., M.K., V.K. and N.V. The authors contributed equally to this work. All authors have read and agreed to the published version of the manuscript.

Funding: The work was supported by grant of the Russian Science Foundation (Project No-18-12-00213).

Acknowledgments: We express our gratitude to Andrea Addazi, Konstantin Belotsky, Maxim Bezuglov, Timur Bikbaev, Vladimir Korchagin, Antonino Marciano, Andrey Mayorov and Egor Tretiakov for useful discussions.

Conflicts of Interest: The authors declare no conflict of interest.

References

1. Linde, A.D. *Particle Physics and Inflationary Cosmology*; Harwood: Chur, Switzerland, 1990.
2. Kolb, E.W.; Turner, M.S. *The Early Universe*; Addison-Wesley: Boston, MA, USA, 1990.
3. Gorbunov, D.S.; Rubakov, V.A. *Introduction to the Theory of the Early Universe. Cosmological Perturbations and Inflationary Theory*; World Scientific: Singapore, 2011.
4. Gorbunov, D.S.; Rubakov, V.A. *Introduction to the Theory of the Early Universe: Hot Big Bang Theory*; World Scientific: Singapore, 2011.
5. Khlopov, M.Y. *Cosmoparticle Physics*; World Scientific: Singapore, 1999.
6. Khlopov, M.Y. *Fundamentals of Cosmoparticle Physics*; CISP-Springer: Cambridge, UK, 2012.
7. Khlopov, M. Cosmological Reflection of Particle Symmetry. *Symmetry* **2016**, *8*, 81. [CrossRef]
8. Khlopov, M. Fundamental particle structure in the cosmological dark matter. *Int. J. Mod. Phys. A* **2013**, *28*, 1330042
9. Khlopov, M.Y. Removing the conspiracy of BSM physics and BSM cosmology. *Int. J. Mod. Phys. D* **2019**, *28*, 1941012. [CrossRef]
10. Beylin, V.; Khlopov, M.; Kuksa, V.; Volchanskiy, N. Hadronic and Hadron-Like Physics of Dark Matter. *Symmetry* **2019**, *11*, 587. [CrossRef]
11. Aprile, E.; Aalbers, J.; Agostini, F.; Alfonsi, M.; Amaro, F.D.; Anthony, M.; Arneodo, F.; Barrow, P.; Baudis, L.; Bauermeister, B.; et al. (XENON Collab). First Dark Matter Search Results from the XENON1 Experiment. *Phys. Rev. Lett.* **2017**, *119*, 181301. [CrossRef] [PubMed]
12. Huo, R.; Matsumoto, S.; Tsai, J.L.S.; Yanagida, T.T. A scenario of heavy but visible baryonic dark matter. *JHEP* **2016**, *1609*, 162. [CrossRef]
13. Luca, V.; Mitridate, A.; Redi, M.; Smirnov, J.; Strumia, A. Colored Dark Matter. *Phys. Rev. D* **2018**, *97*, 115024. [CrossRef]

14. Beylin, V.; Kuksa, V. Dark Matter in the Standard Model Extension with Singlet Quark. *Adv. High Energy Phys.* **2018**, *18*, 8670954. [CrossRef]
15. Beylin, V.; Kuksa, V. Possibility of hadronic dark matter. *Int. J. Mod. Phys. D* **2019**, *28*, 1941001. [CrossRef]
16. Bazhutov, Y.N.; Vereshkov, G.M.; Kuksa, V.I. Experimental and Theoretical Premises of New Stable Hadron Existence. *Int. J. Mod. Phys. A* **2017**, *2*, 1759188. [CrossRef]
17. Khlopov, M.Y. Physics of dark matter im the light of dark atoms. *Mod. Phys. Lett. A* **2011**, *26*, 2823–2839. [CrossRef]
18. Khlopov, M.Y. Dark Atoms and Puzzles of Dark Matter Searches. *Int. J. Mod. Phys. A* **2014**, *29*, 1443002. [CrossRef]
19. Cudell, J.R.; Khlopov, M. Dark atoms with nuclear shell: A status review. *Int. J. Mod. Phys. D* **2015**, *24*, 1545007. [CrossRef]
20. Tanabashi, M.; Hagiwara, K.; Hikasa, K.; Nakamura, K.; Sumino, Y.; Takahashi, F.; Tanaka, J.; Agashe, K.; Aielli, G.; Amsler, C.; et al. The Review of Particle Physics. *Phys. Rev. D* **2018**, *98*, 030001. [CrossRef]
21. Cacciapaglia, G.; Pica, C.; Sannino, F. Fundamental Composite Dynamics: A Review. *Phys. Rept.* **2020**, *877*, 1–70. [CrossRef]
22. Csáki, C.; Grojean, C.; Terning, J. Alternatives to an Elementary Higgs. *Rev. Mod. Phys.* **2016**, *88*, 045001. [CrossRef]
23. DeGrand, T. Lattice tests of beyond Standard Model dynamics. *Rev. Mod. Phys.* **2016**, *88*, 015001. [CrossRef]
24. Hill, C.T.; Simmons, E.H. Strong Dynamics and Electroweak Symmetry Breaking. *Phys. Rep.* **2003**, *381*, 235–402. [CrossRef]
25. Bellazzini, B.; Csáki, C.; Serra, J. Composite Higgses. *Eur. Phys. J. C* **2014**, *74*, 2766. [CrossRef]
26. Panico, G.; Wulzer, A. *The Composite Nambu-Goldstone Higgs*; Springer: Berlin/Heidelberg, Germany, 2016; Volume 913. [CrossRef]
27. Schmaltz, M.; Tucker-Smith, D. Little Higgs review. *Ann. Rev. Nucl. Part. Sci.* **2005**, *55*, 229–270. [CrossRef]
28. Chen, M.C. Models of little Higgs and electroweak precision tests. *Mod. Phys. Lett. A* **2006**, *21*, 621–638. [CrossRef]
29. Perelstein, M. Little Higgs models and their phenomenology. *Prog. Part. Nucl. Phys.* **2007**, *58*, 247–291. [CrossRef]
30. Kilic, C.; Okui, T.; Sundrum, R. Vector-like Confinement at the LHC. *J. High Energy Phys.* **2010**, *2010*, 18. [CrossRef]
31. Pasechnik, R.; Beylin, V.; Kuksa, V.; Vereshkov, G. Chiral-symmetric technicolor with standard model Higgs boson. *Phys. Rev. D* **2013**, *88*, 075009. [CrossRef]
32. Pasechnik, R.; Beylin, V.; Kuksa, V.; Vereshkov, G. Vector-like technineutron Dark Matter: Is a QCD-type Technicolor ruled out by XENON100? *Eur. Phys. J. C* **2014**, *74*, 2728. [CrossRef]
33. Lebiedowicz, P.; Pasechnik, R.; Szczurek, A. Search for technipions in exclusive production of diphotons with large invariant masses at the LHC. *Nucl. Phys. B* **2014**, *881*, 288–308. [CrossRef]
34. Pasechnik, R.; Beylin, V.; Kuksa, V.; Vereshkov, G. Composite scalar Dark Matter from vector-like $SU(2)$ confinement. *Int. J. Mod. Phys. A* **2016**, *31*, 1650036. [CrossRef]
35. Beylin, V.; Bezuglov, M.; Kuksa, V. Analysis of scalar Dark Matter in a minimal vector-like Standard Model extension. *Int. J. Mod. Phys. A* **2017**, *32*, 1750042. [CrossRef]
36. Antipin, O.; Redi, M. The half-composite two Higgs doublet model and the relaxion. *J. High Energy Phys.* **2015**, *2015*, 1–16. [CrossRef]
37. Agugliaro, A.; Antipin, O.; Becciolini, D.; De Curtis, S.; Redi, M. UV-complete composite Higgs models. *Phys. Rev. D* **2017**, *95*, 035019. [CrossRef]
38. Barducci, D.; De Curtis, S.; Redi, M.; Tesi, A. An almost elementary Higgs: Theory and practice. *J. High Energy Phys.* **2018**, *2018*, 17. [CrossRef]
39. Antipin, O.; Redi, M.; Strumia, A. Dynamical generation of the weak and Dark Matter scales from strong interactions. *J. High Energy Phys.* **2015**, *2015*, 157. [CrossRef]
40. Mitridate, A.; Redi, M.; Smirnov, J.; Strumia, A. Dark matter as a weakly coupled dark baryon. *J. High Energy Phys.* **2017**, *2017*, 210. [CrossRef]
41. Appelquist, T.; Brower, R.C.; Buchoff, M.I.; Fleming, G.T.; Jin, X.Y.; Kiskis, J.; Kribs, G.D.; Neil, E.T.; Osborn, J.C.; Rebbi, C.; et al. Stealth dark matter: Dark scalar baryons through the Higgs portal. *Phys. Rev. D* **2015**, *92*, 075030. [CrossRef]

42. Cai, C.; Cacciapaglia, G.; Zhang, H.H. Vacuum alignment in a composite 2HDM. *J. High Energy Phys.* **2019**, *2019*, 130. [CrossRef]
43. Beylin, V.; Bezuglov, M.; Kuksa, V.; Volchanskiy, N. An analysis of a minimal vector-like extension of the Standard Model. *Adv. High Energy Phys.* **2017**, *2017*, 1765340. [CrossRef]
44. Low, I.; Skiba, W.; Tucker-Smith, D. Little Higgses from an antisymmetric condensate. *Phys. Rev. D* **2002**, *66*, 072001. [CrossRef]
45. Csaki, C.; Hubisz, J.; Kribs, G.D.; Meade, P.; Terning, J. Variations of little Higgs models and their electroweak constraints. *Phys. Rev. D* **2003**, *68*, 035009. [CrossRef]
46. Gregoire, T.; Tucker-Smith, D.; Wacker, J.G. What precision electroweak physics says about the SU(6)/Sp(6) little Higgs. *Phys. Rev. D* **2004**, *69*, 115008. [CrossRef]
47. Han, Z.; Skiba, W. Little Higgs models and electroweak measurements. *Phys. Rev. D* **2005**, *72*, 035005. [CrossRef]
48. Brown, T.; Frugiuele, C.; Gregoire, T. UV friendly T-parity in the SU(6)/Sp(6) little Higgs model. *J. High Energy Phys.* **2011**, *2011*, 108. [CrossRef]
49. Gopalakrishna, S.; Mukherjee, T.S.; Sadhukhan, S. Status and Prospects of the Two-Higgs-Doublet SU(6)/Sp(6) little-Higgs Model and the Alignment Limit. *Phys. Rev. D* **2016**, *94*, 015034. [CrossRef]
50. Pauli, W. On the conservation of the lepton charge. *Nuovo C.* **1957**, *6*, 204–215. [CrossRef]
51. Gürsey, F. Relation of charge independence and baryon conservation to Pauli's transformation. *Nuovo C.* **1958**, *7*, 411–415. [CrossRef]
52. Vysotskii, M.I.; Kogan, Y.I.; Shifman, M.A. Spontaneous breakdown of chiral symmetry for real fermions and the $N = 2$ supersymmetric Yang–Mills theory. *Sov. J. Nucl. Phys.* **1985**, *42*, 318.
53. Verbaarschot, J. The Supersymmetric Method in Random Matrix Theory and Applications to QCD. In *Proceedings of the 35th Latin American School of Physics on Supersymmetries in Physics and Its Applications (ELAF 2004)*; Bijker, R., Castaños, O., Fernández, D., Morales-Técotl, H., Urrutia, L., Villarreal, C., Eds.; American Institute of Physics: Melville, NY, USA, 2004; Volume 744, pp. 277–362. [CrossRef]
54. Campbell, D.K. Partially conserved axial-vector current and model chiral field theories in nuclear physics. *Phys. Rev. C* **1979**, *19*, 1965–1970. [CrossRef]
55. Dmitrašinović, V.; Myhrer, F. Pion-nucleon scattering and the nucleon Σ term in an extended linear Σ model. *Phys. Rev. C* **2000**, *61*, 025205. [CrossRef]
56. Delorme, J.; Chanfray, G.; Ericson, M. Chiral Lagrangians and quark condensate in nuclei. *Nucl. Phys. A* **1996**, *603*, 239–256. [CrossRef]
57. Gasiorowicz, S.; Geffen, D.A. Effective Lagrangians and Field Algebras with Chiral Symmetry. *Rev. Mod. Phys.* **1969**, *41*, 531–573. [CrossRef]
58. Parganlija, D.; Kovács, P.; Wolf, G.; Giacosa, F.; Rischke, D.H. Meson vacuum phenomenology in a three-flavor linear sigma model with (axial-)vector mesons. *Phys. Rev. D* **2013**, *87*, 014011. [CrossRef]
59. Bai, Y.; Hill, R.J. Weakly interacting stable hidden sector pions. *Phys. Rev. D* **2010**, *82*, 111701. [CrossRef]
60. Antipin, O.; Redi, M.; Strumia, A.; Vigiani, E. Accidental Composite Dark Matter. *J. High Energy Phys.* **2015**, *2015*, 39. [CrossRef]
61. Khlopov, M.Y. Fractionally charged particles and quark confinement. *JETP Lett.* **1981**, *33*, 162.
62. Alexander, S.; Orbach, R. Density of states on fractals: « fractons ». *J. Phys. Lett.* **1982**, *43*, 625. [CrossRef]
63. Zel'dovich, Y.B.; Okun', L.B.; Pikel'ner, S.B. Quarks: Astrophysical and physicalchemical aspects. *Sov. Phys. Uspekhi* **1966**, *8*, 702. [CrossRef]
64. Glashow, S.G. A Sinister Extension of the Standard Model to $SU(3) \times SU(2) \times SU(2) \times U(1)$. In Proceedings of the XI Workshop on Neutrino Telescopes, Venice, Italy, 22–25 February 2005; p. 9.
65. Khlopov, M.Y. Composite dark matter from 4th generation. *JETP Lett.* **2006**, *83*, 1–4. [CrossRef]
66. Goertz, F. Composite Higgs theory. *Proc. Sci.* **2018**. [CrossRef]
67. Sannino, F.; Tuominen, K. Orientifold theory dynamics and symmetry breaking. *Phys. Rev. D* **2005**, *71*, 051901. [CrossRef]
68. Hong, D.K.; Hsu, S.D.H.; Sannino, F. Composite higgs from higher representations. *Phys. Lett. B* **2004**, *597*, 89–93. [CrossRef]
69. Dietrich, D.D.; Sannino, F.; Tuominen, K. Light composite higgs from higher representations versus electroweak precision measurements: Predictions for LHC. *Phys. Rev. D* **2005**, *72*, 055001. [CrossRef]

70. Dietrich, D.D.; Sannino, F.; Tuominen, K. Light composite higgs and precision electroweak measurements on the Z resonance: An update. *Phys. Rev. D* **2006**, *73*, 037701. [CrossRef]
71. Gudnason, S.B.; Kouvaris, C.; Sannino, F. Towards working technicolor: Effective theories and dark matter. *Phys. Rev. D* **2006**, *73*, 115003. [CrossRef]
72. Khlopov, M.Y.; Kouvaris, C. Strong interactive massive particles from a strong coupled theory. *Phys. Rev. D* **2008**, *77*, 065002. [CrossRef]
73. Khlopov, M.Y.; Kouvaris, C. Composite dark matter from a model with composite Higgs boson. *Phys. Rev. D* **2008**, *78*, 065040. [CrossRef]
74. Cirelli, M.; Strumia, A.; Tamburini, M. Cosmology and Astrophysics of Minimal Dark Matter. *Nucl. Phys. B* **2007**, *787*, 152–175. [CrossRef]
75. Kuksa, V.; Beylin, V. Hyperfine Splitting of Excited States of new Heavy Hadrons and Low-energy Interaction of Hadronic Dark Matter with Photons, Nucleons and Leptons. *Universe* **2020**, *6*, 84. [CrossRef]
76. Ebert, D.; Galkin, V.O.; Faustov, R.N. Mass spectrum of orbitally and radially excited heavy-light mesons in the relativistic quark model. *Phys. Rev. D* **1998**, *57*, 5663–5669. [CrossRef]
77. Mutuk, H. Mass Spectra and Decay Constants of Heavy-Light Mesons. *Adv. High Energy Phys.* **2018**, *2018*, 8095653.
78. Vereshkov, G.M.; Kuksa, V.I. $U(1) \times SU(3)$-gauge model of baryon-meson interactions. *Yad. Fiz.* **1991**, *54*, 1700.
79. Beylin, V.A.; Kuksa, V.I. Interaction of Hadronic Dark Matter with Nucleons and Leptons. *Symmetry* **2020**, *12*, 567. [CrossRef]
80. Wechsler, R.H.; Tinker, J.L. The Connection between Galaxies and their Dark Matter Halos. *arXiv* **2018**, arXiv:1804.03097.
81. Khlopov, M. Dark atom solution for puzzles of direct dark matter searches. *IOP Conf. Ser. J. Phys. Conf. Ser.* **2019**, *1312*, 012011. [CrossRef]
82. Khlopov, M.Y. Introduction to the special issue on indirect dark matter searches. *Mod. Phys. Lett. A* **2014**, *29*, 1402001. [CrossRef]
83. Cirelli, M. Dark Matter Indirect searches: Phenomenological and theoretical aspects. *J. Phys. Conf. Ser.* **2013**, *447*, 012006. [CrossRef]
84. Roszkowski, L.; Sessolo, E.M.; Trojanowski, S. WIMP dark matter candidates and searches—Current status and future prospects. *Rep. Prog. Phys.* **2018**, *81*, 066201. [CrossRef]
85. Gaskins, J.M. A review of indirect searches for particle dark matter. *Contemp. Phys.* **2016**, *57*, 496–525. [CrossRef]
86. Arcadi, G.; Dutra, M.; Ghosh, P.; Lindner, M.; Mambrini, Y.; Pierre, M.; Profumo, D.; Queiroz, F.S. The waning of the WIMP? A review of models, searches, and constraints. *Eur. Phys. J. C* **2018**, *78*, 203. [CrossRef]
87. Belotsky, K.M.; Esipova, E.A.; Kamaletdinov, A.K.; Shlepkina, E.S.; Solovyov, M.L. Indirect effects of dark matter. *Int. J. Mod. Phys. D* **2019**, *28*, 1941011. [CrossRef]
88. Gaggero, D.; Valli, M. Impact of Cosmic-Ray Physics on Dark Matter Indirect Searches. *Adv. High Energy Phys.* **2018**, 3010514. [CrossRef]
89. Berezinsky, V.; Dokuchaev, V.; Eroshenko, Y. Small-scale clumps in the galactic halo and dark matter annihilation. *Phys. Rev. D* **2003**, *68*, 103003. [CrossRef]
90. Berezinsky, V.; Dokuchaev, V.; Eroshenko, Y. Remnants of dark matter clumps. *Phys. Rev. D* **2008**, *77*, 083519. [CrossRef]
91. Berezinsky, V.; Dokuchaev, V.; Eroshenko, Y. Formation and internal structure of superdense dark matter clumps and ultracompact minihaloes. *J. Cosmol. Astropart. Phys.* **2013**, *11*, 059. [CrossRef]
92. Belotsky, K.; Kirillov, A.; Khlopov, M. Gamma-ray evidences of the dark matter clumps. *Grav. Cosmol.* **2014**, *20*, 47. [CrossRef]
93. Tasitsiomi, A.; Olinto, A.V. Detectability of neutralino clumps via atmospheric Cherenkov telescopes. *Phys. Rev. D* **2002**, *66*, 083006. [CrossRef]
94. Gabici, S.; Evoli, C.; Gaggero, D.; Lipari, P.; Mertcsh, P.; Orlando, E.; Strong, A.; Vittino, A. The origin of galactic cosmic rays: Challenges to the standard paradigm. *Int. J. Mod. Phys. D* **2019**, *28*, 1930022. [CrossRef]
95. Kachelries, M.; Semikoz, D.V. Cosmic Ray Models. *Nucl. Phys. B* **2019**, *109*, 103710. [CrossRef]
96. Allard, D.; Busca, N.G.; Decerprit, G.; Olinto, A.V.; Parizot, E. Implications of the cosmic ray spectrum for the mass composition at the highest energies. *J. Cosmol. Astropart. Phys.* **2008**, *2008*, 33. [CrossRef]

97. Unger, M.; Farrar, G.R.; Anchordoqui, L.A. Origin of the ankle in the ultrahigh energy cosmic ray spectrum, and of the extragalactic protons below it. *Phys. Rev. D* **2015**, *92*, 123001. [CrossRef]
98. Dova, M.T. Ultra-High Energy Cosmic Rays. In Proceedings of the CERN–Latin-American School of High-Energy Physics, Arequipa, Peru, 6–19 March 2013; CERN-2015-001; CERN: Meyrin, Switzerland, 2015.
99. Torres, D.F.; Anchordoqui, L.A. Astrophysical Origins of Ultrahigh Energy Cosmic Rays. *Rep. Prog. Phys.* **2004**, *67*, 1663–1730. [CrossRef]
100. Bloom, E.D.; Wells, J.D. Multi-GeV photons from electron–dark matter scattering near Active Galactic Nuclei. *Phys. Rev. D* **1998**, *57*, 1299–1302. [CrossRef]
101. Gorchtein, M.; Profumo, S.; Ubaldi, L. Probing Dark Matter with AGN Jets. *Phys. Rev. D* **2010**, *82*, 083514. [CrossRef]
102. Profumo, S.; Ubaldi, L. Cosmic Ray-Dark Matter Scattering: A New Signature of (Asymmetric) Dark Matter in the Gamma Ray Sky. *J. Cosmol. Astropart. Phys.* **2011**, *2011*, 20. [CrossRef]
103. Gorchtein, M.; Profumo, S.; Ubaldi, L. Gamma rays from cosmic-ray proton scattering in AGN jets: The intra-cluster gas vastly outshines dark matter. *J. Cosmol. Astropart. Phys.* **2013**, *2013*, 12.
104. Cappiello, C.V.; Ng, K.C.Y.; Beacom, J.F. Reverse Direct Detection: Cosmic Ray Scattering with Light Dark Matter. *Phys. Rev. D* **2019**, *99*, 063004. [CrossRef]
105. Beylin, V.; Bezuglov, M.; Kuksa, V.; Tretiakov, E.; Yagozinskaya, A. On the scattering of a high-energy cosmic ray electrons off the dark matter. *Int. J. Mod. Phys. A* **2019**, *34*, 1950040. [CrossRef]
106. Beylin, V.; Bezuglov, M.; Kuksa, V.; Tretiakov, E. Quasi-elastic lepton scattering off two-component dark matter in hypercolor model. *Symmetry* **2020**, *12*, 708. [CrossRef]
107. Agashe, K.; Cui, Y.; Necib, L.; Thaler, J. (In)direct Detection of Boosted Dark Matter. *J. Cosmol. Astropart. Phys.* **2014**, *2014*, 062. [CrossRef]
108. Bennett, E.; Hong, D.K.; Lee, J.-W.; David Lin, C.-J.; Lucini, B.; Piai, M.; Vadacchino, D. Meson spectrum of Sp(4) lattice gauge theory with two fundamental Dirac fermions. *arXiv* **2019**, arXiv:1911.00437.
109. Bennett, E.; Hong, D.K.; Lee, J.-W.; David Lin, C.-J.; Lucini, B.; Piai, M.; Vadacchino, D. Sp(4) gauge theories on the lattice: Nf=2 dynamical fundamental fermions. *J. High Energy Phys.* **2019**, *2019*, 53, doi:10.1007/JHEP12(2019)053. [CrossRef]
110. De Roeck, A.; Kim, D.; Moghaddam, Z.C.; Park, J.-C.; Shin, S.; Whitehead, L.H. Probing Energetic Light Dark Matter with Multi-Particle Tracks signatures at DUNE. *arXiv* **2020**, arXiv:2005.08979.
111. Bhattacharya, A.; Gandhi, R.; Gupta, A. The Direct Detection of Boosted Dark Matter at High Energies and PeV events at IceCube. *J. Cosmol. Astropart. Phys.* **2015**, *2015*, 27. [CrossRef]
112. Kopp, J.; Liu, J.; Wang, X.-P. Boosted Dark Matter in IceCube and at the Galactic Center. *J. High Energy Phys.* **2015**, *2015*, 105. [CrossRef]
113. Kim, D.; Park, J.-C.; Shin, S. Dark Matter Collider from Inelastic Boosted Dark Matter. *Phys. Rev. Lett.* **2017**, *119*, 161801. [CrossRef] [PubMed]
114. Giudice, G.F.; Kim, D.; Park, J.-C.; Shin, S. Inelastic Boosted Dark Matter at Direct Detection Experiments. *Phys. Lett. B* **2018**, *780*, 543–552. [CrossRef]
115. Bringmann, T.; Pospelov, M. Novel direct detection constraints on light dark matter. *Phys. Rev. Lett.* **2019**, *122*, 171801. [CrossRef]
116. Heurtier, L.; Kim, D.; Park, J.-C.; Shin, S. Explaining the ANITA Anomaly with Inelastic Boosted Dark Matter. *Phys. Rev. D* **2019**, *100*, 055004. [CrossRef]
117. Ghou, B.; Beacom, J.F. Neutrino-nucleus cross sections for W-boson and trident production. *Phys. Rev. D* **2020**, *101*, 036011.
118. Kuksa, V.I. The convolution formula for a decay rate. *Phys. Lett. B* **2006**, *633*, 545–549. [CrossRef]
119. Kuksa, V.I.; Volchanskiy, N.I. Factorization in the model of unstable particles with continuous masses. *Cent. Eur. J. Phys.* **2013**, *11*, 182–194. [CrossRef]
120. Ryabov, V.A. Ultrahigh-energy neutrinos from astrophysical sources and superheavy particle decays. *Phys. Uspekhi* **2006**, *49*, 9. [CrossRef]
121. Necib, L.; Moon, J.; Wongjirad, T.; Conrad, J.M. Boosted Dark Matter at Neutrino Experiments. *Phys. Rev. D* **2017**, *95*, 075018. [CrossRef]
122. Alhazmi, H.; Kong, K.; Mohlabeng, G.; Park, J.-C. Boosted Dark Matter at the Deep Underground Neutrino Experiment. *J. High Energy Phys.* **2017**, *2017*, 158. [CrossRef]

123. Jaeckel, J.; Yin, W. Boosted Neutrinos and Relativistic Dark Particles as Messengers from Reheating. *arXiv* **2020**, arXiv:2007.15006.
124. Richard, E.; Okumura, K.; Abe, K.; Haga, Y.; Hayato, Y.; Ikeda, M.; Iyogi, K.; Kameda, J.; Kishimoto, Y.; Miura, M.; et al. Measurements of the atmospheric neutrino flux by 430 Super-Kamiokande: Energy spectra, geomagnetic effects, and solar modulation. *Phys. Rev. D* **2016**, *94*, 052001. [CrossRef]
125. Niederhausen, H.; Xu, Y. High Energy Astrophysical Neutrino Flux Measurement Using Neutrino-induced Cascades Observed in 4 Years of IceCube Data. *Int. Cosm. Ray Conf.* **2017**, *301*, 968.
126. Kochanov, A.A.; Morozova, A.D.; Sinegovskaya, T.S.; Sinegovskiy, S.I. Behaviour of the high-energy neutrino flux in the Earth's atmosphere.*Sol. Terr. Phys.* **2015**, *1*, 3–10.
127. Ahlers, M.; Helbing, K.; de los Heros, C.P. Probing particle physics with IceCube. *Eur. Phys. J. C* **2018**, *78*, 924. [CrossRef]
128. Kampert, K.H.; Watson, A.A. Extensive Air Showers and UHECR: A historical review. *arXiv* **2012**, arXiv:1207.4827.
129. Dent, J.B.; Dutta, B.; Newstead, J.L.; Shoemaker, I.M. Bounds on Cosmic Ray-Boosted Dark Matter in Simplified Models and its Corresponding Neutrino-Floor. *Phys. Rev. D* **2020**, *101*, 116007. [CrossRef]
130. Ge, S.-F.; Liu, J.-L.; Yuan, Q.; Zhou, N. Boosted Diurnal Effect of Sub-GeV Dark Matter at Direct Detection Experiment. *arXiv* **2020**, arXiv:2005.09480.
131. Bottai, S.; Giurgola, S. Downward neutrino induced EAS with EUSO detector. *Int. Cosm. Ray Conf.* **2003**, *2*, 1113–1116.
132. Fargion. D. Ultra High Energy Cosmic Rays, Z-Shower and Neutrino Astronomy by Horizontal-Upward Tau Air-Showers. *arXiv* **2003**, arXiv:hep-ph/0306238.
133. Albert, A.; André, M.; Anghinolfi, M.; Anton, G.; Ardid, M.; Aubert, J.-J.; Avgitas, T.; Baret, B.; Barrios-Martí, J.; Basa, S.; et al. An algorithm for the reconstruction of high-energy neutrino induced particle showers and its application to the ANTARES neutrino telescope. *Eur. Phys. J. C* **2017**, *77*, 149. [CrossRef] [PubMed]
134. Zhang, B.-L.; Lei, Z.-H.; Tang, J. Constraints on cosmic-ray boosted DM in CDEX-10. *arXiv* **2020**, arXiv:2008.07116.
135. Beylin, V.; Kuksa, V. On the reactions involving neutrinos and hidden mass particles in hypercolor model. In *Proceedings of the XXIII Int. Workshop "What Comes Beyond the Standard Models?"*; DMFA: Bled, Slovenia, 2020; to be published.
136. Fusco, L.A.; Versari, F. Testing cosmic ray composition models with very large-volume neutrino telescopes. *Eur. Phys. J. Plus* **2020**, *135*, 624. [CrossRef]
137. Rott, C. Status of Dark Matter Searches. *arXiv* **2017**, arXiv:1712.00666.
138. Bai, X.; Bi, B.; Bi, J.; Cao, Z.; Chen, S.Z.; Chen, Y.; Chiavassa, A.; Cui, X.H.; Dai, Z.G.; della Volpe, D.; et al. The large high altitude air shower observatory (LHAASO) science white paper. *arXiv* **2019**, arXiv:1905.02773.

Publisher's Note: MDPI stays neutral with regard to jurisdictional claims in published maps and institutional affiliations.

© 2020 by the authors. Licensee MDPI, Basel, Switzerland. This article is an open access article distributed under the terms and conditions of the Creative Commons Attribution (CC BY) license (http://creativecommons.org/licenses/by/4.0/).

Article

Invisible QCD as Dark Energy

Andrea Addazi [1,2,*], Stephon Alexander [3] and Antonino Marcianò [4,5,*]

1 Center for Theoretical Physics, College of Physics Science and Technology, Sichuan University, Chengdu 610065, China
2 INFN Sezione Roma Tor Vergata, I-00133 Rome, Italy
3 Department of Physics, Brown University, Providence, RI 02912, USA; Stephon_Alexander@brown.edu
4 Center for Field Theory and Particle Physics & Department of Physics, Fudan University, Shanghai 200433, China
5 Laboratori Nazionali di Frascati INFN, 54 Via Enrico Fermi, 00044 Frascati (Rome), Italy
* Correspondence: andrea.addazi@lngs.infn.it (A.A.); marciano@fudan.edu.cn (A.M.)

Received: 23 May 2020; Accepted: 28 May 2020; Published: 31 May 2020

Abstract: We account for the late time acceleration of the Universe by extending the Quantum Chromodynamics (QCD) color to a $SU(3)$ invisible sector (IQCD). If the Invisible Chiral symmetry is broken in the early universe, a condensate of dark pions (dpions) and dark gluons (dgluons) forms. The condensate naturally forms due to strong dynamics similar to the Nambu–Jona-Lasinio mechanism. As the Universe evolves from early times to present times the interaction energy between the dgluon and dpion condensate dominates with a negative pressure equation of state and causes late time acceleration. We conclude with a stability analysis of the coupled perturbations of the dark pions and dark gluons.

Keywords: dark energy; non-Abelian gauge theory; condensate

---

1. Introduction

A confluence of cosmological data tell us that the Universe is currently accelerating (see e.g., Refs. [1–3] and references therein). While the data can be explained with a cosmological constant, it is also possible that the Universe is dominated by a form of Dark Energy (DE) with a negative pressure barotropic index $w \sim -1$. Another possibility is that general relativity is modified in the infrared (IR) admitting self accelerating solutions, and such models are still under active research Ref. [4]. In this work we take the position that the Einstein-Hilbert action is the correct IR description of gravity and the cosmological constant is zero, by some yet unknown mechanism. We present a model where DE emerges from an Invisible QCD (IQCD) sector due to the interaction of invisible pions and gluons which were present in the early Universe.

There are two ways to theoretically motivate an Invisible QCD sector: it has been known for a long time that the $E_8 \times E_8$ Heterotic Superstring Theory, both in the weak and strong coupling limit necessarily gives rise to non-abelian gauge theories that do not directly couple to the standard model $SU(3)_c \times SU(2)_W \times U(1)_Y$ gauge group. While String Theory has this property, it is also possible to obtain a dark copy in another way, which corresponds to modifying general relativity by enlarging the spin connection $\omega(e)^{IJ}_{\mu}$ to a larger group $G \sim SU(N)$, which breaks to $SU(2) \times SU(N-2)$ where the $SU(N-2)$ factor is identified with the dark sector. This philosophy was pursued in Refs. [5–9]. Here we take a phenomenological perspective and simply assume that this dark sector exists and focus on the cosmological consequences.

As a topic contiguous to the one treated in this investigation, we shall provided a short review on the most relevant studies addressing a possible relation of QCD to dark energy. For sure relevant to our discussion is the perspective proposed by E. Witten, who has shown that the small value of the cosmological constant could be explained in terms of a vacuum state with unbroken supersymmetry. Specifically, as summarized in Ref. [10], Witten has argued that, while compactifying Calabi-Yau manifolds, the string tension in higher dimension can be invoked to cancel the vacuum expectation value of QCD and other fields, and hence reproduce the observed value of the cosmological constant. An extensive review on the same argument was offered in Ref. [11] by R. Bousso. While an analysis more focused on the the link between QCD condensate and the observed value of the cosmological constant is the one contained in Ref. [12], where the authors studied quark and gluon condensates in QCD, as associated to the internal dynamics of hadrons, and then succeeded in showing that these condensates actually provide a vanishing contribution to the cosmological constant.

Related to an extended invisible/mirror framework is the investigation carried out in Ref. [13], which appeared after the first submission of this work on the electronic archives [14]. The analysis in Ref. [13] has remarkable consequences at the level of the mirror symmetry, which some of us further investigated in Ref. [15]. Specifically, the authors of Ref. [13] explored the possible compensation of the negative contribution, due to the QCD vacuum, to the ground state energy density of the Universe, thanks to the (opposite in sign) contribution that arises from the chromomagnetic gluon condensate in the invisible/mirror QCD sector. Although not directly related to the fate of the QCD vacuum, it is nonetheless worth mentioning the study achieved by some of us in Ref. [16], on a physical mechanism similar to the one we developed here. This mechanism led to a quasi-de Sitter expansion of the Universe, which inspired the analysis of this paper, by considering a fields backreaction (in the expanding Universe) that sources an effective cosmological constant. This is due to the interaction energy among the gauge hypercharge field and the fermion field.

In this model, late time acceleration arises from extending the color sector of QCD to have an "invisible-copy". IQCD has similar quantum field theoretic properties of QCD, in that it is confining in the IR. It is well known that pions arise as Goldstone modes associated to Chiral Symmetry Breaking (CSB), and in turn the microphysical description of CSB, the Nambu–Jona-Lasinio mechanism Ref. [17,18], is a strongly coupled version of superconductivity induced by hard gluon exchange. A key feature of our DE model uses the same physics of CSB in the IQCD sector. During the matter and radiation era, the dark pions and gluons have negligible effects. However, we will show that at late times the interaction energy between the dark pions and gluons become more significant because they remain nearly constant, mimicking an effective cosmological constant. Through the consistency of the coupled field equations, this interaction energy naturally leads to late time acceleration and we find an interesting connection between the scale of CSB and the scale of DE.

2. The Theory

We assume that Chiral symmetry in the IQCD sector is broken at some scale f_D corresponding to the mass of a dark pion[1] (dpion) π_D. At the renormalizable level, such a hidden sector can communicate with the standard model only through gauge interactions: $SU(3)_c$, $SU(2)_W$ or $U(1)_Y$—see e.g., Refs. [19–21]. In this work, for clarity of presentation, we assume no gauge coupling to the standard model. However, this model can easily be extended such that the gauge-confined quarks are coupled vectorially to $SU(2)_W$. The authors of Ref. [21] showed that a dark sector with purely vectorial coupling to the standard model has remarkable universal features. A specific parity symmetry (known as G-parity) acting only on the hidden sector fields is left unbroken and stable weakly interacting dark-pions become a compelling candidate for a dark-matter particle. It would be interesting to revisit the constraints on the coupling to the visible sector that is simultaneously consistent with dark-matter and DE. We find it intriguing that our model has the possibility of connecting late time acceleration to Dark Matter and will pursue this possibility in Ref. [22].

[1] From now on, we will remove the subscript D when referring to dark pions.

The $SU(N)_D$ gauge theory we are considering is assumed to have a behavior similar to QCD. The gauge coupling becomes strong in the IR limit, triggering confinement and chiral symmetry breaking at a scale Λ_D. Below Λ_D, the effective theory is described by "pions" which are pseudo-Nambu–Goldstone bosons (pNGBs) associated with the spontaneously broken global flavor symmetry of the hidden sector.

For concreteness and without loss of generality, we consider the subgroup, $SU(2)_L \times SU(2)_R \to SU(2)_V$ with its gauge field $A^a{}_\mu$, where $a, b = 1,2,3$ and $\mu, \nu = 0,1,2,3$ are for the dark color and space-time indices, respectively. The gauge field strength F is

$$F^a_{\mu\nu} = \partial_\mu A^a{}_\nu - \partial_\nu A^a{}_\mu - g\epsilon^a{}_{bc} A^b{}_\mu A^c{}_\nu, \tag{1}$$

where ϵ_{abc} is the totally antisymmetric Levi-Civita symbol, the structure constant of the $SU(2)$ algebra. We are led to consider the most general gauge invariant action coupled to "dark quarks", with masses m_i and Lagrangian

$$\mathcal{L}_{\rm IQCD} \equiv \mathcal{L}_{\rm A} + \mathcal{L}_{\rm D} = -\frac{1}{4g^2} F^a_{\mu\nu} F_a^{\mu\nu} + \overline{\psi}_i \, (\imath \, D_\mu \gamma^\mu - m_i) \psi_i \,. \tag{2}$$

Here D_μ stands for the covariant derivative with respect to the dark SU(3) sector and the gravitational connection, $\gamma^\mu = \gamma^I \, e^\mu_I$ and the metric field is decomposed in tetrad, namely $g_{\mu\nu} = e^I_\mu \, e^I_\nu$, the inverse of which is denoted as e^μ_I and the internal $SO(3,1)$ indices of which are $I = 1,2...4$. Given that we are working in a system where the dpion forms as a result of CSB, the decay constant f_D is defined through the coupling of the axial current to the dpion. In particular, dpions can be created by the axial isospin currents.

Matrix elements of $\mathcal{J}^{5\,I}_a(x)$ between the vacuum and an on-shell dpion state can be parametrized as (see e.g., Ref. [23])

$$\langle 0 | \mathcal{J}^{5\,I}_a(x) | \pi^b(p) \rangle = -\imath \, \delta^b_a \, p^I \, f_D \, e^{-\imath p \cdot x} \,, \tag{3}$$

where the chiral current is $\mathcal{J}^{5\,I}_a(x) = \overline{\psi}(x) \, \gamma^I \gamma_5 \, \tau_a \, \psi(x)$ and $|\pi(p)\rangle$ is the pion ground-state with normalization $\langle \pi(p') | \pi(p) \rangle = 2(2\pi)^3 p^0 \delta^3(p' - p)$. Relation (3) can be recast in terms of the dpion field as

$$\langle 0 | \mathcal{J}^{5\,I}_a(x) | \pi^b \rangle = f_D \, \delta^b_a \, \partial^I \pi_b(x) \,. \tag{4}$$

We can rotate the expectation value of $\langle 0 | \mathcal{J}^{5\,I}_a(x) | \pi^b \rangle$ within the internal Lorentz indices' space, so to accomplish an explicitly space-like axial vector with vanishing temporal component. The symmetry of the vacuum state on the FLRW background allows to further reduce Equation (4) to a homogenous axial vector:

$$\langle 0 | \mathcal{J}^{5\,I}_a(x) | \pi^b \rangle = f_D^2 \, \delta^i_a \, \pi(t, 0) \,, \tag{5}$$

where $\pi(t) \equiv ||\pi^a(t)||$, with respect to the internal indices. The interaction of the axial current with the gauge field $\mathcal{L}^{int.}_\pi = g \, A^a_\mu \, \mathcal{J}^{5\,\mu}_a$ is therefore consistent with homogenous and isotropic metrics.

The low energy dpion effective Lagrangian reads

$$\mathcal{L}^0_\pi = -\frac{1}{2} \partial_\mu \pi_a \partial^\mu \pi^a + \frac{\lambda}{4} \left(\pi^a \, \pi_a - f_D^2 \right)^2 . \tag{6}$$

Consequently, the total effective Lagrangian reads

$$\begin{aligned} \mathcal{L}_{tot.} &= \mathcal{L}_{\rm GR} + \mathcal{L}_{\rm A} + \mathcal{L}^{int.}_\pi + \mathcal{L}^0_\pi \\ &= M_p^2 \, R - \frac{1}{4} F^a_{\mu\nu} F_a^{\mu\nu} + g \, A^a_\mu \, \mathcal{J}^{5\,\mu}_a + \mathcal{L}^0_{\rm ß}, \end{aligned} \tag{7}$$

in which we have introduced the reduced Planck mass as $M_p^2 = (8\pi G)^{-1}$. Quark fields have been integrated out in the path integral in order to get the effective action. The interaction term $\mathcal{L}_\pi^{int.} = g\, A^a_\mu\, \mathcal{J}_a^{5\,\mu}$, which entails parity violations of the $SU(2)$ subgroup of the dark sector, preserves renormalizability. The total action is $\mathcal{S}_{tot.} = \int d^4x\, e\, \mathcal{L}_{tot.}$, with e volume density denoting the determinant of the tetrad e^I_μ.

3. Field Equations

Solutions to the field equations that are consistent with a FLRW background can be recovered once a rotationally invariant configuration for the gauge field has been implemented:

$$A^a{}_\mu = \begin{cases} a(t)\, \phi(t)\, \delta^a_i\,, & \mu = i\,, \\ 0\,, & \mu = 0\,. \end{cases} \tag{8}$$

Thanks to Equations (4) and (8), the energy-momentum tensor of the theory is isotropic and homogenous. In particular, the energy-momentum tensor associated to the interaction Lagrangian yields the remarkable feature of having a barotropic index $w = -1$, since energy and pressure densities respectively read

$$\begin{aligned} \rho_{\mathbf{AJ}} &= 3\, g\, f_D^2\, \phi(t)\, \pi(t), \\ -P_{\mathbf{AJ}} &= 3\, g\, f_D^2\, \phi(t)\, \pi(t), \end{aligned}$$

g denoting above the absolute vale of the coupling constant. This naturally leads towards a de Sitter accelerating phase of the Universe, as soon as the interaction term becomes dominant.

In the rotationally symmetric configuration Equation (8) the gauge field-strength's components simplify and read $F^a_{0i} = \partial_t(\phi(t)a(t)\delta^a_i)$ and $F^a_{ij} = -g\,\epsilon^a_{ij}(\phi(t)a(t))^2$, having specified our system in co-moving coordinates $ds^2 = dt^2 - a^2(t)\, d\vec{x}^2$. Using (8), the total gauge Lagrangian becomes

$$\mathcal{L}^{\mathbf{A}} + \mathcal{L}^{\text{int.}} = \frac{1}{2\,g^2}\left(3\,(\dot{\phi} + H\phi)^2 - 3\,g^2\phi^4\right) + 3g\phi\, \bar{J}(a), \tag{9}$$

where $\bar{J}(a) \equiv f_D^2\, \pi(t)$. From $\frac{1}{a^3}\frac{\partial}{\partial t}(a^3 \frac{\partial \mathcal{L}}{\partial \dot{\phi}}) = \frac{\partial \mathcal{L}}{\partial \phi}$, we obtain the equation of motion for ϕ, which captures the dynamics of $A^a{}_\mu$ through Equation (8), namely

$$\ddot{\phi} + 3H\dot{\phi} + (2H^2 + \dot{H})\,\phi + 2\,g^2\phi^3 - g\,\bar{J}(a) = 0\,. \tag{10}$$

The equation of motion for the dpion field is recovered varying Equation (6), within the assumption of spatial homogeneity. This is plausible, since a previous inflationary epoch of the universe can smooth out the dpion field. In the next section, we show that the dpion field remains homogenous against perturbations.

Using the decomposition in a homogeneous absolute value (in the internal space) times a space-dependent unit vector, *i.e.* $\pi_a = ||\pi_a||\, n_a = \pi(t)\, n_a(x)$, we recover for the pion field

$$\ddot{\pi} + 3H\dot{\pi} + \lambda\, \pi(\pi^2 - f_D^2) - 3\, g f_D^2 \phi = 0\,. \tag{11}$$

To gain some insight as to why we might expect to see late time acceleration, consider the slow roll regime of the dpion field, which is obtained by neglecting the acceleration term. In this approximation, when the dpion field exhibits an inverse scaling with time, $\pi = \pi_0\, a^{-1}(t)$, the equation of motion reduces to

$$3H\,\frac{\dot{a}}{a^2} = \frac{\lambda}{a}\left(\frac{\pi_0^2}{a^2} - f_D^2\right). \tag{12}$$

Solving this latter results in a power law acceleration of the Universe, namely $H(t) \simeq t^{-1}$, provided that[2] $\pi(t_0) = \pi_0 \gg f_D$ and, as customary when taking into account cosmological scalar fields, the slow roll condition holds: $3H\dot{\pi} \simeq V' \gg \ddot{\pi}$. When the interaction term $\mathcal{L}_{\rm int} = -g\,\mathfrak{E}(t)\, f_D^2\, \mathfrak{B}(t)$ between the dpion and the gauge field is considered, we will see that this term persists to have a nearly constant energy density yielding a negative pressure equation of state. Finally, it is straightforward to show that a slightly different behavior in the time dependence of the dpion, *i.e.* $\pi = \pi_0\, a^{-n}(t)$ with $n > 0$, would yield the same late time-behavior $H(t) \simeq t^{-1}$.

Late time acceleration is recovered when the gauge field asymptotically evolves in time as the scale factor, and the pion field approaches the constant value $\pi \simeq f_D$. Below we show how these self consistent solutions to the equations of motion for the pion and gauge field can be recovered, by working in comoving coordinates and assuming the expansion of the universe to be given by de Sitter phase. Finally, given the non-linearities in the coupled differential equations, we pursue a numerical analysis of the full system of equations.

First we consider the energy density ρ and the pressure P for our low energy effective system that emerges from the dark sector, namely

$$\rho = \rho_{\rm YM} + \rho_{\rm AJ}\,, \qquad P = \frac{1}{3}\rho_{\rm YM} - \rho_{\rm AJ}, \tag{13}$$

where

$$\rho_{\rm YM} = \frac{3}{2}(\dot{\phi} + H\phi)^2 + \frac{3}{2}g^2\phi^4\,, \qquad \rho_{\rm AJ} = 3\,g\,\phi\,\tilde{J}(a)\,, \tag{14}$$

and recall that the Friedmann equations are given by

$$\begin{aligned} H^2\, M_p^2 &= \frac{1}{2}(\dot{\phi} + H\phi)^2 + \frac{1}{2}g^2\phi^4 + g\,\phi\tilde{J}(a) \\ &\quad + \frac{1}{6}\dot{\pi}^2 + \frac{\lambda}{12}\left(\pi^2 - f_D^2\right)^2, \\ (\dot{H} + H^2)\, M_p^2 &= -\frac{1}{2}(\dot{\phi} + H\phi)^2 - \frac{1}{2}g^2\phi^4 + g\,\phi\,\tilde{J}(a) \\ &\quad - \frac{1}{3}\dot{\pi}^2 + \frac{\lambda}{12}(\pi^2 - f_D^2)^2\,. \end{aligned} \tag{15}$$

Having derived the differential equations that govern the dynamical system, we can now proceed to solve it.

4. Field Dynamics

Unlike usual gauge field theories, where the gauge fields dilute during cosmic expansion, the coupling of the gauge field to the dquark current leads to a growth of its homogenous component. This can be understood from the inspection of the equations of motion on the FLRW background. The growth of the gauge field will generically occur as it scales with $a(t)$, while the interaction energy $\rho_{AJ} = g\,\phi\,\tilde{J}(a)$ remains nearly constant at late times. We can then get ahead with our purpose of solving the full dynamical system for the fields involved, so as to plot the evolution of barotropic index

2 Notice that assuming a very large initial value of the pion field may actually induce to consider the tower of all the higher dimensional operators. Nevertheless, taking into account these terms would just strengthen our argument, since the crucial ingredient that ensures the quasi-de Sitter solution at the cosmological level is the mass of the pion. Thus any other higher order term would imply a power law time evolution at earlier cosmological times. Even more, at very high energy the very same concept of pion ground-state looses meaning, and the asymptotic behaviour for the pion field that is sustaining the accelerated phase of the Universe shall no longer be considered. Bearing in mind that our intent in this section is only to show the power law behavior of the background at earlier cosmological times, we avoid further comments.

$$
\begin{aligned}
w = \frac{P}{\rho} = \frac{P_{\text{AJ}} + P_{\text{YM}} + P_{\pi}}{\rho_{\text{AJ}} + \rho_{\text{YM}} + \rho_{\pi}} = \\
= \frac{\frac{1}{2}(\dot{\phi} + H\phi)^2 + \frac{1}{2}g^2\phi^4 - 3g\phi \bar{J}(a) + \frac{1}{2}\dot{\pi}^2 - \frac{\lambda}{4}(\pi^2 - f_D^2)^2}{\frac{3}{2}(\dot{\phi} + H\phi)^2 + \frac{3}{2}g^2\phi^4 + 3g\phi \bar{J}(a) + \frac{1}{2}\dot{\pi}^2 + \frac{\lambda}{4}(\pi^2 - f_D^2)^2}.
\end{aligned}
\tag{16}
$$

Under customary assumption, we are able to solve for the coupled system of differential equations in the configuration space $\{\phi(t), \pi(t)\}$, and to find solutions consistent with a de Sitter expanding phase. We assume in Equation (11) the slow roll condition for π. Furthermore, we assume that at any time energy densities are dominated by the coupling of the gauge field to the axial current, $\rho_\pi \ll \rho_{\text{AJ}}$ and $\rho_{\text{YM}} \ll \rho_{\text{AJ}}$, and show later that these assumptions are consistent with the solutions obtained. In this heuristic analysis, the dynamical system is analytically solved imposing initial conditions at recombination[3]. From Equations (11) and (15) we find, imposing slow roll conditions on π, and then considering the late-time dominant contribution to the solution,

$$
\pi(t) \simeq f_D \left[1 - c_0 \exp\left(-\frac{2\lambda f_D^2 t}{3H} \right) \right]^{-\frac{1}{2}}, \tag{17}
$$

which asymptotically reaches the value $\pi_\infty = f_D$, and in which c_0 is an initial constant. We can then solve for the gauge field, and within a similar assumption on its time derivatives than the condition imposed for $\pi(t)$ we find

$$
\phi(t) \simeq \frac{f_D}{(2g)^{\frac{1}{3}}} \left[1 - c_0 \exp\left(-\frac{2\lambda f_D^2 t}{3H} \right) \right]^{-\frac{1}{6}}. \tag{18}
$$

Both solutions Equayion (17) and (18) are monotonically decreasing and converge asymptotically towards values that are proportional to f_D; thus their product conspire to provide an accelerating solution well approximated by a de Sitter phase, the effective cosmological constant of which assumes the asymptotic value

$$
\Lambda \simeq \frac{f_D^4}{M_p^2}. \tag{19}
$$

Supernovae data, which entail at current times $H \simeq 10^{-42}$ GeV, are consistent with the asymptotic value for the gauge field $A \simeq f_D \simeq 10^{-3}$ eV, the coupling constant g having been assumed to be order unity. This suggests a fascinating conclusion: cosmic acceleration is the consequence of the CSB in the dark sector, since it occurs at the same scale of energy, *i.e.* $M_{DE} \simeq f_D$. We remark that our heuristic analysis has ben focusing only on the radiation energy density of the dark $SU(3)$ sector, with the aim of showing how the evolution of the dark energy component — namely, the interaction energy density among the pion condensate field and the dark radiation — would differ from that one of radiation, rather than dealing with the observed amount of (visible and dark) matter of the Universe.

Conservation of the energy-momentum tensor is easily checked, and the attractor behavior of the solutions is also recovered. Indeed, specifying the numerical values $g = \lambda = 0.1$, the system of differential equations provides the unique fixed point:

$$
(\phi_f, \pi_f, H_f) = (2.17 \cdot 10^{-3}\text{eV}, \, 2.04 \cdot 10^{-3}\text{eV}, 1.08 \cdot 10^{-35} eV).
$$

---

[3] Integration over the dark quark degrees of freedom may provide higher-order derivative terms in the dark gauge sector, exactly as it happens for the Euler-Heisenberg effective action in QED. Higher order derivative terms then provide corrections to the radiation energy-density redshift dependence, which are subdominant in the asymptotic time limit. In our case, these corrections will be further suppressed by powers of the masses of the heaviest dark quarks present in our picture and powers of the coupling constants.

Upon linearization, the first order dynamical system recast in term of derivatives in the variable $N = \ln a(t)$, provides a matrix the eigenvalues of which have negative real components. This ensures that the fixed point is an attractor, and that the system asymptotically converges towards a de Sitter phase. A detailed analysis of the linearization of the first order dynamical system, and of the related eigenvalues of the matrix that is hence recovered, is provided in Ref. [22].

A more cogent numerical analysis corroborates the analytical investigations reported above. Related plots are shown in Figure 1, which makes evident the transition from a radiation dominated epoch (at the recombination $z \simeq 1100$) to our present time ($z = 0$) dominated by DE. Coupling with Standard Model matter would affect the time-evolution of w at $z \gg 10$, without changing its qualitative behaviour and the asymptotic value at recombination epoch.

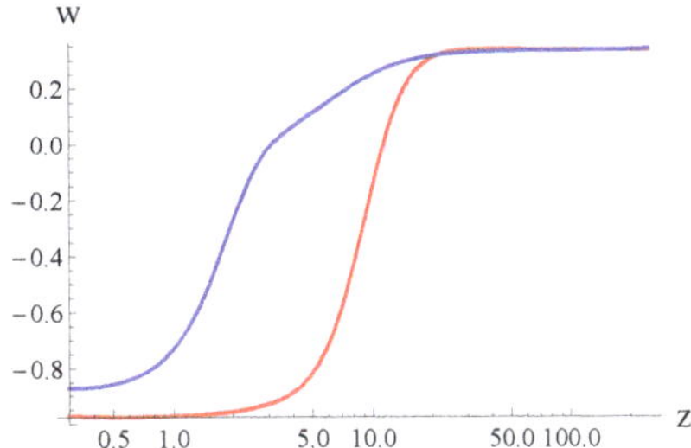

Figure 1. Plot of w against the redshift z, from current time up to recombination. For the blue line, the coupled system of non-linear differential Equations (10), (11) and (15) has been solved numerically for $g = 10^{-1}$ and $\lambda = 1$, and the initial conditions on the functions $H(0) = 10^{-42}$ GeV and $\phi(0) \simeq \pi(0) \simeq f_D = 10^{-3}$ eV, and on their derivatives $\phi'(0) \simeq \pi'(0) = 10^{-5}\,(\text{eV})^2$. Transition from DE to (dark sector) radiation happens for $z \simeq 2$ in the blue line. The plotted red line entails a coupling constant g, the value of which is one order of magnitude smaller than in the case of the blue line.

Notice that in our toy-model, which provides a mechanism for an effective dynamical cosmological constant, the strong energy condition is clearly violated Ref. [24]. Indeed, the (predominant, with respect to the other contributions) interaction energy density $\rho_{\mathbf{AJ}}$ equals the opposite sign of the (predominant, with respect to the other contributions) interaction pressure density $p_{\mathbf{AJ}}$. Nonetheless, the null energy condition is not violated, the interaction energy density providing a positive contribution to the total energy density.

Finally, we wish to emphasize the relevance of the value of the coupling constant g, as already outlined in Figure 1. At this purpose, considering the renormalization group flow of the coupling constant g will provide improvement of the toy-model we present here, and shall be considered in forthcoming more refined and realistic analyses. On a different foot, the relevance of the running of the coupling constant was also studied for QCD in Ref. [15]. Nonetheless, differently that in this latter investigation, we do not expect here that novelties in the attempt to reproduce the cosmological dynamics of the accelerating Universe will be provided by the emergence of a discrete symmetry, but rather that the absolute value of the coupling constant will affect the precise determination of the transition epoch to the dark energy cosmological behavior.

5. Perturbation Analysis

The analysis of perturbations can be developed on the FLRW background by implementing the choice of conformal coordinates $g_{\mu\nu} = a(\eta)^2\,\eta_{\mu\nu}$. Varying action Equation (7), we recover the gauge field equation of motion:

$$\eta^{\mu\rho}(\partial_\mu F^a_{\rho\nu} + g\,\epsilon^a{}_{bc}\,A^b_\mu F^c_{\rho\nu}) = g\,a^4 \mathcal{J}^{5\,a}_\nu\,.$$

From the above equation, the gauge field spatial perturbations immediately follow:

$$\begin{aligned}&\square\delta A_i^a+g\left(\epsilon^a{}_{bc}A_j^b\,\partial^j\delta A_i^c+\epsilon^a{}_{bc}\,\delta A_j^b\,\partial^j A_i^c\right)+\\&+g^2\left(\delta A_b^k A_{[k}^b A_{i]}^a+A_b^k\delta A_{[k}^b A_{i]}^a+A_b^k A_{[k}^b\delta A_{i]}^a\right)=\\&=g\,f_D\,a(\eta)^4\;\partial_i\delta\pi^a(\vec{x},\eta)\,.\end{aligned} \tag{20}$$

The time component perturbations (denoted by $' = d/d\eta$) of the gauge field are found to be:

$$\square\,\delta A_0^a+g\epsilon^a{}_{bc}A_k^b\,\partial^k\delta A_0^c-g^2A_b^i\delta A_{[0}^b\,A_{i]}^a=g\,a^4\,\delta\pi^{a\prime}(\vec{x},\eta)\,.$$

As stated above while using co-moving coordinates, the background solutions for $A_j^a(\eta)$ are subjected to the gauge $A_j^a(\eta)=A(\eta)\,\delta_j^a$, which reduces the spatial perturbation equation to

$$\begin{aligned}&\square\,\delta A_i^a(\vec{x},\eta)+g\,A(\eta)\,[\,\nabla\wedge\delta\vec{A}(\vec{x},\eta)]_i^a+\\&+g^2\left[2A^2(\eta)\,\mathrm{Tr}[\delta A_i^a(\vec{x},\eta)]\,\delta_i^a\right]=g\,a(\eta)^4\;\partial_i\delta\pi^a(\vec{x},\eta)\,,\end{aligned} \tag{21}$$

$$\begin{aligned}&\square\,\delta A_0^a(\vec{x},\eta)+g\,A(\eta)\epsilon^a{}_{bc}\partial^b\delta A_0^c(\vec{x},\eta)+\\&+2g^2A(\eta)^2\delta A_0^a(\vec{x},\eta)=g\,a(\eta)^4\;\delta\pi^{a\prime}(\vec{x},\eta)\,.\end{aligned} \tag{22}$$

Taking the trace of the gauge-field perturbations and writing $\delta A_i^a=\mathrm{Tr}[\delta A_i^a(\vec{x},\eta)]\,\delta_i^a/3$ yield

$$\begin{aligned}&\square\,\mathrm{Tr}[\delta A_i^a(\vec{x},\eta)]+6g^2A^2(\eta)\,\mathrm{Tr}[\delta A_i^a(\vec{x},\eta)]\\&\quad=g\,a(\eta)^4\;\mathrm{Tr}[\partial_i\delta\pi^a(\vec{x},\eta)]\,.\end{aligned} \tag{23}$$

Implementing the same gauge, we recover the equation for the perturbations to the dpion field,

$$\begin{aligned}&\tfrac{\square}{a}\delta\pi^a(\vec{x},\eta)+2\tfrac{a'}{a^2}\delta\pi^{a\prime}(\vec{x},\eta)+\lambda a\,\delta\pi^a(\vec{x},\eta)\,(3\,\pi(\eta)^2-f_D^2)\\&\quad=-gf_D\,\left(\partial^i\delta A_i^a(\vec{x},\eta)+\delta A_0^{a\prime}(\vec{x},\eta)\right)\,.\end{aligned} \tag{24}$$

We are interested in seeing if sub-horizon modes of the dpion field, (24), develop instabilities and cluster. Therefore we focus on wavelengths which are either sub-horizon or above the binding energy of the dpions involved, $H\gg|\vec{k}|\gg f_D$, by studying the evolution of plane-waves $\delta\pi(\vec{x},\eta)\simeq\alpha(\eta)\exp{(\imath\vec{k}\cdot\vec{x})}$ solutions to (24). The equation of motion further reduces to $\alpha''+2Ha\,\alpha'+(\vec{k}^2+\lambda\,a^2f_D^2)\simeq 0$, solutions of which are superpositions of spherical Bessel functions of the first and second type,

$$\begin{aligned}\alpha(\eta)&=\alpha_1\,j_\nu\left(\frac{k}{Ha(\eta)}\right)+\alpha_2\,y_\nu\left(\frac{k}{Ha(\eta)}\right)\,,\\ \nu&=\frac{-H\pm\sqrt{H^2-4\lambda f_D^2}}{2H}\,,\end{aligned}$$

in which $k=|\vec{k}|$, and are convergent to zero at late times for a proper choice of the initial conditions, $\alpha_1\in\mathbb{R}$ and $\alpha_2=0$. Indeed, assuming $\lambda\ll 1$ we see that sub-horizon modes are oscillatory and bounded over all the time axis. This behavior mimics the behavior of super-horizon modes of scalar fields during inflation and is a consequence of the acceleration of space-time. Zero modes are constant and not evolving in time.

Finally, perturbations of the gauge field decrease exponentially (in comoving time), *i.e.* show the conformal time behavior

$$\mathrm{Tr}[\widetilde{\delta A}_i^a(k,\eta)]\simeq 1/a(\eta)\,.$$

We therefore conclude that all sub-horizon perturbations are suppressed. In future work we will perform a full perturbation analysis taking into account metric perturbations Ref. [25].

6. Conclusions and Discussion

We have provided a model of late time acceleration from minimal assumptions, in that aside from instantiating a dark non-abelian copy, we have not introduced any new physics. In fact, we have employed the well known physics of the Nambu–Jona-Lasinio mechanism of chiral symmetry breaking in ordinary QCD applied to IQCD, which is well motivated from string theory. Late time acceleration emerges from the interaction between gravity, a chiral condensate and an invisible gluon that fills the universe today. A preliminary perturbation analysis shows that DE does not cluster on sub-horizon modes.

Finally, we leave to detailed investigations (see e.g., Ref. [22]) the analysis of constraints on the coupling to the visible sector, and the eventual behavior as dark-matter of dquark and dpions in this scenario. Indeed we find intriguing that our model has the possibility of connecting late time acceleration to Dark Matter (see e.g., Refs. [26,27]).

Author Contributions: A.A., S.A., A.M., contributed equally to this article. All authors have read and agreed to the published version of the manuscript.

Funding: S.A. is supported by the US Department of Energy under grant DE-SC0010386. A.M. wishes to acknowledge support by the Shanghai Municipality, through Grant No. KBH1512299, by Fudan University, through Grant No. JJH1512105, and by NSFC, through Grant No. 11875113.

Acknowledgments: We are indebted to David Spergel for enlightening discussions during the early stage of this work. We also thank Robert Caldwell, John Collins, Gia Dvali, Daniel Grinstein and Diego Guadagnoli for their useful feedback.

Conflicts of Interest: The authors declare no conflict of interest.

References

1. Riess, A.G.; et al. [Supernova Search Team Collaboration]. Observational evidence from supernovae for an accelerating universe and a cosmological constant. *Astron. J.* **1998**, *116*, 1009. [CrossRef]
2. Frieman, J.; Turne, M.; Huterer, D. Dark Energy and the Accelerating Universe. *Ann. Rev. Astron. Astrophys.* **2008**, *46*, 385. [CrossRef]
3. Astie, P.; Pain, R. Observational Evidence of the Accelerated Expansion of the Universe. *C. R. Phys.* **2012**, *13*, 521–538.
4. Carroll, S.M.; Duvvuri, V.; Trodden, M.; Turner, M.S. Is cosmic speed-up due to new gravitational physics? *Phys. Rev. D* **2004**, *70*, 043528. [CrossRef]
5. Nesti, F.; Percacci, R. Graviweak Unification. *J. Phys. A* **2008**, *41*, 075405. [CrossRef]
6. Alexander, S.H. Isogravity: Toward an Electroweak and Gravitational Unification. *arXiv* **2007**, arXiv:0706.4481.
7. Smolin, L. The Plebanski action extended to a unification of gravity and Yang-Mills theory. *Phys. Rev. D* **2009**, *80*, 124017. [CrossRef]
8. Alexander, S.; Marcianò, A.; Tacchi, R.A. Towards a Loop Quantum Gravity and Yang-Mills Unification. *Phys. Lett. B* **2012**, *716*, 330–333. [CrossRef]
9. Alexander, S.; Marcianò, A.; Smolin, L. Gravitational origin of the weak interaction's chirality. *Phys. Rev. D* **2014**, *89*, 065017. [CrossRef]
10. Witten, E. The Cosmological constant from the viewpoint of string theory. In *Sources and Detection of Dark Matter and Dark Energy in the Universe*; Springer: Berlin/Heidelberg, Germany, 2001; pp. 27–36.
11. Bousso, R. TASI Lectures on the Cosmological Constant. *Gen. Rel. Grav.* **2008**, *40*, 607–637. [CrossRef]
12. Brodsky, S.J.; Shrock, R. Condensates in Quantum Chromodynamics and the Cosmological Constant. *Proc. Natl. Acad. Sci. USA* **2011**, *108*, 45–50. [CrossRef]
13. Pasechnik, R.; Prokhorov, G.; Teryaev, O. Mirror QCD and Cosmological Constant. *Universe* **2017**, *3*, 43. [CrossRef]
14. Alexander, S.; Marciano, A; Yang, Z. Invisible QCD as Dark Energy. *arXiv* **2016**, arXiv:1602.06557.

15. Addazi, A.; Marcian, A.; Pasechnik, R.; Prokhorov, G. Mirror Symmetry of quantum Yang-Mills vacua and cosmological implications. *Eur. Phys. J. C* **2019**, *79*, 251. [CrossRef]
16. Alexander, S.; Marciano, A.; Spergel, D. Chern-Simons Inflation and Baryogenesis. *J. Cosmol. Astropart. Phys.* **2013**, *2013*, 46. [CrossRef]
17. Nambu, Y.; Jona-Lasinio, G. Dynamical Model of Elementary Particles Based on an Analogy with Superconductivity. I. *Phys. Rev.* **1961**, *122*, 345. [CrossRef]
18. Nambu, Y.; Jona-Lasinio, G. Dynamical Model of Elementary Particles Based on an Analogy with Superconductivity. II. *Phys. Rev.* **1961**, *124*, 246. [CrossRef]
19. Cirelli, M.; Fornengo, N.; Strumia, A. Minimal dark matter. *Nucl. Phys. B* **2006**, *753*, 178. [CrossRef]
20. Kilic, C.; Okui, T.; Sundrum, R. Vectorlike Confinement at the LHC. *J. High Energy Phys.* **2010**, *1002*, 18. [CrossRef]
21. Bai, Y.; Hill, R.J. Weakly Interacting Stable Pions. *Phys. Rev. D* **2010**, *82*, 111701. [CrossRef]
22. Addazi, A.; Marcian, A.; Alexander, S. A Unified picture of Dark Matter and Dark Energy from Invisible QCD. *arXiv* **2016**, arXiv:1603.01853.
23. Peskin, M.E.; Schroeder, D.V. *An Introduction to Quantum Field Theory*; Addison-Wesley: Reading, MA, USA, 1995.
24. Visser, M.; Barcelo, C. Energy conditions and their cosmological implications. In *Cosmo-99*; World Scientific: Singapore, 2000; pp. 99–112.
25. Integrated Sachs-Wolf effect from IQCD Dark Energy and stability of the accelerated solution. Unpublished work.
26. Donà, P.; Marcianò, A.; Zhang, Y.; Antolini, C. Non-Perturbative Yang-Mills Condensate as Dark Energy. *Phys. Rev. D.* **2016**, *93*, 043012. [CrossRef]
27. Addazi, A.; Donà, P.; Marcianò, A. Dynamical interactions of dark energy and dark matter: Yang-Mills condensate and QCD axions. *Chin. Phys. C* **2018**, *42*, 075102. [CrossRef]

© 2020 by the authors. Licensee MDPI, Basel, Switzerland. This article is an open access article distributed under the terms and conditions of the Creative Commons Attribution (CC BY) license (http://creativecommons.org/licenses/by/4.0/).

Article

Dark Side of Weyl Gravity

Petr Jizba [1,*,†], Lesław Rachwał [1,†], Stefano G. Giaccari [2,†] and Jaroslav Kňap [1,†]

1 FNSPE, Czech Technical University in Prague, Břehová 7, 115 19 Praha 1, Czech Republic; leslaw.rachwal@fjfi.cvut.cz (L.R.); knapjaro@fjfi.cvut.cz (J.K.)

2 Department of Sciences, Holon Institute of Technology (HIT), 52 Golomb St., Holon 5810201, Israel; stefanog@hit.ac.il

* Correspondence: p.jizba@fjfi.cvut.cz; Tel.: +420-775-317-307

† These authors contributed equally to this work.

Received: 27 June 2020; Accepted: 4 August 2020; Published: 12 August 2020

Abstract: We address the issue of a dynamical breakdown of scale invariance in quantum Weyl gravity together with related cosmological implications. In the first part, we build on our previous work [*Phys. Rev. D* **2020**, *101*, 044050], where we found a non-trivial renormalization group fixed point in the infrared sector of quantum Weyl gravity. Here, we prove that the ensuing non-Gaussian IR fixed point is renormalization scheme independent. This confirms the feasibility of the analog of asymptotic safety scenario for quantum Weyl gravity in the IR. Some features, including non-analyticity and a lack of autonomy, of the system of β-functions near a turning point of the renormalization group at intermediate energies are also described. We further discuss an extension of the renormalization group analysis to the two-loop level. In particular, we show universal properties of the system of β-functions related to three couplings associated with C^2 (Weyl square), $\mathcal{G}$ (Gauss–Bonnet), and R^2 (Ricci curvature square) terms. Finally, we discuss various technical and conceptual issues associated with the conformal (trace) anomaly and propose possible remedies. In the second part, we analyze physics in the broken phase. In particular, we show that, in the low-energy sector of the broken phase, the theory looks like Starobinsky $f(R)$ gravity with a gravi-cosmological constant that has a negative sign in comparison to the usual matter-induced cosmological constant. We discuss implications for cosmic inflation and highlight a non-trivial relation between Starobinsky's parameter and the gravi-cosmological constant. Salient issues, including possible UV completions of quantum Weyl gravity and the role of the trace anomaly matching, are also discussed.

Keywords: Weyl gravity; renormalization group; inflation; dark energy

---

1. Introduction

It was just two years after the birth of the Yang–Mills (YM) gauge theory [1] when Utiyama in his 1956 seminal paper [2] recognized the similarity between gravity and YM fields, and started a new research line known as "gauge theories of gravity" [3–5]. The central focus of such a program was based on gauging Lorentz, Poincaré, or de-Sitter groups and this, in turn, has led to gravity theories (including Einstein–Hilbert, Einstein–Cartan theories, etc.) that despite their formal appeal share the same fate with Einstein's general relativity. Namely, such theories have dimensionful couplings and hence they are perturbatively non-renormalizable when quantized. More recently, the deep relationship between gauge and gravity theories has further been explored in the light of the holographic principle (as realized, for example, by the AdS/CFT correspondence [6]) and the "gravity = gauge × gauge" principle (as embodied, for example, in the Ben–Carrasco–Johansson color-kinematic correspondence [7]).

Both Lorentz, Poincaré, and de-Sitter groups are subgroups of the 15-parameter conformal group $O(4,2)$, which is the group of spacetime transformations that leave invariant the null interval. It is known that YM theories have a number of desirable features when quantized and this, in part, comes from their conformal invariance. Indeed, the conformal invariance of the YM action immediately implies that no massive (or dimensionful) coupling is present and so the theory is (power-counting) renormalizable. Conformal invariance is also instrumental in finding self-dual YM instantons [8,9]. One might thus expect that by gauging the conformal group one obtains gravity theory that would inherit on a quantum level some of the alluring traits of the YM theory, such as renormalizability, asymptotic freedom, non-trivial topological configurations, etc.

In 1920, Rudolf Bach proposed an action based on the square of the Weyl tensor $C_{\mu\nu\alpha\beta}$ where the Weyl tensor itself is an invariant under local re-scalings of the metric [10]. In 1977, Kaku, Nieuwenhuizen, and Townsend [11] showed that Bach's action is the action of the gauge theory of the conformal group provided that conformal boosts are gauged by means of a non-propagating gauge field. In other words, Bach's theory of gravity may be regarded as the gauge theory of the conformal group. In fact, Ref. [11] was apparently the first paper that explicitly referred to Bach's action as Weyl gravity (WG)—a terminology that we utilize throughout this paper.

As anticipated, quantum Weyl gravity (QWG) has proved to have a number of desirable features that are shared together with quantum YM theory. In particular, as in the YM theory, the coupling constant α is dimensionless, and this makes the theory perturbatively renormalizable. In accordance with analogy with YM instantons, QWG also has gravitational instantons that encode information about a non-perturbative vacuum structure of QWG [12,13]. Particularly intriguing is a parallel between QWG and the $SU(3)$ YM gauge theory of the strong force (QCD). Some speculation about an analogy between quadratic gravity (including QWG) and QCD (which extends an old analogy between general relativity and the chiral Lagrangian of QCD) has occurred already before [14,15], but QWG is very particular in this respect. Similarly to QCD, the gravitational interaction in QWG exhibits an antiscreening behavior at high energies on the account of the negative β-function. There are also strong indications that the ensuing UV fixed point of QWG is Gaussian [16,17] (as in QCD) and that the dynamical breakdown of the scale (Weyl) symmetry in QWG [16] in IR might be compared to the confinement–deconfinement phase transition in QCD where it is anticipated that a mass gap (effective gluon mass) develops in the confining phase [18]. In addition, certain aspects of the QCD functional integral and its measure are also shared by QWG [19].

Nevertheless, QWG is typically not considered as a viable candidate for a fundamental theory of quantum gravity for at least two reasons: (i) its perturbative spectrum contains ghosts (namely a massless spin-2 dipole ghost), and (ii) its β-function at the one-loop level is non-zero, which implies a non-vanishing conformal (trace) anomaly. Since, in QWG, one has conformal symmetry in a gauged (local) version, the appearance of the anomaly is typically considered disastrous for the construction of the quantum theory because it signals the absence of the symmetry defining the theory on the quantum level [20]. This should be contrasted with quite benign conformal anomaly that arises in YM theories.

The problem (i) is not specific only to QWG, but it is shared by all higher derivative gravity (HDG) theories. In these theories, the unitarity is in danger because there are perturbative states with negative kinetic energy or negative mass square parameter—so-called ghosts or tachyonic states, respectively [21]. For QWG, the problem can be most directly seen from the propagator of the metric-field fluctuations around a flat background. The tensor structure projects out spin-2 and spin-1 degrees of freedom. The propagator in the spin-2 sector exhibits one massless pole for normal graviton and another massless pole with negative residue. The tachyonic states can be easily eliminated from the spectrum of the theory by properly shifting the vacuum state. There are two possible interpretations of the massless ghost pole, which depend on the prescription of the $i\varepsilon$ term. One can view it either as a state of negative norm or a state of negative energy. In both cases, the optical theorem implies that the S-matrix based on the ensuing perturbation theory is inevitably non-unitary. Obviously, both ghosts and tachyons in HDG theories are undesirable and various approaches have been invoked to remove them (or their effects) from the observable predictions of the theory: Lee–Wick prescription [22,23], fakeons [24–26], perturbative expansion around true vacuum state [16], non-perturbative numerical methods [27–30], ghost instabilities [31–33], non-Hermitian PT-symmetric quantum gravity [34,35], etc. One might even entertain the idea that unitarity in quantum gravity is not a fundamental concept [36–38]. Thus far, none of the proposed solutions has solved the problem conclusively. Moreover, for QWG, one could question whether perturbative non-unitarity is really an issue for an asymptotically free theory whose IR degrees of freedom are probably different from the ones in the UV.

Instead of debating various attitudes that can be taken toward the ghost/unitarity issue, our aim here is more modest. We will explore, via functional renormalization group (FRG) [39–43], the low-energy phenomenology of the QWG and see whether it can provide a realistic cosmology and what role (if any) is played by ghost fields. We start from the UV fixed point (FP) where the potential quantum gravity has an exact scale-invariance, so that only causal structure of events is relevant. This is a minimalistic assumption about quantum gravity in UV. In order not to invoke any unjustified structure, we consider only metric-field based gravity without any matter field. The UV FP in question might be, for instance, one of the critical points in a series of putative phase transitions that the Universe underwent in the very early (pre-inflationary) period of its evolution. Out of many scale-invariant HDG candidate theories, we choose to work with the simplest one, namely the theory that has only one single coupling constant. The latter corresponds to the QWG theory. Precisely at the UV FP, the QWG has exact local scale-invariance. A consistency of the entire scheme implies that the UV fixed point must be Gaussian [16]. Existence of the Gaussian UV FP for QWG was also conjectured in earlier works [17,44]. We start with this premise and let the theory flow toward IR energy scales. In the close vicinity of the UV FP, the Weyl symmetry in the renormalized action is still preserved as only the Gauss–Bonnet (GB) term is generated. Corrections explicitly violating Weyl symmetry, such as the R^2 term, are generated only at the second (or higher) loop order.

In the vicinity of the UV FP, we choose a truncation ansatz for the effective action that will be further used to set up the FRG flow equation. Our truncation prescription is directly dictated by the one-loop effective action. We further enhance this by incorporating into the FRG equation two non-perturbative effects, namely threshold phenomena and the effect of graviton anomalous dimension. One may also wonder about how a consistent quantum theory can emerge when the action is problematic at tree-level (ghost problem). Nevertheless, the RG flow analysis reveals that by the inclusion of the two non-perturbative effects the quantum theory yields a sensible IR FP. Indeed, by solving the RG flow equation algebraically for β-functions β_C and β_{GB}, we show that there exists a non-Gaussian IR fixed point where both β-functions simultaneously disappear. Aforementioned IR FP represents a critical point after which the (global) scale-invariance is broken. This is reflected through the presence of a composite order-parameter field of the Hubbard–Stratonovich type, which in the broken phase acquires a non-trivial vacuum expectation value.

We map the broken-phase effective action on a two-field hybrid inflationary model that in its low-energy phase approaches the Starobinsky $f(R)$ model with a non-trivial gravi-cosmological constant. The requirement that Einstein's R term in the low energy actions must have a coupling constant $1/2\kappa^2$ ties up the values of Starobinsky's inflation parameter ξ and the gravi-cosmological constant Λ. This fixes the symmetry-breakdown scale for QWG to be at about the GUT inflationary scale. Moreover, the existence of a regime where gravity is approximately scale invariant (fixed-point regime and departure of the RG flow from it) provides a simple and natural interpretation for the nearly-scale-invariance of the power spectrum of temperature fluctuations in the Cosmic Microwave Background radiation.

As for the conformal anomaly, typical imprints of it are new terms generated in the functional-integral action via higher loop corrections. In particular, loop corrections will generate the Weyl-symmetry violating R^2 term in the action. There are various possible ways how one can deal with conformal anomaly. For instance, one might embed QWG in $\mathcal{N} = 4$ conformal supergravity found by Fradkin and Tseytlin in 1985 [45,46] (which is known to be unique anomaly-free theory containing Weyl gravity in its bosonic sector) or in Witten–Berkovits twistor superstring theory [47]. One should then consider QWG as the low-energy limit of such UV-finite models. In these cases, QWG is confined to the purely bosonic sector of spin-2 and spin-1 fluctuations where the conformal symmetry is only softly violated, or one could embed QWG in recently found perturbatively UV-finite quantum gravitational theories [48,49] considered as an extension of superrenormalizable higher derivative theories. Yet another option is to take seriously also theories with anomalous conformal symmetry similarly as done, for instance, in string theory with strings propagating in non-critical spacetime dimensions [50]. Here, we will employ yet another scenario, in particular, we will take advantage of the fact that the symmetry-breaking R^2 term is presumably generated only at the two-loop order (so that to one loop we do not observe the appearance of the ensuing anomalous conformal mode). On the other hand, our FRG approach will show that the IR FP (for the two involved couplings) appears already at the enhanced one-loop level and hence the prospective observational consequences of the trace anomaly do not take over before the Weyl symmetry is dynamically broken. In the broken phase, the trace anomaly is not anymore a signal of inconsistency since the theory there is not scale-invariant to begin with and hence there is no reason why the ensuing energy momentum tensor should be traceless.

The word *dark* in the title of this paper refers to two different things. First, it is related to conceptual and technical issues that plague QWG and that make it discouraging or *dark* in eyes of many practitioners. Second, the appearance of the dynamical gravi-cosmological constant in the broken phase of QWG can be associated with ensuing *dark* energy. It is the purpose of this work to demonstrate that despite the aforementioned problems, QWG may serve as a healthy theoretical setup for the UV-model building of phenomenologically viable quantum theory of gravity with pertinent cosmological implications.

Our paper is organized as follows: in the next Section 2, we discuss some fundamentals of both classical and quantum Weyl gravity. In particular, we highlight a formal similarity of the WG with nonabelian Yang–Mills theories and stress some of dissimilarities and potential problems met during quantization. Section 3 is dedicated to the construction of the FRG flow equation for the QWG in the one-loop enhanced scheme, and with this tool we analyze in Section 4 the running of the β-functions associated with the Weyl tensor square and Gauss–Bonnet terms. In particular, we demonstrate that, apart from the IR–stable fixed point that is reached at a zero-value of the running scale, the RG flow also exhibits a non-trivial bouncing behavior in the vicinity of the IR fixed point. The issue of conformal anomaly of QWG is discussed in Section 5. There we point out various technical and conceptual issued associated with the conformal anomaly in QWG and propose possible remedies. In Section 6, we first employ a Hubbard–Stratonovich (HS) transformation, which introduces a non-dynamical *spurion* scalar field without spoiling the particle spectrum and (perturbative) renormalizability. After the dynamical breakdown of the Weyl symmetry, the HS field acquires a non-trivial vacuum expectation value and gets radiatively generated gradient (kinetic) term. If QWG has any physical relevance, then its effective action in the broken phase must contain an Einstein–Hilbert term. This is shown in the second part of Section 6. We further demonstrate that in the broken phase the corresponding one-loop effective action consists (in the Einstein frame) of two scalar fields—scalaron and dynamical Hubbard–Stratonovich field that interact via derivative coupling. The resulting low-energy behavior in the broken phase can be identified with Starobinsky's $f(R)$-model (SM) with a gravi-cosmological constant (dark side of Weyl gravity) that has a negative sign in comparison to the usual matter-induced cosmological constant. After this, we discuss anomaly matching conditions between symmetric and broken phase of QWG. A brief summary of results and related discussions are provided in Section 7.

2. Some Fundamentals of Quantum Weyl Gravity

2.1. Classical Weyl Gravity

The WG is a pure metric theory that is invariant not only under the action of the diffeomorphism group, but also under Weyl rescaling of the metric tensor by the local smooth functions $\Omega(x)$:

$g_{\mu\nu}(x) \to \Omega^2(x)g_{\mu\nu}(x)$. The simplest WG action functional in four spacetime dimensions that is both diffeomorphism and Weyl-invariant has the form [10,51,52]

$$S = -\frac{1}{4\alpha^2}\int d^4x\sqrt{|g|}\,C_{\mu\nu\rho\sigma}C^{\mu\nu\rho\sigma}, \tag{1}$$

where $C_{\mu\nu\rho\sigma}$ is the *Weyl tensor* which can be written as

$$C_{\mu\nu\rho\sigma} = R_{\mu\nu\rho\sigma} - \left(g_{\mu[\rho}R_{\sigma]\nu} - g_{\nu[\rho}R_{\sigma]\mu}\right) + \tfrac{1}{3}R\,g_{\mu[\rho}g_{\sigma]\nu}, \tag{2}$$

with $R_{\mu\nu\rho\sigma}$ being the Riemann curvature tensor, $R_{\mu\rho} = R_{\mu\nu\rho}{}^{\nu}$ the Ricci tensor, and $R = R_{\mu}{}^{\mu}$ the scalar curvature. Throughout the text, we employ the time-like metric signature $(+,-,-,-)$ whenever pseudo-Riemannian (Lorentzian) manifolds are considered. The dimensionless coupling constant α is conventionally chosen so as to mimic the YM action. On the other hand, in order to make connection with the usual RG practice, it will be more convenient to consider the inverse of the coupling α^2. In the following, we denote the coupling in Equation (1) as ω_C, by employing the prescription $\omega_C \equiv 1/(4\alpha^2)$.

Henceforth, we employ the following notation: for the square of the Riemann tensor contracted naturally that is $R_{\mu\nu\rho\sigma}R^{\mu\nu\rho\sigma}$, we use the symbol $R^2_{\mu\nu\rho\sigma}$, the square of the Ricci tensor $R_{\mu\nu}R^{\mu\nu}$ we denote by simply $R^2_{\mu\nu}$, the square of the Ricci curvature scalar is always R^2, while for the Weyl tensor square $C_{\mu\nu\rho\sigma}C^{\mu\nu\rho\sigma}$ we employ a shorthand and schematic notation C^2. When the latter is treated as a local invariant (not under volume integral) in $d = 4$ dimensions, one finds the following expansion of the C^2 invariant into standard invariants quadratic in curvature

$$C^2 = R^2_{\mu\nu\rho\sigma} - 2R^2_{\mu\nu} + \tfrac{1}{3}R^2. \tag{3}$$

We will also need another important combination of the quadratic curvature invariants, namely Gauss–Bonnet (GB) term

$$\mathcal{G} = R^2_{\mu\nu\rho\sigma} - 4R^2_{\mu\nu} + R^2, \tag{4}$$

which in $d = 4$ is the integrand of the Euler–Poincaré invariant [53]

$$\chi = \frac{1}{32\pi^2}\int d^4x\sqrt{|g|}\,\mathcal{G}. \tag{5}$$

In order to underline the similarity with nonabelian YM theories, we use the Riemann curvature tensor so that it is related to the usual general relativistic one by

$$R_{\mu\nu\lambda}{}^{\kappa} = R^{\kappa}{}_{\mu\nu\lambda}|_{\text{genrel}}. \tag{6}$$

Note that due to skew and interchange symmetries of the Rieman tensor we have

$$R_{\lambda\nu} = R_{\lambda\nu}|_{\text{genrel}} \quad\Rightarrow\quad R = R|_{\text{genrel}}. \tag{7}$$

In addition, also

$$C_{\lambda\mu\nu\kappa}C^{\lambda\mu\nu\kappa} = C_{\lambda\mu\nu\kappa}C^{\lambda\mu\nu\kappa}|_{\text{genrel}}, \tag{8}$$

trivially holds. In particular, in our convention, we consider both Christoffel symbols and curvature tensors as 4×4 matrices; $R_{\mu\nu\lambda}{}^{\kappa} \equiv \{R_{\mu\nu}\}_{\lambda}{}^{\kappa}$ and $\Gamma_{\nu\lambda}{}^{\kappa} \equiv \{\Gamma_{\nu}\}_{\lambda}{}^{\kappa}$. With this, one can write

$$R_{\mu\nu\lambda}{}^{\kappa} = \{\partial_{\mu}\Gamma_{\nu} - \partial_{\nu}\Gamma_{\mu} - [\Gamma_{\mu},\Gamma_{\nu}]\}_{\lambda}{}^{\kappa}, \tag{9}$$

which is clearly analogous to the relation

$$F^a_{\mu\nu} = \{\partial_\mu A_\nu - \partial_\nu A_\mu - ig[A_\mu, A_\nu]\}^a , \tag{10}$$

for the field strength in nonabelian YM theories with A_μ representing the gauge field. A natural geometrical language for the description of this analogy is provided by the fiber-bundle theory. While $\Gamma_{\nu\lambda}{}^\kappa$ represents connection in the frame bundle, A^a_ν plays the same role in the ensuing principal bundle. Similarly, $R_{\mu\nu\lambda}{}^\kappa$ quantifies anholonomy in a parallel transport around a closed infinitesimal loop in the frame bundle, whereas $F^a_{\mu\nu}$ quantifies the anholonomy in the principal bundle. Although the fibre-bundle theory provides the most adequate language for the parallelism between Riemannian-geometry-based gravity and nonabelian YM theories, we will not pursue this point any further here.

With the help of the Chern–Gauss–Bonnet theorem, one can cast the Weyl action S into equivalent form (modulo topological term)

$$S = -\frac{1}{2\alpha^2}\int d^4x\sqrt{|g|}\left(R^2_{\mu\nu} - \tfrac{1}{3}R^2\right) . \tag{11}$$

It should be stressed that although the omitted topological term is clearly not important on the classical level, it is relevant on the quantum level where summation over distinct topologies should be considered (see following subsection). However, even when one stays on topologies with a fixed Euler–Poincaré invariant, the renormalization will inevitably generate (already at one loop) the GB term with a running coupling constant (see, e.g., Section 3).

In passing, we note also that both (1) and (11) are Weyl-invariant only in $d = 4$ dimensions. In fact, under the Weyl transformation $g_{\mu\nu} \to \Omega^2 g_{\mu\nu}$ and the densitized C^2 transforms as

$$\sqrt{|g|}C^2 \to \Omega^{d-4}\sqrt{|g|}C^2 , \tag{12}$$

in general dimension d of spacetime, while $\sqrt{|g|}\,\mathcal{G}$ supplies topological invariant only in $d = 4$. This is particularly important to bear in mind during the quantization where (similarly as in the Yang–Mills theories), one should choose such a regularization method that preserves the local gauge symmetry of the underlying Lagrangian and thus does not introduce any unwanted symmetry breaking terms. For this reason, one should preferentially rely on fixed-dimension renormalization and avoid, e.g., dimensional regularization. This is the strategy we will pursue also in this paper.

Variation of S with respect to the metric yields the field equation of motion (EOM) known as the Bach vacuum equation:

$$[2C^{\mu\lambda\nu\kappa}{}_{;\lambda;\kappa} - C^{\mu\lambda\nu\kappa}R_{\lambda\kappa}] \equiv B^{\mu\nu} = 0 , \tag{13}$$

where $B^{\mu\nu}$ is the *Bach tensor* (trace-free tensor of rank 2) and "$;\alpha$" denotes the usual covariant derivative (with Levi–Civita connection). We remind that the form (13) of the EOM is specific only to $d = 4$. Moreover, apart from being traceless, the Bach tensor is also divergence-free ($B^{\mu\nu};\mu = 0$), which is a consequence of diffeomorphism symmetry. Because Bach tensor results from variational derivative of the action S with respect to symmetric metric tensor $g_{\mu\nu}$, it must also be symmetric ($B^{\nu\mu} = B^{\mu\nu}$). When for a given background $B^{\mu\nu} = 0$ (i.e., given backgound is a classical vacuum solution of Weyl gravity), then we say that it is Bach–flat. More general discussion of classical singularity-free solutions in WG can be found in [54–56].

2.2. Quantization

One can formally quantize WG by emulating the strategy known from quantum field theory, i.e., by introducing a functional integral ($\hbar = c = 1$)

$$Z = \sum_i \int_{\mathcal{M}_i} \mathcal{D}g_{\mu\nu} e^{iS} . \tag{14}$$

Here, $\mathcal{D}g_{\mu\nu}$ denotes the functional-integral measure whose proper treatment involves the Faddeev–Popov gauge fixing of the gauge symmetry Diff $\times$ Weyl$(\mathcal{M}_i)$ plus ensuing Faddeev–Popov (FP) determinant [45,57]. As for local factors $[-\det g_{\mu\nu}(x)]^w$ in the measure, we choose to work with DeWitt convention [58,59]: $w = (d-4)(d+1)/8$. Clearly, when the fixed-dimension renormalization scheme in $d = 4$ is employed, the local factor does not contribute. Strictly speaking, for a full-fledged quantization program, one should consider generator of correlation functions $Z[J_{\mu\nu}]$, i.e., functional of a source field that is coupled to a dynamical metric field. Such object would contain information on all correlations functions, but, unfortunately, it is far beyond our present computational capability. Fortunately, for our purposes, i.e., for the computation of enhanced one-loop effective action the use of quantum partition function (14) will suffice.

The sum in (14) is a sum over four-topologies, that is, the sum over topologically distinct manifolds $\mathcal{M}_i$ (analogous to the sum over genera in string theory or sum over homotopically inequivalent vacua in the Yang–Mills theory) which can potentially contain topological phase factors, e.g., the Euler–Poincaré characteristic of $\mathcal{M}_i$, cf. Ref. [60]. It should be stressed that the sum over four-topologies is a problematic concept since four-manifolds are generally un-classifiable—i.e., there is no algorithm that can determine whether two arbitrary four-manifolds are homeomorphic. On the other hand, simply connected compact topological four-manifolds are classifiable in terms of *Casson handles* [61], which can be applied in functional integrals in Euclidean gravity. For simplicity, we will further assume that all global topological effects can be ignored, so, in particular, we assume that our space $\mathcal{M}_i$ is compact and its tangent bundle is topologically trivial.

To avoid issues related to renormalization of non-physical sectors (i.e., Faddeev–Popov ghosts and longitudinal components of the metric field), it will be convenient in our forthcoming reasonings to employ the York decomposition of the metric fluctuations $h_{\mu\nu}$ defined as

$$g_{\mu\nu} = g_{\mu\nu}^{(0)} + h_{\mu\nu} \tag{15}$$

where we have denoted the background metric as $g_{\mu\nu}^{(0)}$. The York decomposition is then implemented in two steps [44]. In the first step, we rewrite the metric fluctuations as

$$h_{\mu\nu} = \bar{h}_{\mu\nu} + \tfrac{1}{4} g_{\mu\nu} h , \tag{16}$$

where h is a trace part of $h_{\mu\nu}$ and $\bar{h}_{\mu\nu}$ is the corresponding traceless part. More specifically,

$$g^{(0)\mu\nu}\bar{h}_{\mu\nu} = \bar{h}_\mu{}^\mu = 0, \qquad h = g^{(0)\mu\nu}h_{\mu\nu} = h_\mu{}^\mu . \tag{17}$$

We will always tacitly assume that the Lorentz indices are raised or lowered via the background metric, i.e., via $g^{(0)\mu\nu}$ or $g_{(0)\mu\nu}$, respectively. In addition, all covariant derivatives ∇_μ below will be understood as taken with respect to the background metric. In the following, the operator $\Box$ will denote the so-called Bochner Laplacian operator [44], i.e., the covariant operator defined as $\Box \equiv \nabla^\mu \nabla_\mu$.

In the second step, we decompose the traceless part into the transverse, traceless tensor $\bar{h}_{\mu\nu}^\perp$ and to parts carrying the longitudinal (i.e., unphysical) degrees of freedom, namely

$$\bar{h}_{\mu\nu} = \bar{h}_{\mu\nu}^\perp + \nabla_\mu \eta_\nu^\perp + \nabla_\nu \eta_\mu^\perp + \nabla_\mu \nabla_\nu \sigma - \tfrac{1}{4} g_{\mu\nu} \Box \sigma . \tag{18}$$

These mixed-longitudinal (and traceless) parts are written in terms of an arbitrary transverse vector field $\eta_\mu^\perp$ and a scalar (trace) degree of freedom σ. The last fields must satisfy the usual conditions of transversality and tracelessness, i.e.,

$$\nabla^\mu \bar{h}_{\mu\nu}^\perp = 0, \quad \nabla^\mu \eta_\mu^\perp = 0, \quad \bar{h}^{\perp\mu}_{\ \mu} = 0. \tag{19}$$

The true propagating degrees of field in QWG are associated with the transverse and traceless field $\bar{h}_{\mu\nu}^\perp \equiv h_{\mu\nu}^{TT}$. Indeed, from the second variation of the Weyl action expanded around a generic background, it can be seen that $\bar{h}_{\mu\nu}^\perp$ is the only field component that propagates on quantum level. The vector field $\eta_\mu^\perp$ and two scalar fields h and σ completely drop out from the expansion due to diffeomorphism and conformal invariance, respectively [62]. Some explicit examples will be given in Section 4.

3. Exact RG Flow for Quantum Weyl Gravity

For convenience sake, our subsequent reasonings will be done in an $d = 4$ Euclidean space dimensions as this is a typical framework in which the FRG treatment is done. By performing Wick rotation from Minkowski space to Euclidean space, the question of the resulting metric signature arises. When one does, in a standard way, only the change of the time coordinate $t \to -it_E$, (where t_E is the name of the first coordinate in the Euclidean characterization of space) the resulting signature of the metric of space is completely negative, that is $(-,-,-,-)$. It seems natural to define the corresponding GR-covariant d'Alembert operator as $\Box_E = -\nabla^\mu\nabla_\mu$, where the generalization to curved Euclidean space is done by using Bochner Laplacian. However, in all formulas that follow, we find more convenient to use the following definition in the Euclidean signature $\Box = \nabla^\mu\nabla_\mu$. We also remark that this last operator $\Box$, if analyzed on the flat space background, has negative semi-definite spectrum. We will also use a definition of the covariant Euclidean box operator (covariant Laplacian) $\Delta = \Box = -\Box_E$ and this last operator in the Euclidean flat space case has a spectrum which is characterized by $-k^2$, the $d = 4$ Euclidean negative square of a 4-momentum vector k_μ. Accordingly, the signature of the metric in Euclidean space will be taken to be $(+,+,+,+)$.

The aim of this section is to explore the IR behavior of the QWG by starting from the presumed UV FP where the QWG is exact. Existence of such a UV FP was self-consistently checked in Ref. [16]. To this end, we will solve the FRG flow equation [39–42] for the effective average action Γ_k, which reads

$$\partial_t \Gamma_k = \frac{1}{2} \mathrm{Tr} \left[\partial_t R_k (\Gamma^{(2)} + R_k)^{-1} \right]. \tag{20}$$

The IR-cutoff R_k suppresses the contribution of modes with small eigenvalues of the covariant Laplacian $-\Delta \ll k^2$, while the factor $\partial_t R_k$ removes contributions from large eigenvalues of $-\Delta \gg k^2$. In this way, the loop integrals are both IR- and UV-finite [63]. The second variational derivative of the effective action—$\Gamma^{(2)}$ depends on the background metric $g_{\mu\nu}^{(0)}$, which is the argument of the running effective action Γ_k, while k is the running energy (momentum) scale or the momentum of a mode in the Fourier space. We also employ the notational convention $\partial_t = k\partial_k$.

Ideally, Equation (20) would require calculation of the full resummed and RG-invariant effective action. It is, however, difficult to proceed analytically in this way so we make ourselves content here with the conventional procedure, according to which one should employ some well motivated ansatz for the effective action. In particular, in order to evaluate the RHS of Equation (20), we employ the (Euclidean) effective action in the enhanced one-loop scheme. By the enhanced one-loop scheme, we mean one-loop effective action in which also effects of the anomalous dimension and threshold phenomena are included. Corresponding truncation will thus inevitably go beyond the usual polynomial ansatz. On the other hand, for the LHS of (20), we project the flow on the subspace of the three invariants containing precisely four derivatives of the metric (Equations (3) and (4) and R^2 invariant). The reason why we consider effective action on the RHS being different from the effective action on the LHS is dictated by technical convenience. Namely, the RHS acts as source for the RG flow, while the LHS contains the desired structure of the effective action that is appropriate for the extraction of the β-functions.

Let us now briefly describe the basic steps that are used in solving the FRG flow equation. Further technical details as well as necessary derivations can be found in our recent paper [16] and in its supplemental material [64]. On the other hand, our final results will be discussed in more detail in the following section.

The first step is the construction of the one-loop partition function for QWG. It is important here to choose a convenient class of backgrounds and use a background field method. This method is quite standard and widely used in higher derivative gravitational theories and in the FRG context it was first pioneered by Benedetti et al. in Ref. [65]. We selected maximally symmetric spaces (MSS) and Ricci-flat backgrounds, knowing that both of them are also Bach-flat, so they are classical exact solutions to the Bach equation (i.e., vacuum equations in WG) [66]. The partition function is computed from action phrased in terms of York degrees of freedom (recall that metric fluctuations are decomposed into a transverse-traceless part, trace part, and gauge part). A clear advantage of phrasing action in terms of the York decomposition is that the Hessian from the Bach action (that is needed for one-loop effective action) contains neither gauge part nor trace modes due to Weyl symmetry. Therefore, the functional integration over the gauge and trace degrees of freedom can be performed trivially. When quantized, one must, of course, define the functional integral measure, so as to not overcount physical degrees of freedom. This is done via FP procedure, i.e., by Faddeev–Popov gauge fixing of the gauge symmetry Diff $\times$ Weyl plus FP determinant. We must also change variables (to York ones) under the functional integral and include corresponding Jacobian. Finally one must also exclude the contribution from zero modes, which are unwanted. As a double-check, one can verify that all these one-loop partition functions provide six propagating degrees of freedom in QWG as expected. More explicit technical exposition can be found in Refs. [16,44,62].

In the second step, we analyze the general β-functional of the theory considered on our general backgrounds (MSS and Ricci-flat). We ask and address the question about information which can be extracted on these backgrounds about the β-functions of involved couplings. Due to various relations between curvature tensors and between square of tensors, we find that we are able to extract only two combinations of couplings: either $\beta_R + \frac{1}{6}\beta_{\mathrm{GB}}$ on MSS, or $\beta_C + \beta_{\mathrm{GB}}$ on Ricci-flat background. It can be also checked that the usage of general Einstein spaces as on-shell backgrounds does not improve on this situation. However, in a general theory to the quadratic order in curvatures, we may set up an ansatz for the effective average action Γ built out of three couplings ω_C, ω_{GB}, and ω_R corresponding to three quadratic invariants C^2 term, $\mathcal{G} = \mathrm{GB}$ term, and R^2 term [67]. We address this potential inconsistency of the RG system (three couplings and only two extractable β-functions) in the final step of our method of solving the FRG.

The most important part consists of building and solving the FRG flow equation. In Ref. [16], we found a novel form of the FRG flow equation based on the expressions for factors appearing in the quantum partition function of the theory. These factors mimic the simple scalar two-derivative kinetic sectors of the theory; however, they may appear both in the numerator and in the denominator of the partition function and with various mass square parameters (which can also be negative). We take into account quantum wave-function renormalization of quantum fields and add to the flow equation the anomalous dimension η for all fields participating in the quantum dynamics at the one-loop level. To do the IR-suppression of modes in each factor, we should add a suitable cutoff kernel function $R_k(z)$. When all these operations are done, the final form of the FRG flow equation reads

$$\partial_t \Gamma_{L,k} = \frac{1}{2} \sum_i \sum_j \pm \operatorname{Tr}_{\phi_i} \left(\frac{(\partial_t R_k - \eta R_k)\, \hat{\mathbb{1}}}{\hat{\Box} + R_k \hat{\mathbb{1}} - Y_{i,j}} \right), \tag{21}$$

where $Y_{i,j}$ are mass square parameters of the modes ϕ_i and the functional trace is done over space of the same modes, which may also contain some traces over internal indices of the fields. The frontal plus/minus sign in (21) originates from the initial position of the factor in the partition function (whether it was in the denominator, or in the numerator of the partition function, respectively).

The final step consists of choosing the truncation ansatz for the action, whose FRG flow we try to determine. To write the RHS of the flow equation, we explicitly reflect the fact that our theory is QWG theory, where we could add a topological GB term, therefore not influencing at all the perturbative one-loop partition function. On the LHS, we could analyze the flow of the general action quadratic in curvatures with the structure of the Lagrangian $\omega_C C^2 + \omega_{\rm GB}\mathcal{G} + \omega_R R^2$. However, in perturbative QWG at the one-loop level, there is a very interesting and special simplification since at this quantum level the R^2 term is not needed, and we know that perturbatively its one-loop β-function β_R vanishes. We utilize this hierarchy of β-functions in QWG and employ consistently the truncation ansatz without the problematic R^2 term. Therefore, we have two combinations of β-functions ($\frac{1}{6}\beta_{\rm GB}$ and $\beta_C + \beta_{\rm GB}$) and two couplings ω_C and $\omega_{\rm GB}$, so the system is consistent and possible to be solved algebraically for the β-functions involved.

This solving we do in the last technical step, where we evaluate all the functional traces (both in internal space and spacetime volume integrals). We perform the traces using the heat kernel method and Barvinsky–Vilkovisky trace technology. In order to have an analytic control over all formulas, we decide to choose a particular form of the cutoff kernel function $R_k \equiv R_k(z) = (k^2 - z)\,\theta(k^2 - z)$ due to Litim [68,69]. The main point of using the Wetterich Equation (20) is to take into account massive modes which slow down the RG flow in the IR regime. We achieve this by adding cutoff kernels R_k in mass-dependent renormalization scheme of Wilsonian character. This lets us obtain one-loop RG-improved expressions for the two β-functions of the theory with all quantum effects due to anomalous dimension η and IR threshold phenomena included.

4. Analysis of β-Functions and RG Fixed Points

We now discuss the system of β-functions of the theory considered at the one-loop level improved by the usage of FRG methods. Based on the computation presented in Ref. [16], the explicit form of β-functions reads

$$\begin{aligned}\beta_{\rm GB} \;=\; \frac{1}{2}(2-\eta)\Bigg[&-\frac{21}{40}\left(1-\frac{\frac{2}{3}\Lambda}{k^2}\right)^{-1} + \frac{9}{40}\left(1-\frac{\frac{4}{3}\Lambda}{k^2}\right)^{-1} - \frac{179}{45}\left(1+\frac{\Lambda}{k^2}\right)^{-1} \\ &- \frac{59}{90}\left(1+\frac{\frac{1}{3}\Lambda}{k^2}\right)^{-1} + \frac{479}{360}\left(1+\frac{2\Lambda}{k^2}\right)^{-1} - \frac{269}{360}\left(1+\frac{\frac{4}{3}\Lambda}{k^2}\right)^{-1}\Bigg], \end{aligned} \tag{22}$$

and

$$\beta_C + \beta_{\rm GB} = \frac{2-\eta}{2}\frac{137}{60}, \tag{23}$$

with the anomalous dimension of the graviton field given by

$$\eta = -\frac{1}{\omega_C}\beta_C, \tag{24}$$

to the one-loop level of accuracy. By $\omega_C = \omega_C(k)$, we here denote a running coupling parameter in front of the C^2 term in the action (1)—the so-called Weyl coupling. It is important to notice that the β-functions were computed by the background field method in two distinct Bach-flat backgrounds specified above. This is a standard technique pioneered by Benedetti et al. [65]. In spite of this, the β-functions are in general background dependent in the IR regime, we should emphasize that the existence of FP of RG and some other properties (related to critical exponents, the dimensionality of the critical surface or to anomalous dimensions of some CFT operators) are universal and independent of the background chosen. These are indeed the aspects we will be interested in, so that as far as they are concerned we can ignore the explicit dependence of the β-functions on the background spacetime.

The above two β-functions, β_C and β_{GB}, follow from the FRG with the truncation ansatz motivated by the one-loop level and with both the threshold phenomena and non-trivial anomalous dimension of the quantum graviton included. We thus call them one-loop RG-improved. The effects of threshold phenomena are present explicitly only in the expression (22) for β_{GB}. However, due to the combination in (23), the solution for β_C will also inherit these threshold factors. Finally, we observe that the anomalous dimension η enters only multiplicatively in the system of β-functions (22) and (23). This has some simplifying consequences for a search for FP's of the coupled system, both in the UV as well as in the IR regimes.

Let us now discuss the reasons for the presence of threshold phenomena in our system. As it could be seen from the expressions for the one-loop partition functions of the system [16] on MSS,

$$Z^2_{1-\text{loop}} = \frac{\det_1^2\left(\hat{\Box}+\Lambda\hat{\mathbb{1}}\right)\det_1\left(\hat{\Box}+\frac{1}{3}\Lambda\hat{\mathbb{1}}\right)\det_0\left(\hat{\Box}+\frac{4}{3}\Lambda\hat{\mathbb{1}}\right)}{\det_{2T}\left(\hat{\Box}-\frac{2}{3}\Lambda\hat{\mathbb{1}}\right)\det_{2T}\left(\hat{\Box}-\frac{4}{3}\Lambda\hat{\mathbb{1}}\right)\det_0\left(\hat{\Box}+2\Lambda\hat{\mathbb{1}}\right)}, \tag{25}$$

and on Ricci-flat background

$$Z^2_{1-\text{loop}} = \frac{\det_1^3\hat{\Box}\,\det_0^2\hat{\Box}}{\det_2^2\left(\hat{\Box}-2\hat{C}\right)} \tag{26}$$

the box-kinetic operator of all quantum modes is shifted only in the case of MSS background[1]. This is the reason to produce IR thresholds. The shift by a matrix of a Weyl tensor $\hat{C}$ on the Ricci-flat background does not generate any threshold because of the tracelessness of the Weyl tensor. These shifts in the factors in the partition function (25) on MSS backgrounds are analogous to massive modes in standard QFT. Their role is to effectively slow down the RG flow in the IR regime since there the quantum fields become heavy (with mass).

We can further simplify the system of Equations (22) and (23) for the two β-functions. In particular, we do not wish to solve explicitly the system (22) and (23). We just concentrate on the corresponding FP's. This is a much simpler task as we can solve the system of β-functions algebraically. This gives

$$\beta_C = \frac{b-\mathcal{X}}{1+y(\mathcal{X}-b)}, \quad \beta_{GB} = \frac{\mathcal{X}}{1+y(\mathcal{X}-b)}, \tag{27}$$

where

$$\begin{aligned}\mathcal{X} = & -\frac{21}{40}\left(1-\frac{\frac{2}{3}\Lambda}{k^2}\right)^{-1}+\frac{9}{40}\left(1-\frac{\frac{4}{3}\Lambda}{k^2}\right)^{-1}-\frac{179}{45}\left(1+\frac{\Lambda}{k^2}\right)^{-1}-\frac{59}{90}\left(1+\frac{\frac{1}{3}\Lambda}{k^2}\right)^{-1}\\ & +\frac{479}{360}\left(1+\frac{2\Lambda}{k^2}\right)^{-1}-\frac{269}{360}\left(1+\frac{\frac{4}{3}\Lambda}{k^2}\right)^{-1},\end{aligned} \tag{28}$$

with $b = 137/60$ and $y = 1/\omega_C$. The origin of the common denominator $1+y(\mathcal{X}-b)$ is entirely due to the inclusion of the graviton's anomalous dimension η. When η is neglected, the latter is unity. This is the regime in which the Weyl coupling ω_C is big ($\omega_C \to \infty$, so $y \to 0$). This corresponds to

[1] Let us stress once more that both (25) and (26) were obtained by first rewriting the action in York variables and then following the standard Faddeev–Popov quantization procedure which involves the computation of the determinants for the ghost fields associated with diffeomorphism and conformal invariance.

the perturbative regime of the theory (in terms of α_C). When one decides to neglect these common denominators, one gets simplified expressions,

$$\beta_C = b - \mathcal{X}, \quad \beta_{GB} = \mathcal{X}, \tag{29}$$

which are already sufficient to shed light on the issue of existence and character of FP's of the FRG flow. The equations in (29) still include the effects of threshold phenomena. When we neglect the threshold phenomena in our description, then the system of β-functions acquires the form of one-loop perturbative system as derived in [45] for QWG in dimensional regularization scheme. Actually, all these threshold phenomena are contained in the expression called $\mathcal{X}$ above. When one takes the limit Λ/k^2 to zero, then all threshold factors are indeed removed, and the expression $\mathcal{X}$ reduces to just a number $\beta_{GB}^{FT} = -87/20$.

4.1. Ultraviolet Asymptotic Freedom in All Couplings

When $\kappa = k/\sqrt{|\Lambda|} \gg 1$, the threshold phenomena are completely irrelevant and can be neglected, cf. Equation (22). Irrespectively of the initial values of the couplings $\omega_{C0} = \omega_C(t_0)$ and $\omega_{GB0} = \omega_{GB}(t_0)$ (look also at an analysis in the next paragraph), the leading RG running behavior in the UV regime (for $t \gg 1$) is $\omega_C \sim t\beta_C^{FT}$ and $\omega_{GB} \sim t\beta_{GB}^{FT}$. This signifies that the absolute values of the couplings must necessarily grow in the UV. It might be argued that in the UV regime one can also neglect the effects of the anomalous dimension η, cf. Equation (24), since it is suppressed by big values of the ω_C coupling in the UV. For the UV-running ($t \to +\infty$ equivalent to $k \gg k_0$), it suffices to use only the non-RG-improved one-loop perturbation results (29) from above.

Fradkin and Tseytlin [45] were the first to find that the one-loop β-functions for ω-couplings are constants with values

$$\beta_C = \frac{199}{30} \quad \text{and} \quad \beta_{GB} = -\frac{87}{20}. \tag{30}$$

This leads to an asymptotic freedom at the UV FP for all two couplings as we shall prove below. Since $\beta_C > 0$, it is natural to assume that the initial condition of the flow is such that $\omega_C(t_0) > 0$ and similarly, since $\beta_{GB} < 0$, then $\omega_{GB}(t_0) < 0$. If one chooses the opposite condition, then the RG flow tends to decrease the absolute value of the ω-coupling, the coupling crosses zero, and finally it goes on the other side, where the initial conditions are natural in a sense mentioned above. This is because one-loop RG flow in the UV forces the β-functions to be constants, so the increments of the couplings (positive for ω_C and negative for ω_{GB}) are regular and linear in the UV regime. Hence, in the deep UV (close to the UV FP), we can assume that $\omega_C > 0$ and also that $\omega_{GB} < 0$.

All these arguments are self-consistent and lead to the conclusion that the UV fixed point of RG inevitably exists and realizes the asymptotic freedom (AF) scenario (in much the same way as in non-Abelian gauge theories). Our perturbation analysis is carried out in terms of the coupling $g^2 \propto 1/\omega$, so this and the fact that, in the UV regime, $\omega_C \to +\infty$ bolsters the correctness of our perturbative one-loop results even more. Actually, near the UV Gaussian FP, it is the coupling g (analogous to the YM coupling constant) that goes to zero.

The AF characterization of the FP comes when the RG flow is analyzed in terms of g-like couplings. We define them, taking into account above signs of ω_{C0} and ω_{GB0}, in the following way:

$$\omega_C = \frac{1}{g_C^2} = \frac{1}{4\alpha^2} \quad \text{and} \quad \omega_{GB} = -\frac{1}{g_{GB}^2}, \tag{31}$$

and with the inverse relations

$$g_C = \frac{1}{\sqrt{\omega_C}} \quad \text{and} \quad g_{GB} = \frac{1}{\sqrt{|\omega_{GB}|}} = \frac{1}{\sqrt{-\omega_{GB}}}. \tag{32}$$

In this way, we are sure that both g-couplings are non-negative. Now, one can easily derive by differentiation

$$\beta_C = \partial_t \omega_C = -\frac{2}{g_C^3}\partial_t g_C = -\frac{2}{g_C^3}\beta_{g_C}, \tag{33}$$

and

$$\beta_{GB} = \partial_t \omega_{GB} = \frac{2}{g_{GB}^3}\partial_t g_{GB} = \frac{2}{g_{GB}^3}\beta_{g_{GB}}, \tag{34}$$

(mind the sign in the second equation). Then, the β-functions of these couplings are expressed as

$$\beta_{g_C} = -\frac{1}{2}g_C^3\beta_C \quad \text{and} \quad \beta_{g_{GB}} = \frac{1}{2}g_{GB}^3\beta_{GB} \tag{35}$$

which to one-loop accuracy read

$$\beta_{g_C} = -\frac{199}{60}g_C^3 \quad \text{and} \quad \beta_{g_{GB}} = -\frac{87}{40}g_{GB}^3. \tag{36}$$

For vanishing values of the g-couplings, we find that the above β-functions vanish too, so in this situation we have a trivial Gaussian FP of the RG flow. We also see that, for the positive values of the couplings that is $g_C > 0$ and $g_{GB} > 0$, we have that both β-functions β_{g_C} and $\beta_{g_{GB}}$ are negative which signifies that in the UV we meet a FP (asymptotically free theory) in basically the same way like this happens in QCD.

4.2. Scheme Independence of IR Fixed Point

In the paper [16], the detailed analysis of the situation with FP's in the IR was presented. We found both a turning point of the RG flow at some finite energy scale k and also a true IR FP at $k = 0$, so in the deep IR regime. The genuine IR FP has a non-trivial (non-Gaussian) character and it allows for defining the theory in the non-perturbative way, which is free from any IR-type of divergences and it therefore realizes Weinberg asymptotic safety, but in the infrared. The turning point (TP) of the RG flow is related to the non-analytic behavior of the β-functions in terms of running coupling parameters. This in turn corresponds to a very interesting cosmological bounce scenario, when analyzed from the AdS/CFT correspondence point of view. In this subsection, we discuss some universal features of the TP of RG and of its location, while, in the next Section 4.3, we prove the relation between the non-analytic behavior of β-functions and the bounce happening at finite k.

One can easily see that qualitative features of the RG flow presented above do not depend on the details of the regularization and renormalization procedures. For example, taking a closer look at the plot (cf. Figure 1 from [16]) of the dependence of the β-functions on the energy scale, one can convince oneself that the asymptotics in the UV limit of the flow and an existence of the location of the first zero of the RG flow counting from UV direction are universal. Firstly, we would like to discuss the UV regime of the RG flow. The asymptotics in the UV limit are described by one-loop β-functions computed by Fradkin and Tseytlin for four-dimensional conformal gravity and our exact (or rather FRG-improved) β-functions tend asymptotically to these constant values (if the β-functions for ω-type of couplings are considered). (Here, we also suppressed the contribution from the anomalous dimension η since we know that this goes as some inverse powers of the ω_C coupling and should be accurately included only when higher-loop computation accuracy is required. We have $\eta \ll 1$ and effectively we can take $\eta = 0$, which is a judicious assumption in the UV regime of the flow towards asymptotically free point.) As a matter of fact, we see that the β-function of the ω_C is positive (with the precise value $\beta_C = \frac{199}{30}$), while the β-function of the coupling ω_{GB} is negative (with the precise value $\beta_{GB} = -\frac{87}{20}$). Since in this regime we reach an asymptotically free UV FP, where the g-couplings go to zero (and correspondingly ω-couplings tend to infinite values), any RG flow towards such a FP should coalesce with the perturbative one-loop RG flow for small values of the g-couplings. We also see this as a feature of our improved RG flows. Therefore, the locations of horizontal asymptotes of the RG flow from the right (so for $k \to +\infty$) are universal and scheme-independent.

In the paper [16], we studied the energy evolution of the running couplings towards the IR regime of the flow, concentrating on the IR fixed points of the system. The inclusion of threshold phenomena, which are present in any mass-dependent renormalization, is of crucial importance in our analysis. In fact, if we have studied only the simplified system of β-functions (29), we would not find any interesting behavior of the RG flow in the infrared (similarly to the case of QCD in the IR regime where the coupling grows stronger and gets out of the perturbative regime). We do not find any IR FP in such a simplified scheme. The β-functions from the system (29) for $\mathcal{X} = \text{const}$ are always constant, at any energy scale. To search for some non-trivial behavior in the IR, we must thus include some additional non-perturbative effects. This feature is brought about by our usage of FRG methods and account of decoupling of massive modes in the IR domain.

In order to find the FP's of the system in the IR regime, we must solve equations $\beta_C(k) = 0$ and $\beta_{GB}(k) = 0$. One can see a big simplification here because, in order to find zeros, we do not need to solve the full system (27). Actually, we can completely forget the denominators in (27) and solve only Equation (29), where threshold effects embodied in the factors Λ/k^2 are still taken into account. We also notice that the anomalous dimension η does not influence the locations of the possible IR FP's, within the limits implied by our truncation ansatz used in FRG.

Numerical solutions of the equations $\beta_C(\kappa) = 0$ and $\beta_{GB}(\kappa) = 0$ reveal that they are both satisfied at (approximately) simultaneous values of the rescaled energy scale $\kappa = k/\sqrt{|\Lambda|}$. This is a smoking gun for the fixed point of the RG flow. These zeros are automatically zeros of the exact system (27). Moreover, the location of the zeros is almost identical (up to 2% accuracy) for the couplings ω_C and ω_{GB} for both cases of $\Lambda > 0$ and $\Lambda < 0$. The inclusion of higher-loop effects or extension of our truncation ansatz will make this agreement even stronger, so that in an exact fully non-perturbative theory, the locations of two zeros coalesce into the one unique location of a genuine FP for both couplings, in the infrared regime.

As we remarked above, the non-trivial form of the running arises because we have included the effects of threshold phenomena. Let us, for definiteness, analyze closer the case of the parameter $\Lambda > 0$. It is straightforward to understand the behavior of the exact β-functions β_C and β_{GB} regarding their zeros and ensuing non-trivial FP's in the IR domain. For this, it is important to find the behavior of the MSS one-loop partition function treated as a rational function of the energy scale k. The first pole/zero of the partition function (when one is coming from large values of k) we find at $k^2 = \frac{4}{3}\Lambda$, and this is due to the factor $\det_{2T}\left(\hat{\Box} - \frac{4}{3}\Lambda\hat{\mathbb{1}}\right)$, which is present in the denominator of the partition function (25). This latter implies that this factor appears in the FRG flow equation with the positive coefficient because it was a pole (not a zero) of the partition function. As it was explained earlier, for the contribution of this factor to the Wetterich equation, we need to evaluate the following functional trace

$$\mathrm{Tr}_{2T}\left(\frac{(\partial_t R_k - \eta R_k)\,\hat{\mathbb{1}}}{\hat{\Box} + R_k\hat{\mathbb{1}} - \frac{4}{3}\Lambda\hat{\mathbb{1}}}\right), \tag{37}$$

which in the RG flow is with the coefficient $+1$. For details of the structure of the FRG flow equation, the reader is referred to [16] and the Formula (47) there. Here, we meet the moment where we see the dependence on the cut-off kernel function $R_k = R_k(z)$, thus the dependence on the scheme of renormalization. We will now show that for the qualitative features of the flow important for the existence of non-trivial FP this dependence is immaterial. First, the trace in (37) leads to the expression in terms of the heat-kernel B_4 expansion coefficient of the operator, namely to $\left(1 - \frac{4}{3}\Lambda k^{-2}\right)^{-1} B_4\left(\hat{\Box}_{2T} - \frac{4}{3}\Lambda\hat{\mathbb{1}}_{2T}\right)$, again with positive front coefficient. To get the front factor of the above expression, we used the optimized (Litim) form of the cutoff kernel,

$R_k(z) = (k^2 - z)\theta(k^2 - z)$ [68], and we will comment on other possible choices below. One can check that the B_4 coefficient of such an operator is also positive. Indeed, we find

$$b_4\left(\hat{\square}_{2T} - \frac{4}{3}\Lambda\hat{\mathbb{1}}_{2T}\right) = \frac{3}{5}\Lambda^2, \tag{38}$$

which is valid on MSS background and under the space volume integral (to produce the integrated B_4 coefficient). This implies that in the flow of the action $\partial_t\Gamma_k^L$ (as evaluated on MSS background), we have the term $\left(1 - \frac{4}{3}\Lambda k^{-2}\right)^{-1}\left(\frac{3}{5}\Lambda^2\right)$, again with the positive sign. Hence, the expression for the β-function $\beta_{\rm GB}$, which is the only one that can be read from the situation on MSS background, contains the factor $\frac{9}{40}\left(1 - \frac{4}{3}\Lambda k^{-2}\right)^{-1}$, when we again emphasize the positive sign.

By solving algebraically the linear system of the β-functions (29), we also derive that, consequently, the β-function β_C contains in turn the factor $\frac{9}{40}\left(1 - \frac{4}{3}\Lambda k^{-2}\right)^{-1}$, but with the minus sign. This opposite sign is the aftermath of the form of the equations in the system: from MSS, we derive only $\beta_{\rm GB}$ where the threshold factors reside, while on Ricci-flat background we find only the combination $\beta_C + \beta_{\rm GB}$ without threshold factors. Therefore, the threshold factors are inherited by the β-function β_C solved algebraically but with the minus sign. The appearance of the anomalous dimension η (important for quantitative description of the flow) does not change anything for what regards the zero of the system of β-functions as we remarked before in [16] (provided that it is always negative $\eta \leqslant 0$). The factor $\left(1 - \frac{4}{3}\Lambda k^{-2}\right)^{-1}$ is the first one, which decides about the location of a vertical asymptote of the β-functions β_C and $\beta_{\rm GB}$, as seen on the plot of Figure 1 in [16], when the flow comes from the UV direction. In general, the positions of vertical asymptotes of the flow decide the factors in the partition function (both in the denominator and in the numerator).

Now, one can look at this factor from the broader perspective. It describes the decoupling (due to threshold phenomena) of the massive modes with the mass square parameter on MSS background given by $\frac{4}{3}\Lambda$. The coefficient in front is related to the B_4 coefficient of the corresponding operator and that is why it is positive. In a general renormalization scheme, this coefficient is also positive and the threshold factor $\left(1 - \frac{4}{3}\Lambda k^{-2}\right)^{-1}$ which shows the pole, precisely at $k^2 = \frac{4}{3}\Lambda$, must be present in this form or in another more general one, but still in a form exhibiting the pole in the same place. This is due to the gauge-invariant and universal fact that on MSS background in Weyl conformal gravity we find stable (non-tachyonic) $2T$ modes (spin-2 traceless) with the mass square given by $\frac{4}{3}\Lambda$. In addition, this is the highest mass square parameter in the spectrum of all modes there. In a general framework, the general threshold factor could look like this

$$af\left(\left(1 - \frac{4}{3}\Lambda k^{-2}\right)^{-1}\right), \tag{39}$$

where a is a positive constant and $f(x)$ is some smooth, regular and positive function at $x \geqslant 0$. For the β-function β_C this threshold factor appears as

$$bf\left(\left(1 - \frac{4}{3}\Lambda k^{-2}\right)^{-1}\right), \tag{40}$$

with a constant $b < 0$. This form manifests all universal features that we have discussed above. Now, it is a matter of simple analysis of functions that for the GB term coupling $\omega_{\rm GB}$, if $a > 0$ and $\beta_{\rm GB}\to -\frac{87}{20} < 0$, when $k \to +\infty$, then the running β-function $\beta_{\rm GB}(k)$ must meet a zero for some $\sqrt{\frac{4}{3}\Lambda} < k_E < +\infty$, because $a \cdot \left(1 - \frac{4}{3}\Lambda k^{-2}\right)^{-1} \to +\infty$, when $k \to \sqrt{\frac{4}{3}\Lambda}^{\,+}$. In the latter, the "+" superscript signifies that we approach the respective value from the above. Similarly, for the Weyl term coupling ω_C, if $b < 0$ and $\beta_C\to\frac{199}{30} > 0$, when $k \to +\infty$, then the running β-function $\beta_C(k)$ must meet a zero for some $\sqrt{\frac{4}{3}\Lambda} < k_C < +\infty$, because $b \cdot \left(1 - \frac{4}{3}\Lambda k^{-2}\right)^{-1} \to -\infty$, when $k \to \sqrt{\frac{4}{3}\Lambda}^{+}$. These are the invariant features of the flow and that is why the IR FP (for $k = k_{GB}$ for the ω_{GB} coupling and for $k = k_C$ for the ω_C coupling) is universally present in any renormalization scheme that aims at properly taking into account threshold phenomena. It is obvious that any mass-dependent scheme comes with its own form of the function $f = f(x)$, positive for $x \geqslant 0$, to supply the correct decoupling of heavy, massive modes in the IR regime. (As such a renormalization scheme, we cannot, for example, select a DIMREG because it is mass-independent.)

As we have argued, the form of the function $f(x)$ is irrelevant and the IR TP is scheme-independent. Mathematically, the zeros of threshold factors which appear in the expressions for the FRG-improved β-functions are the results of the continuity of the function $f(x)$ and of the existence of the first vertical asymptote at $k^2 = \frac{4}{3}\Lambda$, and of the way it is approached by β-functions. It is also of crucial importance that the factor $\left(k^2 - \frac{4}{3}\Lambda\right)$ first appeared in the denominator of the partition function (25), so the sign was favorable to enforce the change of the sign of the β_{GB}, when moving from negative universal one-loop value in the UV towards lower energies. This happens also because the vertical asymptote is approached from the right (higher energies) to $+\infty$. The change of the sign of β_{GB} must occur between two regimes: UV (when $\beta_{GB} < 0$), and IR near the first vertical asymptote (when $\beta_{GB} \to +\infty$), so the zero line must be crossed for some energy scale $k = k_{GB}$. A similar conclusion holds for the β_C with inherited threshold phenomena obtained from the system (23). Moreover, it is natural to expect that $k_{GB} \approx k_C$ since the difference can be associated only with higher-loop accuracy error as one can easily see by comparing the numerical values, which are quoted below for $\Lambda > 0$:

$$k_C \approx 1.17709\sqrt{\Lambda} \quad \text{and} \quad k_{GB} \approx 1.19163\sqrt{\Lambda}. \tag{41}$$

Finally, one sees that the values of the energy characterizing the IR TP are very close to the lower bound given by $k = \sqrt{\frac{4}{3}\Lambda} \approx 1.1547\sqrt{\Lambda}$. This means that we must inevitably find a FP in the infrared regime. Therefore, the existence of the IR TP is a universal feature of the exact (improved) RG flow, while the details of its location, slopes, and speeds of approaching the FP, etc., depend on the particular choice of the renormalization scheme (or in FRG terminology on the choice of the cut-off kernel function $R_k(z)$).

4.3. Non-Analyticity Near Turning Point

Let us now observe that the system of β-functions Equations (27) and (28) is not autonomous because the equations depend on the initial conditions of the flow, not only on the actual values of the couplings. In other words, we can see that these flow equations depend explicitly on the RG-time t parameter, or on its exponential version $k = k_0 e^t$. In the autonomous system, which is the case, for instance, for one loop in QED, in the dimensional regularization scheme or in simple momentum subtraction renormalization scheme, one has that $\beta = \beta(\omega)$ are functions of the actual values of couplings only. When one includes threshold phenomena for massive modes, then the autonomy of the system of RG flow equations is typically lost. This can be observed also in our case. This conclusion is based on the comparison of the RG running of the same β-functions, obtained for different initial conditions of the flow. We see that even in the situation where the actual values of the couplings are identical the corresponding β-functions for two such flows are unequal. Viewed differently, we can observe that the same values of the β-functions are attained for different values of the actual couplings ω, so the parameter t must also enter into dependence of the β-functions.

The lack of autonomy of the system of β-functions is the main obstacle against the possibility to express the β-functions in terms of couplings only. We remark that in more standard applications of FRG, FP's of the RG flow are looked for such systems $\beta = \beta(\omega)$ and conditions for FP's are conditions on the values of the couplings attained at the FP. In our case, for a genuine IR FP, we must have the additional condition that $t \to -\infty$. As explained in [16], we were able to find a TP for some finite energy scale and a true IR FP for any value of the couplings ω_C and ω_{GB}. Therefore, in our case, we do not have any condition on the couplings at the FP—they are fully unconstrained, or if analyzed in the space of all possible couplings, we have found not a single-point FP, but a two-dimensional surface of FP's of the RG flow.

Here, we present the analysis near the turning point of the RG flow, where $\beta = 0$, which occurs at $t = t_*$ and with the value of the coupling which is $\omega = \omega_*$ (this value of the coupling ω_* depends on the initial condition of the flow that is on $\omega_0 = \omega(t_0)$ and t_0, while t_* is independent of them). We analyze the general situation for one representative coupling, but we can think of the TP as found in the system of ω_C, ω_{GB} couplings for $\kappa_* \approx 1.18$. In the linear Taylor approximation, due to the regularity of the β-function at the zero point understood as a function of the t variable only, we can generically write $\beta = \mathcal{A}(t - t_*)$ for a constant coefficient $\mathcal{A}$ and this leads to the equation

$$\frac{d\omega}{dt} = \mathcal{A}(t - t_*), \tag{42}$$

which is solved by

$$\omega - \omega_* = \frac{\mathcal{A}}{2}(t - t_*)^2 \tag{43}$$

with the initial condition of ODE that $\omega(t_*) = \omega_*$. The above solution shows that there is a minimum of the coupling for the value ω_* at $t = t_*$ for $\mathcal{A} > 0$ (it is a local maximum when the constant coefficient $\mathcal{A} < 0$). The case $\mathcal{A} > 0$ we meet for the ω_C coupling, while for ω_{GB} we have the opposite behavior (due to (23)), so $\mathcal{A} < 0$ there. For definiteness, here we consider the case $\mathcal{A} > 0$. Inverting the relation in (43), we get that $t - t_* \sim \sqrt{\omega - \omega_*}$, so the linearized β-function in terms of the ω-coupling has the non-analytic form $\beta = \sqrt{2\mathcal{A}(\omega - \omega_*)}$. This signifies that there are two branches of the couplings (before and after the minimum of the ω coupling is reached). Due to the square root involved, at TP, there is a cusp-singularity in the local expression $\beta = \beta(\omega)$. The RG flow of the running coupling $\omega = \omega(t)$ has here a turning point because, when t is decreased, the coupling ω first decreases and reaches a local minimum, to finally start growing again and going through the same values of ω for $t < t_*$. The β-function as function of t smoothly crosses the zero line, while $\beta = \beta(\omega)$ has the cusp-behavior at $\omega = \omega_*$ and the double-valued behavior for $\omega > \omega_*$ with two (initially perfectly) locally symmetric branches of β with opposite signs. One could also consider the stability matrix of the linearized RG flow here at the TP. One of its eigenvalue could be finite $\sqrt{2\mathcal{A}}$, but $\frac{\partial\beta}{\partial\omega}$ at $\omega = \omega_*$ is formally positive infinity on the upper branch, so strictly speaking there are no eigenvalues of the linearized RG flow around this point. On the lower branch, we have the opposite situation with formal negative infinity due to the existence of the cusp. This signals the failure of the linearization of the flow $\beta = \beta(\omega)$ around such a point. The flow is non-analytic at $t = t_*$.

Actually, for the β-function of the GB term, the coefficient $\mathcal{A} < 0$ because the zero of β_{GB} is reached from the other side, so, instead of the local minimum for the coupling ω_C, here there is a local maximum for the coupling ω_{GB}. Moreover, there is again formally a negative/positive infinite eigenvalue of the RG flow (depending on which branch one is moving on), hence the stability matrix cannot be properly defined. When we consider two couplings at the same time at TP, a question arises which coupling (only one) has to be chosen to locally eliminate the t variable from the system of β-functions. We decided to remove t in favor of the ω_C coupling. If the β-function of the GB term, at the common TP (placed conventionally at $\kappa = \kappa_C$), is analyzed as a function of ω_C, not of the additional RG-time t variable, and not of the ω_{GB} coupling, then there is a finite positive off-diagonal value of the linearized RG flow matrix because $\beta_{GB} \sim \omega_C - \omega_{C*}$ with some finite coefficient. This is due to the fact that $\kappa_{GB} > \kappa_C$ and that $\beta_{GB}(\kappa_C) > 0$. Here, at the TP, we have a relation that $\omega_C - \omega_{C*}$ is proportional to $(t - t_*)^2$ with a positive coefficient from (43), hence then the two eigenvalues of the system are zero for the GB coupling and formally positive infinity for the ω_C coupling, and these two couplings formally are exact eigenvectors of the matrix of the RG flow. Again, even in the case of two couplings, the stability matrix cannot be determined.

The reader can easily see that the analysis presented above hinges on the fact that the location of the TP of the RG flow in RG-time coordinate t_* is finite. If formally $t_* \to \pm\infty$, then the initial assumption about the local behavior of the β-function $\beta = \mathcal{A}(t - t_*)$ does not make any sense. Hence, the solution in (43) is not realized in this form. As it is known from the general theory of true FP's of the RG flows, they may only appear at the abstract theoretical RG-scale coordinate $t_* \to \pm\infty$. Then, there are various theoretical ways how the asymptotics of the function $\beta(t) \to 0$ is realized, but, in most of the cases, it is possible to linearize the RG flow near the FP, when the β-functions are expressed entirely via the ω-couplings. At such FP, we generically find that $\beta(\omega) = \mathcal{A}(\omega - \omega_*)$ to the first infinitesimal level. Therefore, the flow is analytic and can be linearized to obtain finite derivative $\frac{\partial\beta}{\partial\omega}$ (in the case when we have many couplings, this is a linear matrix) giving us the stability coefficient. The form of the solutions $\omega = \omega(t)$ near the true FP's is typically exponential in t variable and moreover they do not depend on the particular initial values of the flow $\omega_0 = \omega(t_0)$, hence the system of the β-functions in the UV/IR becomes effectively autonomous. All the above statements about FP's are equivalent to each other and they show a clear distinction from the charasteristic of the RG flow at the stop of the flow which happens at TP's.

In addition, in the case of TP's, the statements about the finiteness of t_*, the linearity of $\beta = \beta(t)$ to the first order in $t - t_*$, the non-analyticity and the square-root-like singularity of the function $\beta = \beta(\omega)$, the lack of autonomy of the system of β-functions, and finally the impossibility to linearize the RG flow near TP and to define the stability matrix there are equivalent. For their derivation, we have not used any additional assumption and this is the reason why in the proof of these equivalences one can easily go both ways. For example, from the non-analytic form of the β-function $\beta = \sqrt{2\mathcal{A}(\omega - \omega_*)}$, one derives the local behavior of the β-function near the TP in t variables: $\beta = \mathcal{A}(t - t_*)$, which makes sense only for t_* finite. This is why all the characteristics of the TP as the special point of the RG flow are tightly related and it differs from a true FP of RG. Similarly, from the AdS/CFT point of view, true FP's of RG (both in IR or in UV) correspond to AdS backgrounds in asymptotic conformal regions of the gravitationally dual bulk spacetime, while the TP's correspond to bounce solutions holding in intermediate finite regions of spacetime characterized by some finite values of the AdS-like radial coordinate.

Following the above distinction between TP's and FP's, in the paper [16], we continued the search for true IR FP's. For this purpose, we used the established fact that at $\kappa \approx 1.18$ we found a common TP of the RG system. We exploited the infinitesimal form of the flow at TP and analytically extended it beyond the TP. We treated the TP as a good point from which we could start a new perturbation calculus driven towards the IR regime. Assuming perturbativity (in different couplings than in the UV FP), we were eventually able to find a genuine IR FP at $t \to -\infty$. We characterized this FP as non-trivial and non-Gaussian and computed the characteristic values of the couplings there, so called ω_{C**} and ω_{GB**}, which revealed to be non-vanishing. Moreover, we found that this IR FP is stable for both perturbation directions given by the couplings ω_C and ω_{GB}. The IR-stable true IR FP is the main result to be used for further cosmological and conformal symmetry breakdown related applications of the QWG theory, see also Section 6.

4.4. Extension of the RG Analysis to Two-Loop Order

In this subsection, we attempt to give a preliminary analysis of the RG system of running coupling parameters in QWG, at the two-loop level. We base our considerations only on algebraic, dimensional-analytic, and combinatorial arguments since a detailed computation of UV-divergences and β-functions at this level is still beyond our computational capabilities. We assume that numbers, we are dealing below with, are generic and they do not vanish, and we discuss the general structure of the RG system. In particular, we touched upon the issue of the "new" β-function β_R, which is expected to be generated first time at the two-loop level. We analyze its suppression compared to other β-functions in the system and establish the hierarchy of them. Moreover, we also look at the universality properties of the β-functions for all three couplings ω_C, ω_{GB} and new ω_R at this level and in this way we strengthten and extend the well known results from the one-loop quantum level to QWG theory at two loops.

As remarked, in the paper [16], the first paper to deal with the divergences issue at the two-loop level in QWG, was the one by Fradkin and Tseytlin from 1984 [70]. However, the computation presented there is not complete, since only a subset of two-loop diagrams is analyzed. Nevertheless, we agree with the authors' conclusion that it is very probable that the R^2 divergence shows up for the first time at the two-loop accuracy. This is in contradiction to the conjecture of 't Hooft and Mannheim [52], who instead expect that the conformal symmetry on the quantum level is so powerful that this non-conformal β-function β_R is vanishing to all orders and also non-perturbatively. This would be true, if the conformality was fully present at the quantum level (not only at the one-loop level, where it forces $\beta_R = 0$). We do not think this to be so, in accordance with [46,70], because of the presence of conformal anomaly, already at the one-loop perturbative level. We discuss more issues related to the conformal anomaly in the special discussion Section 5.

4.4.1. Two-Loop Suppression of β-Functions

Let us remind that the one-loop action for QWG reads

$$S = \int d^4x \sqrt{|g|} \left(\omega_C C^2 + \omega_{\rm GB} \mathcal{G} \right) . \tag{44}$$

This also served us as the truncation ansatz for the effective action Γ that we used for the FRG computation. We re-emphasize that the term R^2 with the coupling ω_{R^2} is consistently not included at the tree-level and in the one-loop motivated RG flow equation because such a term is *not* generated by any quantum correction at the one-loop level. In the original one-loop computation by Fradkin and Tseytlin [45], the quantities which are assumed to be small are $\frac{1}{\omega_C}$ and $\frac{1}{\omega_{\rm GB}}$. The loop expansion is precisely in these quantities that is at the one-loop level we have

$$\beta_C^{(1)} = \beta_C^{\rm FT} \quad \text{and} \quad \beta_{\rm GB}^{(1)} = \beta_{\rm GB}^{\rm FT}, \tag{45}$$

where the coefficients $\beta_C^{\rm FT}$, $\beta_{\rm GB}^{\rm FT}$ are simple numbers (30), while up to the two-loop order we must find

$$\beta_C^{(2)} = \beta_C^{\rm FT} + a_{C,C}^{(2)} \frac{1}{\omega_C} + a_{C,{\rm GB}}^{(2)} \frac{1}{\omega_{\rm GB}}, \tag{46}$$

and

$$\beta_{\rm GB}^{(2)} = \beta_{\rm GB}^{\rm FT} + a_{{\rm GB},C}^{(2)} \frac{1}{\omega_C} + a_{{\rm GD,GD}}^{(2)} \frac{1}{\omega_{\rm GB}}, \tag{47}$$

where the numerical coefficients $a_{C,C}^{(2)}$, $a_{C,{\rm GB}}^{(2)}$, $a_{{\rm GB},C}^{(2)}$, and $a_{{\rm GB,GB}}^{(2)}$ are presently unknown, but it is certain that they do not depend on the couplings ω_C, $\omega_{\rm GB}$. The two-loop form of the RG system presented above is the result of assuming the perturbative expansion in $\frac{1}{\omega_C}$ and $\frac{1}{\omega_{\rm GB}}$ variables.

The UV-divergent part of the effective action at the one-loop level is given schematically by

$$\Gamma^{(1)} = \int d^4x \sqrt{|g|} \left(\beta_C^{(1)} C^2 + \beta_{\rm GB}^{(1)} \mathcal{G} + \beta_R^{(1)} R^2 \right), \tag{48}$$

where we find that to one-loop accuracy we have

$$\beta_R^{(1)} = \beta_R^{\rm FT} = 0, \tag{49}$$

due to (partial) conformal symmetry still preserved at the quantum one-loop level. This fact can be viewed as the one-loop remnant of full conformal symmetry present in the action at the tree-level (44). At the two-loop level, we expect $\beta_R^{(2)}$ not to vanish and be given analogously by

$$\beta_R^{(2)} = \beta_R^{\rm FT} + a_{R,C}^{(2)} \frac{1}{\omega_C} + a_{R,\rm GB}^{(2)} \frac{1}{\omega_{\rm GB}} = a_{R,C}^{(2)} \frac{1}{\omega_C} + a_{R,\rm GB}^{(2)} \frac{1}{\omega_{\rm GB}}, \tag{50}$$

where the coefficients $a_{R,C}^{(2)}, a_{R,\rm GB}^{(2)}$ are presently unknown numbers, whose non-vanishing (even of one of them), if unambiguously computed, would completely prove the conjecture of [70]. We explain that the possible structure term $a_{R,R}^{(2)}\omega_R^{-1}$ is not present since only ω_C and $\omega_{\rm GB}$ are the couplings in the original (and also one-loop level) action. The coupling ω_R has to be introduced (and renormalized) only from the two-loop level only. Actually, the reason for its introduction at the two-loop level action is the presence of the R^2 counterterm in $\Gamma^{(2)}$. We need to absorb such a covariant UV-divergent term and for this we need to include the $\omega_R R^2$ term in the bare action. This also means that, for perturbative computations at the level of three loops and higher, we must use the bare action (44) corrected by the presence of this new term $\omega_R R^2$ with arbitrary coefficient ω_R (however, due to hierarchy and suppression of β-functions, as explained below, we should assume that its value is parametrically smaller than the values of other couplings ω_C and $\omega_{\rm GB}$ present in (44) that is we should use $\omega_R \ll \omega_C, \omega_{\rm GB}$). For full one- and two-loop level quantum computations, we can use the bare action as given in (44) and this is reflected in the results for the β-functions to this accuracy given in (46), (47), and (50). It is conceivable that, if one wants to theoretically go to three-loop expressions for the β-functions of any of the coupling $X, Y = \omega_C, \omega_{\rm GB}$, or ω_R, then the terms with the structure $a_{X,Y,R}^{(3)}\omega_Y^{-1}\omega_R^{-1}$ could appear in $\beta_X^{(3)}$ with non-vanishing coefficients $a_{X,Y,R}^{(3)}$.

The two-loop level UV-divergent part of the effective action then takes the form

$$\Gamma^{(2)} = \int d^4x \sqrt{|g|} \left(\tilde\beta_C^{(2)} C^2 + \tilde\beta_{\rm GB}^{(2)} \mathcal{G} + \tilde\beta_R^{(2)} R^2 \right) \tag{51}$$

so the new term R^2 is generated with the coefficient $\tilde\beta_R^{(2)}$, which is always suppressed by one power of the small coupling $\frac{1}{\omega_C}$ or $\frac{1}{\omega_{\rm GB}}$, as in (50). Compared to the one-loop level action (48), where the counterterms were multiplied by only numerical coefficients $\beta_C^{\rm FT}$ and $\beta_{\rm GB}^{\rm FT}$, this is a suppression by additional power of small coupling. We conclude here that the β-function β_R when it finally shows up at the two-loop level is additionally suppressed with respect to other β-functions in the RG system. This signifies that the hierarchy of the β-functions is evident and the running of the ω_R coupling is very small, and that it was fully consistent to assume to the one-loop accuracy that $\omega_R = 0$. This was the fact that we took advantage of in the truncation ansatz for Γ that we used to model FRG to the one-loop level. Sincerely speaking, the significant R^2 term could be generated in the truncation ansatz Γ, but this does not happen immediately, and it requires a long RG-time since the running of ω_R is very slow. Our truncation ansatz for QWG is therefore internally consistent, at least in a big vicinity of the UV FP of RG. We just remark that we took care of the fact that at the two- and higher-loop level coefficients of UV-divergences are not the same as higher-loop β-functions of couplings (but they are in strict relations) and therefore we decorate the terms in (51) by additional tildes.

If one uses the electric charge-like couplings defined for any coupling ω (from the set ω_C, $\omega_{\rm GB}$, and ω_R) (cf., also the analysis in Section 4.1) by

$$\omega_i = \frac{1}{g_i^2}, \tag{52}$$

then the corresponding β-function reads

$$\beta_i = \beta_{g_i} \frac{d\omega_i}{dg_i} = -2g_i^{-3}\beta_{g_i}. \tag{53}$$

Consequently, we find that, for the one-loop level accuracy

$$\beta_{C}^{(1)} = -2g_{C}^{-3}\beta_{g_C}^{(1)} \quad \text{and} \quad \beta_{\rm GB}^{(1)} = 2g_{\rm GB}^{-3}\beta_{g_{\rm GB}}^{(1)}, \tag{54}$$

and hence $\beta_{g_C}^{(1)} \propto g_C^3$, $\beta_{g_{\rm GB}}^{(1)} \propto g_{\rm GB}^3$, and finally $\beta_{g_R}^{(1)} = 0$. Similarly, based on Equations (46) and (52), up to the two-loop level, we find

$$\beta_{C}^{(2)} = -2g_{C}^{-3}\beta_{g_C}^{(2)} = \beta_{C}^{(1)} + a_{C,C}^{(2)}g_C^2 + a_{C,{\rm GB}}^{(2)}g_{\rm GB}^2, \tag{55}$$

which implies that

$$\beta_{g_C}^{(2)} = -\frac{1}{2}g_C^3\left(\beta_{C}^{(1)} + a_{C,C}^{(2)}g_C^2 + a_{C,{\rm GB}}^{(2)}g_{\rm GB}^2\right) = -\frac{1}{2}\left(g_C^3\beta_{C}^{(1)} + a_{C,C}^{(2)}g_C^5 + a_{C,{\rm GB}}^{(2)}g_C^3g_{\rm GB}^2\right). \tag{56}$$

We repeat verbatim for the GB term coupling:

$$\beta_{g_{\rm GB}}^{(2)} = -\frac{1}{2}\left(g_{\rm GB}^3\beta_{\rm GB}^{(1)} + a_{{\rm GB},C}^{(2)}g_C^2g_{\rm GB}^3 + a_{{\rm GB},{\rm GB}}^{(2)}g_{\rm GB}^5\right), \tag{57}$$

and for the ω_R coupling (remembering that $\beta_{g_R}^{(1)} = 0$):

$$\beta_{g_R}^{(2)} = -\frac{1}{2}\left(a_{R,C}^{(2)}g_C^2g_R^3 + a_{R,{\rm GB}}^{(2)}g_{\rm GB}^2g_R^3\right) = \mathcal{O}\left(g^5\right). \tag{58}$$

We see that at the *leading order* the β-function β_{g_R} is proportional to the fifth power of the g-couplings. Again, compared to the expressions for β_{g_C} and $\beta_{g_{\rm GB}}$, which to the leading order (which is a one-loop order) go like g^3, this is a two-loop suppression. We derive that this suppression is present independently of the form of the couplings used in QWG theory.

When using electric-like-couplings g_i, we are in a comfortable situation in which perturbative calculus is conveniently done in these couplings (compare to a different case for ω-like couplings). For example, we write the one-loop effective action $\Gamma^{(1)} \propto \mathcal{O}(g^0)$ as a perturbation in small couplings g_i to the classical action $S \propto \mathcal{O}(g^{-2})$, and then the genuine corrections at the two-loop level are

$$\Gamma^{(2)} \propto \mathcal{O}(g^2). \tag{59}$$

The coupling of the R^2 term in the effective action Γ for QWG is generated from the level of (at least) two loops and its first coefficient is proportional to g^2, so compared to other terms like C^2 or E present at the one-loop level and multiplied by the coefficients of order g^0, this coefficient is again highly suppressed. We also draw a parallel that, analogously like in QWG considered above, in QCD, the β-functions scale like: $\beta_g^{(1)} \sim g^3$ for the one-loop level and $\beta_g^{(2)} \sim g^5$ for the two-loop level, and similarly the tree-level action scales as $S \propto \mathcal{O}(g^{-2})$, the one-loop level effective action scales as $\Gamma^{(1)} \propto \mathcal{O}(g^0)$, and the two-loop effective action scales as $\Gamma^{(2)} \propto \mathcal{O}(g^2)$, where g is the Yang Mills coupling parameter.

4.4.2. Universality of Two-Loop Order β-Functions in Weyl Gravity

Let us now present a general argument for the universal properties of the RG system of β-functions of all three couplings ω_C, $\omega_{\rm GB}$, and ω_R in QWG at the second loop level. Our exposition is loosely based on Ref. [71]. First, we recall that to the one-loop accuracy all β-functions of perturbative couplings are universal and do not depend on the choice of the renormalization scheme, and are also gauge-fixing independent, if the corresponding couplings are in front of dimension four gauge-invariant terms in the action. Of course, the form of the β-functions depend on the version of the couplings used (whether this is an ω-type coupling, or g-like coupling, or something else) as clearly seen in the formulas from the last subsection. Here, we analyze the change of the β-functions under a general redefinition of the couplings. Such redefinition mimics the change done by using a different regularization or renormalization scheme, or the change of the gauge-fixing conditions. Provided that this change is continuously connected to the point $g_i = 0$ and that it is analytic in the couplings g_i, we find that the type of the coupling is preserved. (We remark that the change between ω-type and g-type couplings is not regular at the point $g = 0$, hence it does not satisfy the above criterion, and as can be clearly seen from expressions in Equations (30) and (36), the corresponding β-functions are different, though they are obviously related to each other.) For the analysis in QWG, we also use the fact that ω_{GB} coupling cannot appear in any Feynman diagram computation, since it multiplies the Gauss–Bonnet term, which is a total derivative in $d = 4$. Hence, there cannot be any dependence on ω_{GB} in any perturbative β-function of the theory. This observation simplifies also the analysis of the previous subsection, where we can effectively set everywhere $1/\omega_{\mathrm{GB}} \to 0$.

The β-function of the R^2 term is expected first to show up at the two-loop level and in the form

$$\beta_R^{(2)} = \frac{d\omega_R}{dt} = \mathfrak{a}\frac{1}{\omega_C}, \tag{60}$$

where $\mathfrak{a} \equiv a_{R,C}^{(2)}$ is some constant coefficient, which has not yet been computed. Effectively, we have here the system of two couplings ω_R and ω_C and three β-functions. (In accordance with the remark made above, we do not have ω_{GB} as the coupling on which these β-functions could depend.) First, the β-function of the coupling ω_C is in its sector universal (it is behaving effectively like the only coupling in the town). By looking at the formula in Equation (46) when we set $a_{C,\mathrm{GB}}^{(2)}$ to zero, we see that this RG sector is identical with a two-loop sector of a theory with only one coupling ω_C. As it is known from [71], two-loop β-function for any QFT system of only one unique coupling is also invariant under the general coupling redefinition transformations.

Now, the question is only about the β-function β_R. We prove below that its two-loop value is also universal. In the two-couplings system (ω_R, ω_C), we are allowed to change couplings only in the following way, written to the leading order ω^0,

$$\omega_R' = \omega_R + A_0 + A\frac{\omega_R}{\omega_C} + B\frac{\omega_C}{\omega_R} + \mathcal{O}\left(\omega^{-1}\right), \tag{61}$$

$$\omega_C' = \omega_C + C_0 + C\frac{\omega_R}{\omega_C} + D\frac{\omega_C}{\omega_R} + \mathcal{O}\left(\omega^{-1}\right). \tag{62}$$

However, there exists a requirement that, in any gauge choice or parametrization method, the β-functions β_R' and β_C', when expressed in terms of primed couplings, cannot depend on ω_R' because there are no Feynman diagrams depending on this coupling (here on ω_R') needed to be considered to the two-loop level of accuracy. (The bare action to the two-loop level contains only one term $\omega_C C^2$.) Hence, one can check that the only possible changes of the couplings are shifts according to:

$$\omega_R' = \omega_R + A_0, \tag{63}$$

$$\omega_C' = \omega_C + C_0, \tag{64}$$

and then, of course, we have that the transformed β-functions read

$$\beta_R' = \beta_R = \mathfrak{a}\frac{1}{\omega_C} = \mathfrak{a}\frac{1}{\omega_C'} + \mathcal{O}\left(\frac{1}{\omega_C'^2}\right), \tag{65}$$

$$\beta_C' = \beta_C = \beta_C^{\mathrm{FT}} + a_{C,C}\frac{1}{\omega_C} = \beta_C^{\mathrm{FT}} + a_{C,C}\frac{1}{\omega_C'} + \mathcal{O}\left(\frac{1}{\omega_C'^2}\right), \tag{66}$$

to the quadratic order in the coupling ω'_C. This proves the universality in the ω_R and ω_C sector of the expressions for the two-loop β-functions of these couplings of the theory.

Similarly, we find the Gauss–Bonnet coupling $\omega_{\rm GB}$, which does not appear in any perturbative Feynman rule of the theory, so none of the β-functions can depend on it. Hence, the only permissible changes of couplings are given by

$$\omega'_C = \omega_C + C_0, \tag{67}$$

$$\omega'_{\rm GB} = \omega_{\rm GB} + D_0. \tag{68}$$

Then, according to (47), the β-function is to two-loop order accuracy

$$\beta^{(2)}_{\rm GB} = \beta^{\rm FT}_{\rm GB} + a^{(2)}_{{\rm GB},C}\frac{1}{\omega_C}, \tag{69}$$

and a transformed β-function is

$$\beta'_{\rm GB} = \beta^{\rm FT}_{\rm GB} + a^{(2)}_{{\rm GB},C}\frac{1}{\omega_C} = \beta^{\rm FT}_{\rm GB} + a^{(2)}_{{\rm GB},C}\frac{1}{\omega'_C} + \mathcal{O}\left(\frac{1}{\omega'^2_C}\right), \tag{70}$$

so again there is a universality of this two-loop expression for $\beta^{(2)}_{\rm GB}$.

In conclusion of this subsection, we note that all three β-functions β_C, $\beta_{\rm GB}$, and β_R are universal to two-loop level of accuracy. This, in particular, means that, if one finds the $\mathfrak{a}$ coefficient non-zero by explicit computation, then the appearance of the R^2 term in the two-loop effective action is unambiguous and universal fact, which cannot be removed by any gauge transformation, change of the scheme of RG, or redefinition of couplings. Ensuing consequences for conformal symmetry in QWG are very significant and intimately related to the conformal anomaly problem.

Basically, all the couplings in QWG at the two-loop level, behave as if they were in separate one-coupling sectors of the couplings' space of the theory. Everything boils down to the fact that the RG system of β-functions at two-loop level depends only on one coupling ω_C. This is a special feature of QWG that, on the two-loop level, we need only to deal with one coupling of the Weyl square term, the GB term will not have any impact perturbatively and only from the third loop we need to include the new coupling ω_R of the R^2 term. This new term $\omega_R R^2$ in the bare action is also heralded by the presence of trace anomaly already at the one-loop level, for which we turn now for discussion.

5. Discussion of the Trace Anomaly Issue

Let us now discuss the issue of the trace (or conformal) anomaly. Firstly, we recall the following fact about trace anomaly in ordinary Yang–Mills gauge theories in $d = 4$ spacetime dimensions. Standard YM theories are described by the action $S_g = -\frac{1}{2}\int d^4x\, {\rm tr}(F^2_{\mu\nu})$. This entails that, on the classical level, the theory is conformally invariant and hence the trace anomaly on this level vanishes. One could compute the trace of the classical energy-momentum tensor (obtained by the Hilbert method of variation with respect to some fiducial metric tensor of any curved background) and then it is found to vanish in agreement with conformal symmetry. On the other hand, if the theory is not very special, then for a generic situation at the quantum (loop) level there is a non-vanishing β-function β of the YM coupling, which is a signal of the presence of trace anomaly. Due to the RG-invariance in the effective action of the YM model at the one-loop level, we have the term

$${\rm tr}\left(F_{\mu\nu}\log\left(\frac{D^2}{\mu^2}\right)F_{\mu\nu}\right), \tag{71}$$

where D^2 is the square of the gauge-covariant derivative operator and μ is this renormalization scale which was also used to put renormalization conditions for fields and compensate $\mu-$dependence of the running dimensionless YM coupling. This RG running is in an invariant manner (and universal to one-loop level) described by the non-vanishing β-function β. One could ask for the conformal properties of the newly generated term (71) in the finite pieces of the effective action. Clearly, it is not even scale-invariant due to the presence of the dimensionful renormalization energy scale μ. Hence, this term cannot be conformally invariant. If one tries to evaluate the trace of the energy-momentum tensor coming from this term, it is definitely non-zero, hence there is a trace anomaly due to quantum effects. The trace was zero on the classical level (as the consequence of the presence and full realization of conformal symmetry), but it is non-zero on the quantum level, so there is an anomaly related to the fact that on quantum level there is no conformal symmetry anymore in this model. As it was elucidated above, this fact is in tight links with the presence of non-vanishing β-functions of the model. To one-loop level, the value of the β-function is universal, but as better observables we could choose scattering amplitudes in this model. While they are constrained by scale-invariance on the tree-level, due to the presence of the term (71) at one-loop level, the scattering amplitudes show behavior which explicitly breaks scale-invariance (and conformal-invariance in particular too). Simply, the amplitudes depend on the energy scale, while at tree-level they do not. For example, the 4-gluon scattering amplitude on tree-level is expressed only by the square of the constant value of the YM coupling at tree-level and there is no any dependence on the incoming gluon energies. After inclusion of the finite corrections, like the term (71), in the effective action, the resulting 4-gluon scattering amplitudes do show dependence on energies, as a consequence of the presence of the scale μ in the term (71). These are the observable results and we just interpret them that the scale-invariance is not present on the quantum level of this model and does not constrain amplitudes anymore. Simply saying, quantum physics of standard pure gauge theories is more complicated and more interesting than just what was on the tree-level constrained by scale-invariance. The amplitudes are more complicated and show more intricate behavior with energy scales as the result of freedom from conformal symmetry. The trace anomaly is not here a problem and can be used to generate some terms in the effective action. Actually, it signals that quantum gauge field theory are more interesting and worth studying. Here, conformal symmetry was never meant to be used as local gauge symmetry hence its lack after inclusion of quantum corrections (due to polarization effects of gluons) is not problematic. We could say that conformal symmetry was an accidental symmetry of the tree-level (classical) generic YM theory and at the quantum level we have seen that it is not there anymore. Since we do not put much of emphasis to this symmetry in gauge field theories, then its loss it is not a big deal like this happens with other accidental symmetries. We can sacrifice easily this conformal invariance and we shall not regret it (although now the computations on the quantum level are much more involved). This is the physics of YM theories, and still it is fully consistent without conformal invariance on the quantum level.

There is a completely different attitude for quantum conformal gravity theories, since there by definition we want to use the conformal symmetry with its fundamental and not accidental role. This fundamental role is signified in that we want to use this symmetry to constrain the form of all possible terms in the gravitational action (only C^2 action is acceptable in $d = 4$ dimensions), to constrain the spectrum of the model, which then is different from a generic spectrum of four-derivative gravitational theory. These things shall not be understood as accidents due to conformal symmetry, they are definite virtues. In addition, the fact that they happen is not surprising but rather demanded and expected from the fundamental role of conformal symmetry in constraining gravitational interactions. We want to use full conformal group as the gauge symmetry group of conformal-gravitational interactions. By gauging a full 15-parameter conformal group, we give the decisive and fundamental role to both the local Poincaré and local conformal symmetries. Only together could they be used to put a control over quantum theory of gravitational interactions.

Let us now return to the issue of the trace (or conformal) anomaly. For the energy-momentum tensor of the total system (matter + gravity), we define the trace in the classical and quantum cases respectively as

$$T = g_{\mu\nu} T^{\mu\nu} \quad \text{or} \quad \langle T \rangle = \langle g_{\mu\nu} \hat{T}^{\mu\nu} \rangle . \tag{72}$$

For quantum conformally invariant theories, we should find that $\langle g_{\mu\nu} \hat{T}^{\mu\nu} \rangle = 0$. By explicit computation on the classical level for conformal models, we find that $T = 0$, but, on the quantum level, one can find $\langle T \rangle \neq 0$, namely by general arguments [72]

$$\langle T \rangle = cC^2 + a\mathcal{G} + a_P P + a_L \Box R , \tag{73}$$

where $P = \epsilon_{\mu\nu\lambda\zeta} R^{\mu\nu}{}_{\rho\sigma} R^{\rho\sigma\lambda\zeta}$ is the parity-odd Pontryagin density, which can be a priori excluded in parity-preserving theories (as QWG is), and the last term is ambiguous as it depends on the renormalization prescription. Coefficients a and c are the central charges that can be directly related to the corresponding β-functions studied in the previous subsections (namely c and a are gauge-dependent for spin $3/2$ and 2 but $c + a$ is not [73,74]). One can find a straightforward relation between the logarithmically divergent part of the quantum effective action and the anomalous terms above. In particular, at one loop, one can check that

$$\langle T \rangle = \frac{1}{(4\pi)^2} b_4 , \tag{74}$$

so that $c = \beta_C^{(1)}$ and $a = \beta_{GB}^{(1)}$. In our case, this anomaly poses a problem: the classical theory (before quantization) was with local conformal symmetry, while quantum theory is without it. Quantum effects break the gauge symmetry of the model on the full quantum level (e.g., conformal analogues of Slavnov–Taylor identities are not satisfied). Consequently, the symmetry is without any power to constrain UV-divergences, Green functions, the form of the quantum corrections to the effective action or scattering amplitudes. Thus, in order to get a better grasp on the trace anomaly issue, it is imperative to use the effective action Γ rather than semiclassical arguments.

First, in pure C^2 gravity, we note that the β-function of Weyl coupling β_C is non-vanishing at the quantum level (starting from first loop). Hence, in the effective action Γ_{fin}, we must have a term

$$\beta_C C \log\left(\frac{\Box}{\mu^2}\right) C , \tag{75}$$

with μ being an arbitrary renormalization scale. This is due to RG-invariance of the total effective action Γ and the β-function β_C. In conclusion, the total action in $d = 4$ is

$$\Gamma_{\text{tot}} = \Gamma_{C^2} + \Gamma_{\text{eff}} \supset \sqrt{g}\left[C^2 + \beta_C C \log\left(\frac{\Box}{\mu^2}\right) C\right] . \tag{76}$$

Relevant observation pertaining to it is that Γ_{tot} is not conformally invariant action in $d = 4$! Definitely, conformal symmetry is not present on the quantum level in Weyl square gravity.

There is the perturbative trace anomaly issue, but the problem is also at non-perturbative level. This means that the theory, despite being perturbatively renormalizable is, however, non-perturbatively non-renormalizable [46]. This is because the radiative correction break Weyl symmetry in a dramatic way. Pure C^2 gravity is anomalous. In other words, there is RG running (towards IR) as explained in this work, but the problem is at high energies (UV-limit) for overall consistency of the model of conformal gravity. It is very important to cancel this anomaly since conformal symmetry appears in a local (gauged) version. Similarly, like in gauge theory models, we have to ascertain that there are no gauge anomalies due to fermions in the matter sector since the presence of such anomaly would be disastrous for the gauge symmetry on the quantum level. Theory would be inconsistent again because of lack of symmetry on the quantum level. In a perturbative vein, there would be pop up perturbative UV-divergences that we will not be able to absorb in any form of counterterms possible in gauge-invariant actions. These counterterms would not be gauge-invariant since on the quantum level there is no any symmetry constraining their form. In addition, if there is not a constraining (or forbidding) symmetry law on the quantum level in QFT, everything that is possible to be generated is generated (according to the QFT's Murphy's law).

The possible trace anomaly resolution is given below. It is known that there exists special matter content coupled to C^2 gravity such that $\beta_C = 0$ (and all other β-functions in the matter sector vanish too) and then the terms $C \log \Box C$ in the effective action Γ are not generated at all. For example, in the case of $\mathcal{N} = 4$ supergravity, Fradkin and Tseytlin [70] showed this can be achieved by coupling it to $\mathcal{N} = 4$ super-YM (SYM) gauge field theory model, which is known to be conformal on flat spacetime. This is brought by example from coupling $\mathcal{N} = 4$ supergravity due to Fradkin and Tseytlin to $\mathcal{N} = 4$ SYM gauge field theory model, which is known to be conformal on flat spacetime. Then, the quantum conformality is present in the coupled model and the anomaly is canceled due to mutual interactions between gravitational and SYM sectors. Both of these sectors are necessary for the successful trace anomaly cancellation. Then, if we have secured the presence of conformal symmetry on the quantum level, we can exploit the possibility to constrain scattering amplitudes, correlation functions, form of the effective action Γ, etc. It is well known that conformal symmetry is omnipotent in constraining the form of the terms in the effective action Γ because, in a fixed dimensionality of spacetime, only a few terms (finite number of them!) are conformally invariant compared with an infinite number of terms which are consistent with gauge symmetry or coordinate diffeomorphism invariance. The theory could be so powerful that the quantum effective action is idempotent with respect to quantization procedure (understood as going from classical tree-level action functional S to the fully quantum effective action functional Γ). Moreover, this also opens the possibility to resolve GR-like singularities using conformal symmetry of some exact solutions of quantum Γ, which now are also exact solutions of the classical theory.

From a different perspective, if there is no conformal (gauged) anomaly on the quantum level, then the theory is endowed with quantum scale-invariance, there are no non-zero β-functions, no RG flow and the theory sits at the UV FP of RG. This is a situation that has to be secured in UV by embedding the quantum Weyl gravity theory in the bigger picture, e.g., in the $\mathcal{N} = 4$ supergravity or in a twistor superstring theory of Berkovits and Witten. Another possibility is that the theory reaches in the UV regime a non-trivial (non-Gaussian) FP of RG. With the additional conditions that the dimension of the critical surface on which putative UV FP lies is finite, this realizes the Weinberg's Asymptotic Safety scenario (AS). In these circumstances, the UV-divergence problem is avoided and there are no β-functions (even on the non-perturbative level) since the theory sits at the FP. There is no RG flow and scale-invariance of the situation is enhanced to full conformal invariance and the UV-action that describes such a FP is an action of conformal field theory (CFT). There is a whole discipline of studies of CFT's but little is known about CFT when the gravity is quantum and dynamical. However, this is a desired CFT that describes the quantum theory of conformal gravitational interactions at very high energies. As with any other CFT to fully describe it, we would need to give the set of primary conformal operators and their anomalous dimensions (conformal weights). This constitutes the set of CFT data. Based on it, we could describe any correlation function within this gravitational CFT. To get away from the FP and start RG flow at lower energies, some deformation operator must be added to such CFT. That is, we think that, by deforming CFT by adding an operator which is not conformally invariant, we start a non-trivial RG flow away from the UV FP. In addition, then following the flow, we can reach even the domain of low-energy physics.

As it was pointed out long ago, the pure C^2 theory may reveal to be non-renormalizable from the level of two loops onward. This is due to the presence of Weyl anomaly [46]. Therefore, some authors claim that because of the non-zero trace anomaly C^2 gravity does not survive quantization and hence it is inconsistent on the quantum level until the conformal anomaly (CA) is made vanish. Its most direct effect may be the presence of the R^2 term in the divergent part of the effective action. (Such term does not show up at one-loop, but it is expected at the two-loop level because all symmetries that could forbid its presence there are not realized on the quantum level). The UV-divergences which could be covariantly collected in the term R^2 are not absorbable by the counterterms of the original Weyl theory and that is why the theory may show up to be perturbatively non-renormalizable, when a definite answer will be given to the presence or not of the R^2 term on the two-loop level [70].

We emphasized that, in our work [16], we worked in the fixed-dimension regularization scheme in $d = 4$ because this should not produce any spurious (e.g., dimension dependent) Weyl-symmetry violating terms in the effective action. Thus, we confined ourselves to the cutoff scheme at fixed dimensionality of spacetime. In this way, we do not discard or avoid the trace anomaly (which indeed shows up already at the one-loop level), and the statement of its existence is scheme-independent. Strictly speaking, the onset of the trace anomaly is renormalization prescription independent because the β-function is zero only at the fixed point and the existence of the fixed point is indeed renormalization prescription independent. On the other hand, the actual non-zero value of the trace of the energy-momentum (i.e., trace anomaly) is renormalization prescription dependent, since the β-function is universal only to the second order in the coupling constant. The one-loop value is scheme-independent but from two loops on we expect that such dependence will start to develop. The conformal anomaly is proportional to the β-functions of the model. In a different vein, these β-functions can be read off from perturbative UV-divergences of the effective action, hence the problem of trace anomaly is a problem rooted in the UV regime or, in other words, in an UV-completion of the theory.

At the UV FP which is asymptotically free for pure C^2 gravity, the conformal anomaly is still there. The β-functions for omega-type couplings are not vanishing at this FP (couplings multiplying directly invariants like C^2 or $\mathcal{G}$). Only β-functions in alpha-type couplings (similar to electric charge-like couplings) vanish, but the anomaly in terms of ω_C and ω_{GB} is non-zero even in the UV FP.

Let us summarize here our approach towards the quantization of C^2 theory and RG running towards the IR limit. We start in the UV fixed point where theory is exactly Weyl-invariant (just a single dimensionless coupling) and the renormalization group flow deforms the theory (i.e., induces other terms) as the theory flows towards IR fixed point. It turns out that the theory has a non-trivial (non-Gaussian) fixed point in IR and hence it is asymptotically safe in IR or, in another words, it is non-perturbatively renormalizable in IR. This AS is in accordance with Weinberg definition [75], but this time for the IR, rather than UV sector. Our situation with the issue of trace anomaly is slightly different since we have asymptotic safety in IR. Thus, we do not assume that theory is valid at all energy scales down to IR. It very quickly flows to the IR FP that is so close to the UV FP that the R^2 term does not even have enough (RG-)time to appear in the effective action (it does not appear there at one-loop level), and after IR FP is reached we do not have anymore scale invariance. In other words, we surpass the problem of perturbative non-renormalizability by non-perturbative renormalizability in the IR regime. In passing, we should stress that the presented IR AS scenario is conceptually very different from the conventional UV AS known from Einstein [75] and higher-derivative gravity [17,76,77].

One may wonder whether following RG trajectory towards lower energy one ends up at a reasonable point (be it Gaussian-like or Banks–Zaks FP) or diverge. In the latter case, the theory would need some kind of IR-completion or protection against infrared problems and IR-divergences. For this, if we have AS in IR, then the theory is non-perturbatively renormalizable and thus solves all such problems near IR FP.

As already mentioned, the problems related to conformal anomaly are essentially related to the UV regime. One might thus guess that having AS in IR FP does not a priori help in solving UV problems. For this, we need to provide evidence for a separate UV FP. In addition, it is obvious that looking locally in the parameter space the existence of IR FP does not imply anything for the existence of any UV FP. One possible scenario that can allow for inferring possible existence of UV FP for QWG is to embed the theory in a broader context of more fundamental (a better behaved) theories like twistor string theory or $\mathcal{N} = 4$ conformal supergravity. In particular, the bosonic C^2 sector that we consider here can be understand as a truncation of the anomaly-free $\mathcal{N} = 4$ supergravity of Fradkin and Tseytlin [45]. The super $\mathcal{N} = 4$ case of Fradkin and Tsetylin was later seen to provide the exceptional solution to the issue of trace anomaly [45] (however, it seems that there is still a unitarity problem for this model). It is the latter theory which is in the UV without any problem, there is no conformal anomaly, and no non-trivial β-functions. This theory is perturbatively renormalizable and UV-finite and hence there are not any problems with quantization. In our previous work [16], we consider the IR version of QWG when we reduce and consistently integrate out degrees of freedom and we study only the bosonic spin-2 sector. From a different perspective, QWG appears as the low energy limit of twistor string theory where there are not apparent problems with conformal anomaly. The conformal anomaly is simply a problem of the UV regime, but we assume that somehow this problem is solved there and we start with the theory, which is consistently reduced, in the intermediate energies. We assume that such UV FP exists and somehow addresses the anomaly issue, and then we try to see if such a postulate is consistent. At the same time, there is a subtle difference with what some people typically mean by a UV problem of conformal gravity. We believe that most practitioners just start perturbatively from IR FP (hence assume Gaussian or Banks–Zaks IR FP) and deduce the problem in UV by increasing energies in the RG flows. Natural questions arise: what about if one cannot even start perturbatively from IR as in our case (and also, e.g., in QCD). What then one can say about UV FP? We offer a plausible scenario that is logically consistent and reasonably well motivated. We assume that UV FP is just a full-fledged critical point in a series of hypothetical phase transitions that the Universe has undergone in its very early stage. It might be FP that descends from the spontaneous symmetry breaking (SSB) phase of twistor string theory or $\mathcal{N} = 4$ conformal supergravity. In any case, SSB FP has exact conformal symmetry and the simplest theory in $d = 4$ that lives at that critical manifold is QWG (only one coupling ω_C). Then, we may hypothesize about UV-completion in a form of string theory or something similar but clearly we do not have to do this now, if we have some strong evidence for UV FP and for solving some UV problems there. Moreover, if someone would like to see the RG evolution from IR FP towards UV, then he would probably need to go through horrible technical problems that would be even more complicated than in QCD. We are more like physicists, who start with asympotically free (deconfined) QCD and then deduce non-perturbative (confining) IR phase. This way is far simpler than going vice versa.

It should be stressed that when using FRG we do not a priori assume that there is any IR FP. This is a clear advantage of FRG approach. We just chose the truncation ansatz, which in our case is motivated by perturbative one-loop results. If we had known the higher-loop (or even better the full non-perturbative) effective action, then we could have utilized its form and set up a more informed truncation ansatz, which in turn could have provided more reliable (or more refined) results.

6. Physics beyond an IR Critical Point

In this section, we study the physics of the 2nd order phase transition that is associated with the IR FP of QWG discussed in Section 4. The characteristic (in fact defining) feature of 2nd order phase transitions is the existence of the order parameter field that has zero vacuum expectation value (VEV) in the symmetric phase and non-zero VEV in the broken phase. A typical method used in this context is the theory of effective potentials. This approach will be presented in the following two subsections: In Section 6.3, we will see that one can harness the conformal anomaly to say something more on the structure of an actual critical point and corresponding phases in its vicinity.

6.1. Composite Hubbard–Stratonovich Field as an Order-Parameter Field, Emergent Starobinsky Model

Because we require that our theory should induce Einstein action in the low-energy limit, we rewrite the R^2-part in the Bach action (11) with the help of the Hubbard–Stratonovich transformation [78–80] as

$$\exp(iS_{R^2}) \equiv \exp\left(\frac{i}{6\alpha^2}\int d^4x\sqrt{|g|}\,R^2\right) = \int \mathcal{D}\sigma \exp\left[-i\int d^4x\sqrt{|g|}\left(\frac{\sigma R}{2} + \frac{3}{32\omega_C}\sigma^2\right)\right]. \quad (77)$$

It is clear that the HS transformation (77) is nothing but a simple identity based on a functional Gaussian integral. Although the auxiliary HS field $\sigma(x)$ does not have a bare kinetic term, one might expect that, due to loop corrections, the renormalized action will develop in the IR regime a gradient term, which then allows for identifying the HS boson with a bona fide propagating mode. This mechanism is well-known from condensed matter theory [81,82] and particle physics [83–86]. A typical example is obtained when the BCS superconductivity is reduced to its low-energy effective level. There the HS boson coincides with the disordered field whose dynamics is described via the Ginzburg–Landau equation [87].

The σ field in (77) can be separated into a background field $\bar{\sigma} \equiv \langle\sigma\rangle$ corresponding to VEV of σ plus fluctuations $\delta\sigma$. Since $\bar{\sigma}$ is dimensional, it must be zero in the case when the theory is (globally) Weyl-invariant. On the other hand, when the Weyl symmetry is broken, $\bar{\sigma}$ develops a non-zero value. Thus, the σ field plays the role of the order-parameter field. With the benefit of hindsight, we further introduce an arbitrary (hyperbolic) mixing angle $\vartheta \in \mathbb{R}$ and make the following splitting

$$S_{R^2} = S_{R^2}\cosh^2\vartheta - S_{R^2}\sinh^2\vartheta\,. \quad (78)$$

If we now apply the HS transformation only to the $S_{R^2}\sinh^2\vartheta$ part of the action (11), we obtain

$$S_{R^2} = \int d^4x\sqrt{|g|}\left[-\frac{\sigma}{2}R + \frac{2\omega_C\cosh^2\vartheta}{3}R^2 + \frac{3}{32\omega_C\sinh^2\vartheta}\sigma^2\right], \quad (79)$$

and hence the full action (1) can be rewritten (modulo topological term) as

$$S = \int d^4x\sqrt{|g|}\left[-2\omega_C R_{\mu\nu}^2 - \frac{\sigma}{2}R + \frac{2\omega_C\cosh^2\vartheta}{3}R^2 + \frac{3}{32\omega_C\sinh^2\vartheta}\sigma^2\right]. \quad (80)$$

By employing the Weyl symmetry, one can formally generate a gradient term for the σ field already on the level of a bare-action. In fact, if we perform the Weyl rescaling of the form $g_{\mu\nu} = |\sigma|^{-1}\tilde{g}_{\mu\nu}$ we get the gradient term

$$-\sqrt{|\tilde{g}|}\,\frac{3}{2}\frac{\sigma\Box\sigma}{\sigma^2} + \ldots, \quad (81)$$

(dots refer to higher order derivative terms of the σ field). It should be, however, clear that such a gradient term cannot define genuine propagating mode since it depends on the conformal scaling and hence it is a gauge dependent concept. This is actually consistent with the fact that the WG has no propagating scalar degree of freedom on the bare-action level. On the other hand, when the scale symmetry is broken, then the σ-field will be trapped in a particular broken phase with specific kinetic as well as potential term, cf. Equation (87).

Although the full theory described by the action S is manifestly independent of the mixing angle ϑ, truncation of the perturbation series at a finite loop order will inevitably destroy this independence. The optimal result can be reached via the principle of minimal sensitivity [88–90], which posits that if a perturbation theory depends on some unphysical parameter (e.g., $\sinh^2\vartheta$ in our case) the best result is achieved if each perturbation order has the weakest possible dependence on the parameter ϑ.

As a result, the value of ϑ is determined at each loop order from the vanishing of the corresponding derivative of effective action.

Let us now see that the VEV $\bar{\sigma}$ indeed acquires a non-zero value at low enough energies. Since the corresponding phase transition must be of 2nd order ($\bar{\sigma}$ must change continuously from zero in symmetric phase to non-zero in broken phase), this is tantamount to showing that the IR FP must exist and that it is associated with the (global) scale-symmetry breakdown. To this end, we employ the standard effective-action methodology in the mean field approximation. In particular, we replace in (80) all σ fields with their VEV $\bar{\sigma}$ and compute the effective potential by integrating over metric-field fluctuations. This can be done again in the framework of the York decomposition discussed in Section 2.2. In order to simplify our computations, we use as the background the Minkowski flat spacetime. This in turn means that our computations serve only as proof of the principle that one should expect a phase transition in QWG with interesting cosmological implications in its broken phase. For more precise cosmological analysis, one should, of course, use cosmologically more pertinent classes of Bach-flat backgrounds.

As discussed in [78], the one loop-effective potential acquires the form

$$\begin{aligned} V_{\text{eff}}(\bar{\sigma},\vartheta) \;&=\; -\frac{9}{4^4\pi^2\omega_C^2}\frac{\bar{\sigma}^2}{(4\sinh^2\vartheta+1)^2}\left[\log\left(\frac{3\bar{\sigma}}{2\omega_C(4\sinh^2\vartheta+1)\mu^2}\right)-\frac{3}{2}\right] \\ &\quad+\frac{3\bar{\sigma}^2}{4^3\pi^2\omega_C^2}\left[\log\left(\frac{\bar{\sigma}}{2\omega_C\mu^2}\right)-\frac{3}{2}\right]-\frac{3\bar{\sigma}^2}{4^2\omega_C\sinh^2\vartheta}\,, \end{aligned} \tag{82}$$

and this result is true both in dimensional [78] and ζ-function [91] renormalization scheme. Here, μ is the renormalization scale.

The corresponding VEV $\bar{\sigma}$ is obtained through minimizing V_{eff} which gives

$$\bar{\sigma}(\vartheta) \;=\; 2\omega_C\mu^2\exp\left[\frac{3\sinh^2\vartheta\log\left(\frac{3}{4\sinh^2\vartheta+1}\right)+16\pi^2\omega_C(4\sinh^2\vartheta+1)^2}{\sinh^2\vartheta(32\sinh^4\vartheta+16\sinh^2\vartheta-1)}+1\right]. \tag{83}$$

We might note that the value $\bar{\sigma}=0$ is *not* a local minimum for V_{eff} when $\sinh^2\vartheta > (\sqrt{6}-2)/8$ because in such a case $V_{\text{eff}}<0$, while for $\bar{\sigma}=0$ one has $V_{\text{eff}}=0$. In order to see whether a situation with $\bar{\sigma}\neq 0$ can be realized, we employ the principle of minimal sensitivity, namely we require that

$$0 \;=\; \frac{dV_{\text{eff}}(\bar{\sigma}(\vartheta),\vartheta)}{d\sinh^2\vartheta} \;=\; \frac{\partial\bar{\sigma}(\vartheta)}{\partial\sinh^2\vartheta}\frac{\partial V_{\text{eff}}}{\partial\bar{\sigma}}+\frac{\partial V_{\text{eff}}}{\partial\sinh^2\vartheta} \;=\; \frac{\partial V_{\text{eff}}}{\partial\sinh^2\vartheta}\,. \tag{84}$$

This equation admits two branches of real solutions [78]. The branch that corresponds to the symmetric phase ($\bar{\sigma}=0$) corresponds to small values of $\sinh^2\vartheta$ and can be approximately written as $\sinh^2\vartheta = 0.02592337-0.0000197\alpha^2+\mathcal{O}(\alpha^2)$. The broken-phase branch ($\bar{\sigma}\neq 0$) corresponds to large values of $\sinh^2\vartheta$ with actual value depending on ω_C. Consequently, we can rewrite to order $\mathcal{O}(1/\sinh^4\vartheta)$ Equation (83) as

$$\bar{\sigma}(\vartheta) \;=\; 2\omega_C\mu^2\exp\left[1+\frac{8\pi^2\omega_C}{\sinh^2\vartheta(\omega_C)}\right]. \tag{85}$$

In particular, for any value of the dimensionless coupling ω_C, we can choose the renormalization scale μ, in such a way that $\bar{\sigma}\sim 1/\kappa^2$, which in turn will guarantee phenomenologically correct gravitational forces at long distances. Consequently, Newton's constant κ^2 is dynamically generated. Owing to the last term in (79)–(80), the appearance of a cosmological constant in the low-energy limit of the broken phase is also a consequence of QWG. In the following, it will be convenient to rescale $\sigma\mapsto\sigma/\kappa^2$ so that $\bar{\sigma}\sim 1$ and the field itself is dimensionless.

As already mentioned in Section 2.2, the local scale symmetry dictates that the scalar degree of freedom must decouple from the on-shell spectrum of QWG. When the conformal symmetry is broken the scalar field reappears in action through a radiatively induced gradient term of the HS field σ. The explicit form of the kinetic term can be decided from the momentum-dependent part of the σ-field self-energy Σ_σ. In Ref. [78], it was shown that the corresponding leading order gradient term is of the form $\frac{1}{2\kappa^2\bar{\sigma}}\partial_\mu\sigma\partial^\mu\sigma$. By assuming that in the broken phase a cosmologically relevant metric is that of the Friedmann–Robertson–Walker (FRW), then, modulo a topological term, the additional constraint

$$\int d^4x\sqrt{|g|}3R_{\mu\nu}^2 = \int d^4x\sqrt{|g|}R^2, \tag{86}$$

holds due to a conformal flatness of the FRW metric [92]. It was argued in [78] that from (11) and (79) one obtains in the broken phase the steepest descent (or WKB) gravitational action of the form

$$S_{b.p.,\sigma} = -\frac{1}{2\kappa^2}\int d^4x\sqrt{|g|}\left(\sigma R - \xi^2R^2 - \frac{(\partial_\mu\sigma)^2}{\bar{\sigma}} - 2\Lambda\sigma^2\right), \tag{87}$$

In the long-wave limit, one can neglect fluctuations of the σ field and consider that σ is basically described by its VEV, i.e., $\sigma = \bar{\sigma}$. This yields in the broken-phase the mean-field effective action

$$S_{b.p.,\bar{\sigma}} = -\frac{1}{2\kappa^2}\int d^4x\sqrt{|g|}(R - \xi^2R^2 - 2\Lambda), \tag{88}$$

By comparing (88) with (80), we obtain that

$$\Lambda = \frac{3}{32\omega_C\kappa^2\sinh^2\vartheta}, \quad \xi^2 = \frac{4\omega_C\kappa^2\sinh^2\vartheta}{3} \quad \Rightarrow \quad \Lambda = \frac{1}{8\xi^2}, \tag{89}$$

The action (88) is nothing but the Starobinsky action with the cosmological constant. In the Starobinsky model (SM), the linear Einstein term determines the long-wavelength behavior while the R^2-term dominates short distances and drives inflation [93], which is followed by the gravitational reheating with the decrease of R^2 [94]. In phenomenological cosmology, the SM represents metric gravity with a curvature-driven inflation. In particular, it does not contain any fundamental scalar field that could play the role of an inflaton field, even though a scalar field/inflaton formally appears when transforming the SM to the Einstein frame [95]. We stress that the cosmological constant Λ is entirely of a geometric origin (it descends from the QWG) and it enters with the opposite sign in comparison with the usual matter-sector induced (i.e., de Sitter) cosmological constant. In the following, we will call such gravitation-induced cosmological constant as *gravi cosmological* constant and it represents what we call the *dark side of QWG*.

In this connection, the following point deserves mentioning. Since the conformal symmetry prohibits the existence of a (scale-full) cosmological constant, the gravi-cosmological constant must correspond to a scale at which the conformal symmetry breaks, which in turn determines the cut-off scale of the σ-field. The magnitude of ξ in the SM is closely linked to the scale of inflation [46]. Using the values relevant for the Cosmic Microwave Background radiation (CMB) with 50–60 e-foldings, the Planck data [96,97] require that $\xi \sim 10^{-13}$ GeV^{-1} or equivalently $\xi/\kappa \sim 10^5$. The vacuum energy density is thus $\rho_\Lambda \equiv \Lambda/\kappa^2 \sim 10^{-10}(10^{18}$ GeV$)^4$, which corresponds to a zero-point energy density of a scalaron with an ultraviolet cut-off at about 10^{15}–10^{16} GeV. This coincides with a range of the GUT inflationary scale. For compatibility with an inflationary-induced large structure formation, the conformal symmetry should be broken before (or during) inflation. We will further discuss this issue in the following section.

6.2. Broken Phase QWG and Hybrid Inflation

By now, it is well recognized that the scalar cosmological perturbations are (nearly) scale-invariant with the value of the spectral index $n_s = 0.97$, which is tantalizingly close to 1—exact scale invariance [78,98]. The combination of data from Planck and BICEP2/Keck Array BK15 [97] tightens the upper bound on the tensor-to-scalar ratio to obtain: $r < 0.056$ at 95% CL. These include, e.g., the SM, the non-minimally coupled model ($\propto \phi^2 R/2$) with a $V(\phi) \propto \phi^4$-potential, inflation model based on a Higgs field and the so-called universal attractor models [97].

In the previous subsection, we have seen that the long-wave limit of QWG approaches in the broken phase the SM. It might thus be interesting to see what inflationary cosmology can be deduced from the action (87). In particular, we would like to see to that what extends the predictions of CMB perturbations based on the Lagrangian (87) is compatible with Planck and BICEP2/Keck Array BK15 cosmological data.

To this end, one can set up for $S_{b.p.,\sigma}$ a dual description in terms of a non-minimally coupled auxiliary scalar field λ with the action [94,95]

$$S_{\{\sigma,\lambda\},J} \;=\; -\frac{1}{\kappa^2}\int d^4x\,\sqrt{|g|}\left(\frac{\sigma+2\xi\lambda}{2}R+\frac{\lambda^2}{2}-\frac{(\partial_\mu\sigma)^2}{2\bar{\sigma}}-\Lambda\sigma^2\right). \tag{90}$$

This is basically an HS-transformed $S_{b.p.,\sigma}$ with λ being the HS-field. To analyze (90), we choose to switch from the Jordan frame to the Einstein frame [94] where the curvature R enters without a non-minimally coupled fields σ and λ. This is obtained via rescaling: $g_{\mu\nu} \mapsto (\sigma+2\xi\lambda)^{-1}g_{\mu\nu}$, giving

$$\begin{aligned} S_{\{\sigma,\lambda\},E} \;=\; & -\frac{1}{\kappa^2}\int d^4x\sqrt{|g|}\left[\frac{\tilde{R}}{2}-\frac{3\xi^2(\partial_\mu\lambda)^2}{(\sigma+2\xi\lambda)^2}-\frac{3\xi(\partial_\mu\lambda)(\partial^\mu\sigma)}{(\sigma+2\xi\lambda)^2}-\frac{(\partial_\mu\sigma)^2}{2\bar{\lambda}(\sigma+2\xi\lambda)}\right.\\ & \left. -\;\frac{3(\partial_\mu\sigma)^2}{4(\sigma+2\xi\lambda)^2}+\frac{\lambda^2}{2(\sigma+2\xi\lambda)^2}-\frac{\Lambda\sigma^2}{(\sigma+2\xi\lambda)^2}\right]. \end{aligned} \tag{91}$$

The above metric rescaling is valid only for $(\sigma+2\xi\lambda) > 0$. The action (91) can be brought into a diagonal form if we pass from fields $\{\sigma,\lambda\}$ to $\{\sigma,\psi\}$ where the new field ψ is obtained via the redefinition $\lambda = [\exp(\sqrt{2/3}|\psi|)-\sigma]/(2\xi)$. In terms of ψ, the action reads

$$S_{\psi,E} = -\frac{1}{\kappa^2}\int d^4x\sqrt{|g|}\left[\frac{\tilde{R}}{2}-\frac{1}{2}(\partial_\mu\psi)^2+V(\psi,\sigma)\;-e^{-\sqrt{2/3}|\psi|}\,\frac{(\partial_\mu\sigma)^2}{2\bar{\sigma}}\right], \tag{92}$$

where $V(\psi,\sigma) = \frac{1}{8\xi^2}\left(1-2\sigma e^{-\sqrt{2/3}|\psi|}\right)$. The strength of σ-field oscillations is controlled by the size of a coefficient in front of the σ-gradient term, i.e., $e^{-\sqrt{2/3}|\psi|}/\kappa^2$. In particular, the local squared amplitude of the σ-field fluctuations on energy scale $\mathcal{E}$ is of order

$$\delta\sigma^2(x) \;=\; \langle(\sigma(x)-\bar{\sigma})^2\rangle \;\sim\; \kappa^2\mathcal{E}^2 e^{\sqrt{2/3}|\psi(x)|}\,. \tag{93}$$

At large values of the dimensionless scalar field ψ, i.e., at values of the dimensionful field $\tilde{\psi} = \psi/\kappa$ that are large compared to the Planck scale, the gradient coefficient is very small and σ-field rapidly fluctuates. Assuming that QWG was broken before the beginning of inflation, then after a brief period of fierce oscillations the σ-fluctuations are strongly damped at around the value $\psi \lesssim 10$ (i.e., $\tilde{\psi} \lesssim 10m_p$). Indeed, if the fluctuations $\sqrt{\delta\sigma^2(x)}$ are smaller that 10^{-17}, i.e., smaller than the GUT inflationary scale, then

$$\kappa\exp\left(\frac{1}{2}\sqrt{\frac{2}{3}}|\psi(x)|\right) \;\lesssim\; 10^{-17} \quad\Rightarrow\quad \tilde{\psi} \;\lesssim\; 10m_p\,. \tag{94}$$

From that period on, the σ-field settles at its potential minimum at $\bar{\sigma} = 1$. Note that $V(\psi, \bar{\sigma}) \leq 1/(8\xi^2) \ll m_p^2$, which is a necessary condition for a successful inflation. At values of $\tilde{\psi} \sim 10m_p$, the potential $V(\psi, \bar{\sigma})$ is sufficiently flat to produce the phenomenologically acceptable slow-roll inflation, with the inflaton field ψ. From the inflationary potential in the Einstein frame, we can infer that this case represents hybrid inflation, in the sense that the inflationary potential is due to non-zero $\bar{\sigma}$. Using the slow-roll parameters

$$\epsilon = \frac{1}{2}m_p^2 \left(\frac{\partial_\psi V(\psi,\bar{\sigma})}{V(\psi,\bar{\sigma})}\right)^2, \quad \eta = m_p^2 \frac{\partial_\psi^2 V(\psi,\bar{\sigma})}{V(\psi,\bar{\sigma})}, \tag{95}$$

($\partial_\psi \equiv \partial/\partial\psi$) one can write down the tensor-to-scalar ratio r and the spectral index n_s in the slow-role approximation as [97]

$$r = 16\epsilon, \quad n_s = 1 - 6\epsilon + 2\eta\,. \tag{96}$$

In terms of the number N of e-folds left to the end of inflation,

$$N = -\kappa^2 \int_\psi^{\psi_f} d\psi \frac{V(\psi,\bar{\sigma})}{\partial_\psi V(\psi,\bar{\sigma})} \approx \frac{3}{4\bar{\sigma}} e^{\sqrt{2/3}|\psi|}, \tag{97}$$

(ψ_f represents the values of the inflaton at the end of inflation, i.e., when $e^{-\sqrt{2/3}|\psi|} \sim 1$) one gets

$$n_s \approx 1 - \frac{2}{N}, \quad r \approx \frac{12}{N^2}, \tag{98}$$

which for $N = 50 \div 60$ (i.e., values relevant for the CMB) is remarkably consistent with Planck data [96,97].

While during the inflation, the σ-field is constant (due to a large coefficient in front of the gradient term), allowing a *large-valued* inflaton field to descend slowly from the potential plateau, inflation ends gradually when σ regains its canonical kinetic term, and a *small-valued* inflaton field picks up kinetic energy. From (92), the dominant interaction channel at small $|\psi|$ is $(\partial_\mu\sigma)^2|\psi|$, hence the vacuum energy density stored in the inflaton field is transferred to the σ field via inflaton decay $\psi \to \sigma + \sigma$ (reheating), possibly preceded by a non-perturbative stage (preheating). Note also that the gravi-cosmological constant Λ that was instrumental in setting the inflaton potential in (92) has the *opposite* sign when compared with ordinary (matter-sector induced) cosmological constant.

An inflationary scheme discussed can be naturally incorporated in a broader theoretical context of "conformal inflation" model, which has been the subject of much recent investigation [99–102]. Setting the inflation period around the time of Weyl-symmetry breakdown brings about a number of attractive features, including: natural justification for inflation models with a plateau, technically accessible inflationary correlators, specific predictions for the B-mode vorticity fluctuations in CMB power spectra, etc. Let us also notice that the existence of a single scalar field with cutoff at around the GUT scale and coupled to broken phase of QWG (e.g., our σ or GUT Higgs field) would contribute with a *positive* zero-point energy that could offset gravi-cosmological Λ and leave behind a small observable cosmological constant. This could provide a viable mechanism for breaking down 120 orders of magnitude difference that presently exists between theoretical predictions and astronomical observations of Λ. Consequently, the dark side of QWG would not be ultimately so dark.

6.3. Some Reflections on Anomaly Matching

We have seen that the effective potential approach provides a powerful method for discussing physics in both broken and unbroken phase. Unfortunately, the physics in the immediate vicinity of the critical point is poorly grasped by this approach. The point is that, due to the appearance of long-range correlations, perturbative approaches are quite unreliable. It might seem rather surprising that the physics in the vicinity of the critical point might be for CFT's discussed in terms of conformal anomaly. The reason for this is basically twofold. First, conformal anomaly has an universal structure phrased in terms of two central charges (see Section 5). Second, there is a powerful theorem [103] that ensures that there is an anomaly matching across the critical point of the 2nd order phase transition.

In Section 4, we have considered an RG flow of QWG, which ends up in an IR FP that is associated with a CFT. Thus, one could wonder how this IR FP should be associated with low-energy gravity where the scale symmetry is absent. One intriguing possibility is that the IR CFT has a moduli space of vacua, such that in all of them but the one at the "origin" conformal (or scale) invariance is spontaneously broken. Such examples are known, expecially in superconformal field theories [104]. This allows for identifying a broken phase of the theory where the analytic structure of the amplitudes is not constrained by conformal invariance, but still the numerical value of the trace anomalies can be possibly matched (as proven, e.g., in Ref. [103] for CFT's with global conformal invariance). This poses strong constraints on the broken phase theory that is in spirit very similarly to what happens in *'t Hooft anomaly matching* [105]. In particular, requiring for the two cental charges

$$c_{<} = c_{>} \qquad a_{<} = a_{> } \tag{99}$$

("$>$" refers to *physical* energy scales above the FP in question and "$<$" refers to *physical* energy scales below the FP[2]) it was shown to be possible to determine the couplings of the action for the dilaton which emerges in the broken phase as (potentially composite) Goldstone boson related to the spontaneously or dynamically broken conformal symmetry [103]. Actually, in a supersymmetric scenario where the trace anomaly is in the same supermultiplet as R-symmetry charge; this boils down to the usual 't Hooft anomaly matching condition for which the generating functional of amplitudes accounting for anomalies is well understood. This could make it possible to understand the IR regime in the broken phase in a quite novel way. We leave this interesting and important topic for our future investigation.

7. Conclusions

In this paper, based on the approach of functional renormalization group, we have mainly investigated the IR physics of the quantum Weyl gravity, which turns out to be surprisingly rich and interesting, in particular in connection with the structure of its IR FP and related cosmological implications.

An important point that we have utilized here was the fact that at one loop the R^2-infinities of the effective action are absent so that we do not observe the appearance of the anomalous conformal mode. We point out, however, that the validity of this result goes beyond the usual one-loop order of the perturbation expansion since it was derived in an enhanced one-loop scheme where the effects of both anomalous dimensions and the threshold phenomena were also taken into account. Although it is expected that such situation will change at two loops (contrary to chiral anomalies, which are one-loop exact), one can still view this fact as an indication of a specific feature of the model that could potentially persist even to higher loops, where conclusive calculations are still missing. In our FRG approach, the IR FP (for the two involved couplings) develops already at the enhanced one-loop level and hence the prospective observational consequences of the trace anomaly do not take over before the Weyl symmetry is dynamically broken. Furthermore, we proved the suppression of β-functions at two-loop order and the fact that the possible presence of the R^2-infinity at two loops, albeit unambiguous, will modify the β-functions system only at three loops. We can thus judiciously assume that the results obtained in the enhanced one-loop scheme (e.g., existence of IR FP, bounce behavior, etc.) are quite robust and valid even when higher-loop orders would be included.

---

2 Here, we should stress that *physical scale* is not the same as the *running scale* in the RG flow. The running RG-flow scale is always re-parameterized so that it reaches its limiting value ($\pm\infty$) at the FP. In physics, the critical point that is represented by an RG FP always happens at some finite energy scale—phase transition scale.

A problem that typically besets higher-derivative gravity theories is the unitarity issue. For QWG, this is commonly phrased in terms of spin-2 ghost field. As already mentioned in the Introduction, there exist various remedies to this issue in the literature. If unresolved, the ghost problem prevents QWG from being a possible UV completion of Einstein's general relativity. However, this conclusion is based solely on the analysis of a classical Hamiltonian or on the structure of tree-level propagators. Basically, for any physical process that explicitly manifests the ghost problem, there is usually an implicit assumption that the perturbative analysis reflects the true physical spectrum. It is quite instructive to compare QWG with QCD. The latter is also renormalizable and asymptotically free theory, but, in this case, one is accustomed to the fact that the physical spectrum bears no resemblance to the perturbative degrees of freedom. In particular, the gluon is not in the physical spectrum. This is a consequence of confinement, but it is also understood more directly in terms of the behavior of the full gluon propagator. Essentially, there is an IR suppression of the propagator that is sufficient to remove the gluon pole. Similarly, we have seen that the FRG analysis of QWG shows that the IR FP is non-Gaussian; hence, QWG theory is non-perturbatively renormalizable in IR.

The presence of ghosts in the linearized spectrum and the quantum breaking of conformal invariance (trace anomaly) can, however, be seen as an indication that QWG should be understood in the context of a more fundamental theory. One such theory could be the $d = 4$ twistor string [106,107], where superconformal symmetry in spacetime is explicitly preserved contrary to ordinary string theory. In fact, it was found in Ref. [47] that its spectrum includes to the one of the $\mathcal{N} = 4$ superconformal gravity, which is made up of 2 $\mathcal{N} = 4$ graviton multiplets and 4 gravitini supermultiplets. One of the graviton multiplets is in the non-standard ghost-like sector. The gravitini supermultiplets contain the 15 gauge bosons of the local $SU(4)_R$ symmetry. The $SU(4)_R$ symmetry is potentially anomalous because its coupling is chiral and, as it is gauged, its cancellation requires the dimension of the gauge group G be 4. Thus, one gets the conclusion that $G = SU(2) \times U(1)$ or $U(1)^4$. An analogous condition is consistent with, but not implied by worldsheet arguments based on the worldhsheet conformal anomaly. Apart from the $SU(4)^3_R$, also spacetime conformal anomaly is possible and it has actually been argued that $\mathcal{N} = 4$ superconformal gravity can be made UV-finite and therefore anomaly-free by coupling it to exactly four $\mathcal{N} = 4$ super Maxwell multiplets [70]. The computation was carried out at one-loop order, but the result should hold at all orders as the β-function in $\mathcal{N} > 1$ CSG and conformal anomaly of super YM may receive contributions only from one loop. In the case of pure extended supergravity, it would actually be quite interesting to perform the analysis of functional RG in a situation where a nontrivial IR FP at one-loop order would be an exact result and where also the absence of an R^2 would be granted at any perturbative order. It is a known fact that, in $\mathcal{N} > 1$ CSG, a non-renormalization theorem based on formal superspace arguments implies UV-finiteness from two loops on [108], so that the R^2-infinity at two loops could not be generated. Whereas the presence of the trace anomaly could itself affect the argument, it is still possible to conjecture that the R^2-term could not exist in theories with $\mathcal{N} \geqslant 3$ [45]. In the case of $\mathcal{N} = 4$, one-loop exactness is also a consequence of the fact that all these anomalies are related by supersymmetry, and vanish when the $SU(4)^3_R$ anomaly does. As the latter is a chiral one-loop exact anomaly, its cancellation should entail the cancellation of conformal anomalies and other anomalies to all order. The version of $\mathcal{N} = 4$ CSG that was actually found in the context of twistor-string is a "non-minimal" one where the 4-derivative complex scalar ϕ couples to the Weyl graviton through a term of the kind $f(\phi)\left(C_{\mu\nu\rho\sigma}\right)^2$. Imposing a manifest $SU(1,1) \simeq SL(2,R)$ invariance requires f to be a constant so that one can recover the "minimal" version of CSG. It is quite interesting that only recently explicit actions for these theories have been constructed [109–111] making an RG flow analysis viable. The anomaly-free theory described above could be related to the Gaussian UV FP of QWG. We also would like to point out that the role of superconformal symmetry in cosmological applications has found large resonance in recent years [112,113] and in this context the relation between higher derivative terms and emergent order-parameter fields as discussed in Section 6.1 has also been highlighted (cf., e.g., Ref. [114]). It would, of course, be very interesting to investigate whether a non-Gaussian IR FP can also be identified through RG flow analysis in such scenarios. We would like however to point out that, once we are able to construct a model which contemplates QWG at some intermediate energy, we think the study presented in this paper (conducted through RG flow methods) can find interesting applications in a lot of physically viable models of QG.

In passing, we note that trace anomalies have also been recently discussed as valuable instruments to follow the violation of conformal symmetry along the RG flow. In particular, in four dimensions, any such flow can be reinterpreted in terms of a spontaneously or dynamically broken conformal symmetry. This has led to a deeper understanding of the a-theorem [103,115], and it would be definitely interesting to understand also the RG flow of QWG in this framework.

Apart from these more conceptual issues, we also discussed a cosmology that is implied by the broken phase of QWG. In the one-loop approaximation, we have been able to map the broken-phase effective action on a two-field hybrid inflationary model that, in its low-energy phase, approaches the Starobinsky $f(R)$ model with a gravi-cosmological constant. In particular, the inflationary potential obtained contains two scalar fields that interact via derivative coupling: scalaron ψ and Hubbard–Stratonovich order-parameter field σ (basically dilaton). The scalaron appears when we transform the broken-phase effective action to the Einstein frame and due to its slow-role potential it plays the role of an inflaton. The Hubbard–Stratonovich scalar, on the other hand, results from the dynamical transmutation of the spurion HS field and it mediates the inflationary potential. The derivative interaction between the two scalars provides a viable mechanism for a reheating scenario and graceful exit. The requirement that Einstein's R term in the low energy actions must have a coupling constant $1/2\kappa^2$ ties up the values of Starobinsky's inflation parameter ξ and the gravi-cosmological constant Λ. This in turn fixes the symmetry-breakdown scale for QWG to be at about the GUT inflationary scale. Moreover, the existence of a regime where gravity is approximately scale invariant (fixed-point regime and departure of the RG flow from it) provides a simple and natural interpretation for the nearly-scale-invariant power spectrum of temperature fluctuations in the CMB. In this connection, we have seen that the physics in the vicinity of the critical point (and, in particular, in the broken phase) can be analyzed in terms of conformal anomaly matching. Since trace anomaly is a fundamental property of QWG, this is clearly worth addressing. The study on simple Bach-flat backgrounds could be instrumental in understanding the importance of the trace anomaly for the early Universe cosmology, e.g., in setting up a specific cosmologically relevant inflationary potential. This adds up to the list of topics indicating that QWG with its rich spectrum of cosmological predictions could be considered as a viable theory of quantum gravity.

It should be stressed that, for simplicity's sake, our cosmological considerations were done in flat Minkowski background. In order to get more refined (and also more realistic) information on the inflationary potential, we should perform explicit computations of the broken-phase effective potential, in a non-trivial but cosmologically pertinent background, namely on MSS. These backgrounds are tailor-made for treatment of inflation and, in addition, they can provide us with some further guidance for the possible resolution of the cosmological constant problem on de Sitter spacetime. Work along these lines is presently in progress.

Author Contributions: P.J., L.R., S.G.G. and J.K. all jointly discussed, conceived and wrote the manuscript. All authors have read and agreed to the published version of the manuscript.

Funding: P.J. was supported by the Czech Science Foundation Grant No. 19-16066S. J.K. was supported by the Grant Agency of the Czech Technical University in Prague, Grant No. SGS19/183/OHK4/3T/14. S.G. was supported by the Israel Science Foundation (ISF), Grant No. 244/17.

Acknowledgments: The authors are grateful to Leonardo Modesto and Philip Mannheim for fruitful discussions.

Conflicts of Interest: The authors declare no conflict of interest.

Abbreviations

The following abbreviations are used in this manuscript:

FNSPE	Faculty of Nuclear Sciences and Physical Engineering
WG	Weyl Gravity
YM	Yang–Mills
QWG	Quantum Weyl Gravity
RG	Renormalization group
FRG	Functional Renormalization Group
FP	Fixed Point
TP	Turning Point
GB	Gauss–Bonnet
HS	Hubbard–Stratonovich
UV	Ultraviolet
IR	Infrared
MSS	Maximally Symmetric Spaces
CSG	Conformal Supergravity
SYM	Super–Yang–Mills
HDG	Higher Derivative Gravity
QG	Quantum Gravity
CA	Conformal Anomaly
AS	Asymptotic Safety
CFT	Conformal Field Theory
SM	Starobinsky Model
QFT	Quantum Field Theory
RHS	Right Hand Side
LHS	Left Hand Side
QCD	Quantum Chromodynamics
QED	Quantum Electrodynamics
AF	Asymptotic Freedom
AdS	Anti-de Sitter
EOM	Equation of Motion
DIMREG	Dimensional Regularization
ODE	Ordinary Differential Equation
FT	Fradkin–Tseytlin
GR	General Relativity
SSB	Spontaneous Symmetry Breaking
FRW	Friedmann–Robertson–Walker
CMB	Cosmic Microwave Background
VEV	Vacuum Expectation Value
GUT	Grand Unified Theory

References

1. Yang, C.N.; Mills, R. Conservation of Isotopic Spin and Isotopic Gauge Invariance. *Phys. Rev.* **1954**, *96*, 191. [CrossRef]
2. Utiyama, R. Invariant theoretical interpretation of interaction. *Phys. Rev.* **1956**, *101*, 1597. [CrossRef]
3. Kibble, T.W.B. Lorentz invariance and the gravitational field. *J. Math. Phys.* **1961**, *2*, 212. [CrossRef]
4. Néeman, Y.; Regge, T. Gauge Theory of Gravity and Supergravity on a Group Manifold. *Nuovo Cimento* **1978**, *1N5*, 1.
5. Ivanov, E.A.; Niederle, J. Gauge Formulation of Gravitation Theories. 1. The Poincare, De Sitter and Conformal Cases. *Phys. Rev. D* **1982**, *25*, 976. [CrossRef]
6. Maldacena, J.M. The Large N limit of superconformal field theories and supergravity. *Int. J. Theor. Phys.* **1999** *38*, 1113. [CrossRef]

7. Borsten, L. Gravity as the square of gauge theory: A review. *La Rivista del Nuovo Cimento* **2020**, *43*, 97. [CrossRef]
8. Taubes, C.H. A framework for Morse Theory for the Yang–Mills functional. *Invent. Math.* **1988**, *94*, 327. [CrossRef]
9. Taubes, C.H. Self-dual Yang-Mills connections on non-self-dual 4-manifolds. *J. Differ. Geom.* **1982**, *17*, 139. [CrossRef]
10. Bach, R. Zur Weylschen Relativitätstheorie und der Weylschen Erweiterung des Krümmungstensorbegriffs. *Math. Z.* **1921**, *9*, 110. [CrossRef]
11. Kaku, M.; Townsend, P.; Nieuwenhuizen, P.V. Gauge Theory of the Conformal and Superconformal Group. *Phys. Lett. B* **1977**, *69*, 304. [CrossRef]
12. Strominger, A.; Horowitz, G.T.; Perry, M.J. Instantons in Conformal Gravity. *Nucl. Phys. B* **1984**, *238*, 653. [CrossRef]
13. Hartnoll, S.A.; Policastro, G. Spacetime foam in twistor string theory. *Adv. Theor. Math. Phys.* **2006**, *10*, 181. [CrossRef]
14. Hill, T.C. Conjecture on the physical implications of the scale anomaly. *arXiv* **2005**, arXiv:hep-ph/0510177.
15. Maggiore, M. Dark energy and dimensional transmutation in R^2 gravity. *arXiv* **2005**, arXiv:1506.06217.
16. Jizba, P.; Rachwał, L.; Kňap, J. Infrared behavior of Weyl Gravity: Functional Renormalization Group approach. *Phys. Rev. D* **2020**, *101*, 044050. [CrossRef]
17. Codello, A.; Percacci, R. Fixed points of higher derivative gravity. *Phys. Rev. Lett.* **2006**, *97*, 221301. [CrossRef]
18. Cornwall, J.M. Dynamical Mass Generation in Continuum QCD. *Phys. Rev. D* **1982**, *26*, 1453. [CrossRef]
19. Holdom, B.; Ren, R. QCD analogy for quantum gravity. *Phys. Rev. D* **2016**, *93*, 124030. [CrossRef]
20. Asorey, M.; Gorbar, E.V.; Shapiro, I.L. Universality and Ambiguities of the Conformal Anomaly. *Class. Quant. Grav.* **2003**, *21*, 163. [CrossRef]
21. Boulware, D.G.; Deser, S. Can gravitation have a finite range? *Phys. Rev. D* **1972**, *6*, 3368. [CrossRef]
22. Lee, T.D.; Wick, G.C. Negative Metric and the Unitarity of the S Matrix. *Nucl. Phys. B* **1969**, *9*, 209. [CrossRef]
23. Lee, T.D.; Wick, G.C. Finite Theory of Quantum Electrodynamics. *Phys. Rev. D* **1970**, *2*, 1033. [CrossRef]
24. Anselmi, D. Fakeons And Lee-Wick Models. *JHEP* **2018**, *2*, 141. [CrossRef]
25. Anselmi, D. Fakeons, Microcausality And The Classical Limit Of Quantum Gravity. *Class. Q. Grav.* **2019**, *36*, 065010. [CrossRef]
26. Anselmi, D.; Piva, M. Quantum Gravity, Fakeons And Microcausality. *JHEP* **2018**, *11*, 021. [CrossRef]
27. Tkach, V.I. Towards Ghost-Free Gravity and Standard Model. *Mod. Phys. Lett. A* **2012**, *27*, 1250131. [CrossRef]
28. Smilga, A.V. Supersymmetric field theory with benign ghosts. *J. Phys. A* **2014**, *47*, 052001. [CrossRef]
29. Tomboulis, E. 1/N Expansion and Renormalization in Quantum Gravity. *Phys. Lett.* **1977**, *70B*, 361. [CrossRef]
30. Kaku, M. Strong Coupling Approach to the Quantization of Conformal Gravity. *Phys. Rev. D* **1983**, *27*, 2819. [CrossRef]
31. Cusin, G.; de O.Salles, F.; Shapiro, I.L. Tensor instabilities at the end of the Λ CDM universe. *Phys. Rev. D* **2016**, *93*, 044039. [CrossRef]
32. Asorey, M.; Rachwal, L.; Shapiro, I. Unitarity issues in higher derivative field theories. *Galaxies* **2018**, *6*, 23. [CrossRef]
33. Donoghue, J.F.; Menezes, G. Unitarity, stability, and loops of unstable ghosts. *Phys. Rev. D* **2019**, *100*, 105006. [CrossRef]
34. Bender, C.M.; Mannheim, P.D. No-ghost theorem for the fourth-order derivative Pais-Uhlenbeck oscillator model. *Phys. Rev. Lett.* **2008**, *100*, 110402. [CrossRef] [PubMed]
35. Bender, C.M.; Mannheim, P.D. Exactly solvable PT-symmetric Hamiltonian having no Hermitian counterpart. *Phys. Rev. D* **2008**, *78*, 025022. [CrossRef]
36. Hartle, J.B. Unitarity and causality in generalized quantum mechanics for nonchronal space-times. *Phys. Rev. D* **1994**, *49*, 6543. [CrossRef]
37. Politzer, H.D. Simple quantum systems in space-times with closed timelike curves. *Phys. Rev. D* **1992**, *46*, 4470. [CrossRef]
38. Lloyd, S.; Maccone, L.; Garcia-Patron, R.; Giovannetti, V.; Shikano, Y. Quantum mechanics of time travel through post-selected teleportation. *Phys. Rev. D* **2011**, *84*, 025007. [CrossRef]
39. Reuter, M. Nonperturbative evolution equation for quantum gravity. *Phys. Rev. D* **1998**, *57*, 971.

40. Reuter, M.; Wetterich, C. Exact evolution equation for scalar electrodynamics. *Nucl. Phys. B* **1994**, *427*, 291. [CrossRef]
41. Reuter, M.; Wetterich, C. Effective average action for gauge theories and exact evolution equations. *Nucl. Phys. B* **1994**, *417*, 181. [CrossRef]
42. Wetterich. C. Exact evolution equation for the effective potential. *Phys. Lett. B* **1993**, *301*, 90. [CrossRef]
43. Codello, A.; Percacci, R.; Rachwal, L.; Tonero, A. Computing the Effective Action with the Functional Renormalization Group. *Eur. Phys. J. C* **2016**, *76*, no. 4, 226. [CrossRef]
44. Percacci, R. *An Introduction to Covariant QuantumGravity and Asymptotic Safety*; World Scientific: New York, NY, USA, 2017. [CrossRef]
45. Fradkin, E.S.; Tseytlin, A.A. Conformal Supergravity. *Phys. Rept.* **1985**, *119*, 233.
46. Capper, D.M.; Duff, M.J. Conformal Anomalies and the Renormalizability Problem in Quantum Gravity. *Phys. Lett. A* **1975**, *53*, 361. [CrossRef]
47. Berkovits, N.; Witten, E. Conformal supergravity in twistor-string theory. *JHEP* **2004**, *8*, 009. [CrossRef]
48. Modesto, L.; Rachwal, L. Super-renormalizable and finite gravitational theories. *Nucl. Phys. B* **2014**, *889*, 228. [CrossRef]
49. Koshelev, A.S.; Sravan Kumar, K.; Modesto, L.; Rachwal, L. Finite quantum gravity in dS and AdS spacetimes. *Phys. Rev. D* **2018**, *98*, 046007. [CrossRef]
50. Polchinski, J. *String Theory*, 2nd ed.; Cambridge University Press: Cambridge, UK, 1998 [CrossRef]
51. Weyl, H. Reine Infinitesimalgeometrie. *Math. Z.* **2018**, *2*, 384.
52. Mannheim, P.D. Mass Generation, the Cosmological Constant Problem, Conformal Symmetry, and the Higgs Boson. *Prog. Part. Nucl. Phys.* **2017**, *94*, 125. [CrossRef]
53. Nakahara, M. *Geometry, Topology and Physics*; IOP Publishing, Ltd.: Bristol, UK, 1990. [CrossRef]
54. Rachwal, L. Conformal Symmetry in Field Theory and in Quantum Gravity—Review. *Universe* **2018**, *4*, 125.
55. Modesto, L.; Rachwal, L. Finite conformal quantum gravity and spacetime singularities. *J. Phys. Conf. Ser.* **2017**, *942*, 012015. [CrossRef]
56. Modesto, L.; Rachwal, L. Finite Conformal Quantum Gravity and Nonsingular Spacetimes. *arXiv* **2016**, arXiv:1605.04173. [CrossRef]
57. Hamber, H.W. *Quantum Gravitation, The Feynman Path Integral Approach*; Springer: Berlin, Germany, 2009.
58. DeWitt, B.S. *General Relativity: An Einstein Centenary Survey*; Hawking, S.W., Israel, W., Eds.; Cambridge University Press: Cambridge, UK, 1979.
59. DeWitt, B.S. *Relativity, Groups and Topology II*; DeWitt, B.S., Stora, R., Eds.; North-Holland: Amsterdam, The Netherlands, 1984.
60. Carlip, S. Dominant topologies in Euclidean quantum gravity. *Class. Q. Grav.* **1998**, *15*, 2629.
61. Freedman, M.H. The topology of four-dimensional manifolds. *J. Differ. Geom.* **1982**, *17*, 357. [CrossRef]
62. Irakleidou. M.; Lovrekovic, I. Conformal gravity one-loop partition function. *Phys. Rev. D* **2016**, *93*, 104043. [CrossRef]
63. Wetterich, C. Infrared limit of quantum gravity. *Phys. Rev. D* **2018**, *98*, 026028. [CrossRef]
64. Finer Technical Details. Supplemental Material. Available online: http://link.aps.org/supplemental/10.1103/PhysRevD.101.044050 (accessed on 11 August 2020). [CrossRef]
65. Benedetti, D.; Machado, P.F.; Saueressig, F. Taming perturbative divergences in asymptotically safe gravity. *Nucl. Phys. B* **2010**, *824*, 168.
66. Li, Y.D.; Modesto, L.; Rachwal, L. Exact solutions and spacetime singularities in nonlocal gravity. *J. High Energy Phys.* **2015**, *1512*, 173. [CrossRef]
67. Salvio, A. Quadratic Gravity. *Front. Phys.* **2018**, *6*, 77. [CrossRef]
68. Litim, D.F. Optimized renormalization group flows. *Phys. Rev. D* **2001**, *64*, 105007. [CrossRef]
69. Litim, D.F. Optimization of the exact renormalization group. *Phys. Lett. B* **2000**, *486*, 92. [CrossRef]
70. Fradkin, E.S.; Tseytlin, A.A. Conformal Anomaly in Weyl Theory and Anomaly Free Superconformal Theories. *Phys. Lett. B* **1984**, *134B*, 187. [CrossRef]
71. Hooft, G.T. *Under the Spell of the Gauge Principle*; World Scientific Publishing: Singapore, 1994. [CrossRef]
72. Bonora, L.; Pasti, P.; Bregola, M. Weyl Cocycles. *Class. Quant. Grav.* **1986**, *3*, 635.
73. Christensen, S.; Duff, M. Axial and Conformal Anomalies for Arbitrary Spin in Gravity and Supergravity. *Phys. Lett. B* **1978**, *76*, 571. [CrossRef]

74. Christensen, S.; Duff, M. New Gravitational Index Theorems and Supertheorems. *Nucl. Phys. B* **1979**, *154*, 301. [CrossRef]
75. Weinberg, S. Ultraviolet Divergences in Quantum Theories of Gravitation. In *General Relativity: An Einstein Centenary Survey*; Hawking, S.W., Israel,W., Eds.; Cambridge University Press: Cambridge, UK, 1979; Chapter 16, pp. 790–831. [CrossRef]
76. Cai, Y.-F.; Easson, D.A. Asymptotically safe gravity as a scalar-tensor theory and its cosmological implications. *Phys. Rev. D* **2011**, *84*, 103502.
77. Cai, Y.-F.; Easson, D.A. Black holes in an asymptotically safe gravity theory with higher derivatives. *JCAP* **2010**, *9*, 002. [CrossRef]
78. Jizba, P.; Kleinert, H.; Scardigli, F. Inflationary cosmology from quantum Conformal Gravity. *Eur. Phys. J. C* **2015**, *75*, 245. [CrossRef]
79. Hubbard, K. Calculation of partition functions. *Phys. Rev. Lett.* **1959**, *3*, 77. [CrossRef]
80. Stratonovich, R.L. On a Method of Calculating Quantum Distribution Functions. *Soviet Physics Doklady* **1958**, *2*, 416. [CrossRef]
81. Sachdev, S. *Quantum Phase Transitions*; Cambridge University Press: Cambridge, UK, 2011.
82. Altland, A.; Simons, B. *Condensed Matter Field Theory*; Cambridge University Press: Cambridge, UK, 2013.
83. Mathews, P.T.; Salam, A. The Green's functions of quantised fields. *Nuovo Cimento* **1954**, *12*, 563.
84. Mathews, P.T.; Salam, A. Propagators of quantized field. *Nuovo Cimento* **1955**, *2*, 120. [CrossRef]
85. Coleman, S. *Aspects of Symmetry: Selected Erice Lectures*; Cambridge University Press: Cambridge, UK, 1988. [CrossRef]
86. Kleinert, H. On the hadronization of quark theories. In *Understanding the Fundamental Constituents of Matter*; Zichichi, A., Ed.; Plenum Press: New York, NY, USA, 1978; pp. 289–390.
87. Ginzburg, V.L.; Landau, L.D. On the Theory of superconductivity. *J. Exp. Theor. Phys.* **1950**, *20*, 1064.
88. Stevenson, P.M. Optimized Perturbation Theory. *Phys. Rev. D* **1981**, *23*, 2916.
89. Bender, C.M.; Milton, K.A.; Pinsky, S.S.; Simmons, L.M., Jr. A New Perturbative Approach to Nonlinear Problems. *J. Math. Phys.* **1989**, *30*, 1447. [CrossRef]
90. Kleinert, H.; Schulte-Frohlinde, V. *Critical Propertiesof φ^4-Theories*; World Scientific: Singapore, 2001. [CrossRef]
91. Urban, P. Physical Aspects of Conformal Gravity. Master's Thesis, Czech Technical University, Prague, Czech Republic, 2018. Available online: https://physics.fjfi.cvut.cz/publications/mf/2018/dp_mf_18_urban.pdf (accessed on 11 August 2020).
92. Birrell, N.D.; Davies, P.C.W. *Quantum Fields in Curved Space*; Cambridge University Press: Cambridge, UK, 1982.
93. Koshelev, A.S.; Modesto, L.; Rachwal, L.; Starobinsky, A.A. Occurrence of exact R^2 inflation in non-local UV-complete gravity. *J. High Energy Phys.* **2016**, *11*, 067.
94. Capozziello, S.; Faraoni, V. *Beyond Einstein Gravity; A Survey of Gravitational Theories for Cosmology and Astrophysics*; Springer: London, UK, 2011. [CrossRef]
95. Whitt, B. Fourth Order Gravity as General Relativity Plus Matter. *Phys. Lett. B* **1984**, *145*, 176.
96. Ade, P.A.; Aikin, R.W.; Bock, J.J.; Brevik, J.A.; Filippini, J.P.; Ghosh, T.; Hildebrt, S.R.; Hui, H.; Kefeli, S.; Moncelsi, L.; et al. The Keck Array and BICEP2 Collaborations. BICEP2 / Keck Array IX: New bounds on anisotropies of CMB polarization rotation and implications for axionlike particles and primordial magnetic fields. *Phys. Rev. D* **2017**, *96*, 102003. [CrossRef]
97. Akrami, Y.; Arroja, F.; Ashdown, M.; Aumont, J.; Baccigalupi, C.; Ballardini, M.; Banday, A.J.; Barreiro, R.B.; Bartolo, N.; Basak, S.; et al. Planck, Planck 2018 results. X. Constraints on inflation. *arXiv* **2018**, arXiv:1807.06211. [CrossRef]
98. Mannheim, P.D. Cosmological Perturbations in Conformal Gravity. *Phys. Rev. D* **2012**, *85*, 124008.
99. Kallosh, R.; Linde, A. Universality Class in Conformal Inflation. *JCAP* **2013**, *7*, 002. [CrossRef]
100. Costa, R.; Nastase, H. Conformal inflation from the Higgs. *JHEP* **2014**, *6*, 145. [CrossRef]
101. Bars, I.; Steinhardt, P.; Turok, N. Local Conformal Symmetry in Physics and Cosmology. *Phys. Rev. D* **2014**, *89*, 043515. [CrossRef]
102. Rinaldi, M.; Cognola, G.; Vanzo, L.; Zerbini, S. Reconstructing the inflationary $f(R)$ from observations. *JCAP* **2014**, *8*, 015. [CrossRef]

103. Schwimmer, A.; Theisen, S. Spontaneous Breaking of Conformal Invariance and Trace Anomaly Matching. *Nucl. Phys. B* **2011**, *847*, 590. [CrossRef]
104. Fayet, P. Spontaneous Generation of Massive Multiplets and Central Charges in Extended Supersymmetric Theories. *Nucl. Phys. B* **1979**, *149*, 137. [CrossRef]
105. Hooft, G.T. Naturalness, chiral symmetry, and spontaneous chiral symmetry breaking. *NATO Adv. Study Inst. Ser. B Phys.* **1980**, *59*, 135. [CrossRef]
106. Berkovits, N. An Alternative string theory in twistor space for $\mathcal{N} = 4$ superYang–Mills. *Phys. Rev. Lett.* **2004**, *93*, 011601.
107. Witten, E. Perturbative gauge theory as a string theory in twistor space. *Commun. Math. Phys.* **2004**, *252*, 189. [CrossRef]
108. Stelle, K.S.; Townsend, P.K. Vanishing Beta Functions in Extended Supergravities. *Phys. Lett. B* **1982**, *113*, 25–30. [CrossRef]
109. Buchbinder, I.; Pletnev, N.; Tseytlin, A. "Induced" $\mathcal{N} = 4$ conformal supergravity. *Phys. Lett. B* **2012**, *717*, 274. [CrossRef]
110. Ciceri, F.; Sahoo, B. Towards the full $\mathcal{N} = 4$ conformal supergravity action. *JHEP* **2016**, *1*, 059. [CrossRef]
111. Butter, D.; Ciceri, F.; de Wit, B.; Sahoo, B. Construction of all $\mathcal{N} = 4$ conformal supergravities. *Phys. Rev. Lett.* **2017**. *118*, 081602. [CrossRef]
112. Kallosh, R.; Kofman, L.; Linde, A.D.; Van Proeyen, A. Superconformal symmetry, supergravity and cosmology. *Class. Quant. Grav.* **2000**, *17*, 4269. [CrossRef] [PubMed]
113. Ferrara, S.; Kallosh, R.; Linde, A.D.; Marrani, A.; Van Proeyen, A. Superconformal Symmetry, NMSSM, and Inflation. *Phys. Rev. D* **2011**, *83*, 025008. [CrossRef]
114. Cecotti, S.; Kallosh, R. Cosmological Attractor Models and Higher Curvature Supergravity. *JHEP* **2014**, *5*, 114. [CrossRef]
115. Komargodski, Z.; Schwimmer, A. On Renormalization Group Flows in Four Dimensions. *JHEP* **2011** *12*, 099. [CrossRef]

© 2020 by the authors. Licensee MDPI, Basel, Switzerland. This article is an open access article distributed under the terms and conditions of the Creative Commons Attribution (CC BY) license (http://creativecommons.org/licenses/by/4.0/).

Review

An SU(2) Gauge Principle for the Cosmic Microwave Background: Perspectives on the Dark Sector of the Cosmological Model

Ralf Hofmann

Institut für Theoretische Physik, Universität Heidelberg, Philosophenweg 16, 69120 Heidelberg, Germany; r.hofmann@thphys.uni-heidelberg.de

Received: 27 July 2020; Accepted: 20 August 2020; Published: 24 August 2020

Abstract: We review consequences for the radiation and dark sectors of the cosmological model arising from the postulate that the Cosmic Microwave Background (CMB) is governed by an SU(2) rather than a U(1) gauge principle. We also speculate on the possibility of actively assisted structure formation due to the de-percolation of lump-like configurations of condensed ultralight axions with a Peccei–Quinn scale comparable to the Planck mass. The chiral-anomaly induced potential of the axion condensate receives contributions from SU(2)/SU(3) Yang–Mills factors of hierarchically separated scales which act in a screened (reduced) way in confining phases.

Keywords: light scalar fields; axial anomaly; SU(2) Yang–Mills thermodynamics; de-percolation of axionic lumps; cosmological and galactic dark-matter densities

---

1. Introduction

Judged by decently accurate agreement of cosmological parameter values extracted from (i) large-scale structure observing campaigns on galaxy- and shear-correlation functions as well as redshift-space distortions towards a redshift of unity by photometric/spectroscopic surveys (sensitive to Baryonic Accoustic Oscillations (BAO), the evolution of Dark Energy, and non-linear structure growth), for recent, ongoing, and future projects see, e.g., [1–5]; (ii) fits of various CMB angular power spectra based on frequency-band optimised intensity and polarisation data collected by satellite missions [6–8]; and (iii) fits to cosmologically local luminosity-distance redshift data for Supernovae Ia (SNeIa) [9,10] the spatially flat standard Lambda Cold Dark Matter (ΛCDM) cosmological model is a good, robust starting point to address the evolution of our Universe. About twenty years ago, the success of this model has triggered a change in paradigm in accepting a present state of accelerated expansion induced by an essentially dark Universe made of 70% Dark Energy and 25% Dark Matter.

While (i) and (ii) are anchored on comparably large cosmological standard-ruler co-moving distance scales, the sound horizons $r_{s,\mathrm{d}}$ and $r_{s,*}$, which emerge at baryon-velocity freeze-out and recombination, respectively, during the epoch of CMB decoupling and therefore refer to very-high-redshift physics ($z > 1000$), (iii) is based on direct distance ladders linking to host galaxies at redshifts of say, $0.01 < z < 2.3$, which do not significantly depend on the assumed model for Dark Energy [11] and are robust against sample variance, local matter-density fluctuations, and directional bias [12–15].

Thanks to growing data quality and increased data-analysis sophistication to identify SNeIa [16] and SNeII [17] spectroscopically, to establish precise and independent geometric distance indicators [18,19] (e.g., Milky Way parallaxes, Large Magellanic Cloud (LMC) detached eclipsing binaries, and masers in NGC 4258), tightly calibrated period-to-luminosity relations for LMC cepheids and cepheids in SNeIa host galaxies [19], the use of the Tip of the Red Giant Branch (TRGB) [20,21] or the Asymptotic Giant Branch (AGB, Mira) [22] to connect to the distance ladder independently of cepheids, the high value of the basic cosmological parameter H_0 (Hubble expansion rate today) has (see [21], however), over the last decade and within ΛCDM, developed a tension of up to $\sim 4.4\,\sigma$ [19] in SNeIa distance-redshift fits using the LMC geometrically calibrated cepheid distance ladder compared to its low value extracted from statistically and systematically accurate medium-to-high-l CMB angular power spectra [7,8] and the BAO method [3]. Both, the CMB and BAO probe the evolution of small radiation and matter density fluctuations in global cosmology: Fluctuations are triggered by an initial, very-high-redshift (primordial) spectrum of scalar/tensor curvature perturbations which, upon horizon entry, evolve linearly [23] up to late times when structure formation generates non-linear contributions in the galaxy spectra [2,3] and subjects the CMB to gravitational lensing [24].

Recently, a cosmographic way of extracting H_0 through time delays of strongly lensed high-redshift quasars, see e.g., [25,26], almost matches the precision of the presently most accurate SNeIa distance-redshift fits [19]—$H_0 = 71.9^{+2.4}_{-3.0}$ $\mathrm{km\,s^{-1}\,Mpc^{-1}}$ vs. $H_0 = (74.22 \pm 1.82)$ $\mathrm{km\,s^{-1}\,Mpc^{-1}}$— supporting a high local value of the Hubble constant and rendering the local-global tension even more significant [27]. The high local value of $H_0 = (72 \cdots 74)$ $\mathrm{km\,s^{-1}\,Mpc^{-1}}$ [19,27] (compared to the global value of $H_0 = (67.4 \pm 0.5)$ $\mathrm{km\,s^{-1}\,Mpc^{-1}}$ [8]) is stable to sources of systematic uncertainty such as line-of-sight effects, peculiar motion (stellar kinematics), and assumptions made in the lens model. There is good reason to expect that an improved localisation of sources for gravitational-wave emission without an electromagnetic counterpart and the increase of statistics in gravitational-wave events accompanied by photon bursts (standard sirenes) within specific host galaxies will lead to luminosity-distance-high-redshift data producing errors in H_0 comparable of those of [19], independently of any distance-ladder calibration [28], see also [29].

Apart from the Hubble tension, there are smaller local-global but possibly also tensions between BAO and the CMB in other cosmological parameters such as the amplitude of the density fluctuation power spectrum (σ_8) and the matter content (Ω_m), see [30]. Finally, there are persistent large-angle anomalies in the CMB, already seen by the Cosmic Background Explorer (COBE) and strengthened by the Wilkinson Microwave Anisotropy Probe (WMAP) and Planck satellite, whose atoms (a) lack of correlation in the TT two-point function, (b) a rather significant alignment of the low multipoles (p-values of below 0.1%), and (c) a dipolar modulation, which is independent of the multipole alignment (b), indicate a breaking of statistical isotropy at low angular resolution, see [31] for a comprehensive and complete review.

The present work intends to review and discuss a theoretical framework addressing a possibility to resolve the above-sketched situation. The starting point is to subject the CMB to the thermodynamics of an extended gauge principle in replacing the conventional group U(1) by SU(2). Motivated [32] by explaining an excess in CMB line temperature at radio frequencies, see [33] and references therein, this postulate implies a modified temperature-redshift relation which places CMB recombination at a redshift of $\sim$1700 rather than $\sim$1100 and therefore significantly reduces the density of matter at that epoch. To match ΛCDM at low z, one requires a release of Dark Matter from early Dark Energy within the dark ages at a redshift z_p: lumps and vortices, formely tighly correlated within a condensate of ultralight axion particles, de-percolate into independent pressureless (and selfgravitating) solitons due to cosmological expansion, thereby contributing actively to non-linear structure formation at low z. A fit to CMB angular power spectra of a cosmological model, which incorporates these SU(2) features on the perfect-fluid level but neglects low-z radiative effects in SU(2) [34], over-estimates TT power on large angular scales but generates a precise fit to the data for $l \geq 30$. With $z_p \sim 53$ a locally favoured value of $H_0 \sim 74\,\mathrm{km\,s^{-1}\,Mpc^{-1}}$ and a low baryon density $\omega_{b,0} \sim 0.017$ are obtained. Moreover, the conventionally extracted near scale invariance of adiabatic, scalar curvature perturbations comes out to be significantly broken by an infrared enhancement of their power spectrum. Finally, the fitted redshift $\sim$6 of re-ionisation is compatible with the detection of the Gunn-Peterson-trough in the spectra of high-z quasars [35] and distinctly differs from the high value of CMB fits to ΛCDM [7,8].

As of yet, there are loose ends to the SU(2) based scenario. Namely, the physics of de-percolation requires extra initial conditions for matter density fluctuations at z_p. In the absence of a precise modelling of the 'microscopics' of the associated soliton ensembles[1] it is only a guess that these fluctuations instantaneously follow the density fluctuations of primordial Dark Matter as assumed in [34]. Moreover, it is necessary to investigate to what extent profiles of the axion field (lumps of localised energy density) actively seed non-linear structure formation and whether their role in galactic halo formation and baryon accretion meets the observational constraints (Tully–Fisher relation, etc.). Furthermore, radiative effects on thermal photon propagation at low z [36], which are expected to contribute or even explain the above-mentioned atoms (a), (b), and (c) of the CMB large-angle anomalies, see [37], and which could reduce the excess in low-l TT power of [34] to realistic levels, need to be incorporated into the model. They require analyses in terms of Maxwell's multipole-vector formalism [31] and/or other robust and intuitive statistics to characterise multipole alignment and the large-angle suppression of TT.

This review-type paper is organised as follows. In Section 2 we briefly discuss the main distinguishing features between the conventional ΛCDM model and cosmology based on a CMB which obeys deconfining SU(2) Yang–Mills thermodynamics. A presentation of our recent results on angular-power-spectra fits to Planck data is carried out Section 3, including a discussion of the parameters H_0, n_s, σ_8, $z_{\rm re}$, and $\omega_{\rm b,0}$. In Section 4 we interpret the rough characteristics of the Dark Sector employed in [34] to match ΛCDM at low z. In particular, we point out that the value of z_p seems to be consistent with the typical dark-matter densities in the Milky Way. Finally, in Section 5 we sketch what needs to be done to arrive at a solid, observationally well backed-up judgement of whether SU(2)$_{\rm CMB}$ based cosmology (and its extension to SU(2) and SU(3) factors of hierarchically larger Yang–Mills scales including their nonthermal phase transitions) may provide a future paradigm to connect local cosmology with the very early Universe. Throughout the article, super-natural units $\hbar = k_B = c = 1$ are used.

2. SU(2)$_{\rm CMB}$ vs. Conventional CMB Photon Gas in ΛCDM

The introduction of an SU(2) gauge principle for the description of the CMB is motivated theoretically by the fact that the deconfining thermodynamics of such a Yang–Mills theory exhibits a thermal ground state, composed of densely packed (anti)caloron [38] centers with overlapping peripheries [39,40], which breaks SU(2) to U(1) in terms of an adjoint Higgs mechanism [41]. Therefore, the spectrum of excitations consists of one massless gauge mode, which can be identified with the CMB photon, and two massive vector modes of a temperature-dependent mass on tree level: thermal quasi-particle excitations. The interaction between these excitations is feeble [41]. This is exemplified by the one-loop polarisation tensor of the massless mode [36,42,43]. As a function of temperature, polarisation effects peak at about twice the critical temperature T_c for the deconfining-preconfining transition. As a function of increasing photon momentum, there are regimes of radiative screening/ antiscreening, the latter being subject to an exponential fall-off [36]. At the phase boundary ($T \sim T_c$) electric monopoles [41], which occur as isolated and unresolved defects deeply in the deconfining phase, become massless by virtue of screening due to transient dipoles [44] and therefore condense to endow the formely massless gauge mode with a quasiparticle Meissner mass m_γ. This mass rises critically (with mean-field exponent) as T falls below T_c [41]. Both, (i) radiative screening/antiscreening of massless modes and (ii) their Meissner effect are important handles in linking SU(2)$_{\rm CMB}$ to the CMB: while (i) induces large-ange anomalies into the TT correlation [37] and contributes dynamically to the CMB dipole [45,46] (ii) gives rise to a nonthermal spectrum

[1] Lump sizes could well match those of galactic dark-matter halos, see Section 4.

of evanescent modes[2] for frequencies $\omega < m_\gamma$ once T falls below T_c. This theoretical anomaly of the blackbody spectrum in the Rayleigh–Jeans regime can be considered to explain the excess in CMB radio power below 1 GHz, see [33] and references therein, thereby fixing $T_c = 2.725\,\mathrm{K}$ and, as a consequence of $\lambda_c = 13.87 = \frac{2\pi T_c}{\Lambda_{\rm CMB}}$ [41], the Yang–Mills scale of SU(2)$_{\rm CMB}$ to $\Lambda_{\rm CMB} \sim 10^{-4}$ eV [32].

Having discussed the low-frequency deviations of SU(2)$_{\rm CMB}$ from the conventional Rayleigh–Jeans spectrum, which fix the Yang–Mills scale and associate with large-angle anomalies, we would now like to review its implications for the cosmological model. Of paramount importance for the set-up of such a model is the observtion that SU(2)$_{\rm CMB}$ implies a modified temperature (T)-redshift (z) relation for the CMB which is derived from energy conservation of the SU(2)$_{\rm CMB}$ fluid in the deconfining phase in an FLRW universe with scale factor a normalised to unity today [47]. Denoting by $T_c = T_0 = 2.725\,\mathrm{K}$ [32] the present CMB baseline temperature [6] and by $\rho_{{\rm SU(2)}_{\rm CMB}}$ and $P_{{\rm SU(2)}_{\rm CMB}}$ energy density and pressure, respectively, of SU(2)$_{\rm CMB}$, one has

$$a \equiv \frac{1}{z+1} = \exp\left(-\frac{1}{3}\log\left(\frac{s_{{\rm SU(2)}_{\rm CMB}}(T)}{s_{{\rm SU(2)}_{\rm CMB}}(T_0)}\right)\right), \tag{1}$$

where the entropy density $s_{{\rm SU(2)}_{\rm CMB}}$ is defined by

$$s_{{\rm SU(2)}_{\rm CMB}} \equiv \frac{\rho_{{\rm SU(2)}_{\rm CMB}} + P_{{\rm SU(2)}_{\rm CMB}}}{T}. \tag{2}$$

For $T \gg T_0$, Equation (1) simplifies to

$$T = \left(\frac{1}{4}\right)^{1/3} T_0(z+1) \approx 0.63\, T_0(z+1). \tag{3}$$

For arbitrary $T \geq T_0$, a multiplicative deviation $\mathcal{S}(z)$ from linear scaling in $z+1$ can be introduced as

$$\mathcal{S}(z) = \left(\frac{\rho_{{\rm SU(2)}_{\rm CMB}}(z=0) + P_{{\rm SU(2)}_{\rm CMB}}(z=0)}{\rho_{{\rm SU(2)}_{\rm CMB}}(z) + P_{{\rm SU(2)}_{\rm CMB}}(z)}\,\frac{T^4(z)}{T_0^4}\right)^{1/3}. \tag{4}$$

Therefore,

$$T = \mathcal{S}(z)\, T_0(z+1). \tag{5}$$

Figure 1 depicts function $\mathcal{S}(z)$.

Amusingly, the asymptotic T-z relation of Equation (3) also holds for the relation between $\rho_{{\rm SU(2)}_{\rm CMB}}(z)$ and the conventional CMB energy density $\rho_\gamma(z)$ in ΛCDM (the energy density of a thermal U(1) photon gas, using the T-z relation $T = T_0(z+1)$). Namely,

$$\rho_{{\rm SU(2)}_{\rm CMB}}(z) = 4\left(\frac{1}{4}\right)^{4/3}\rho_\gamma(z) = \left(\frac{1}{4}\right)^{1/3}\rho_\gamma(z) \quad (z \gg 1). \tag{6}$$

Therefore, the (gravitating) energy density of the CMB in SU(2)$_{\rm CMB}$ is, at the same redshift $z \gg 1$, by a factor of ~0.63 smaller than that of the ΛCDM model even though there are eight (two plus two times three) gauge-mode polarisations in SU(2)$_{\rm CMB}$ and only two such polarisations in the U(1) photon gas.

---

[2] That the deep Rayleigh–Jeans regime is indeed subject to classical wave propagation is assured by the fact that wavelengths that are greater than the spatial scale $s \equiv \pi T|\phi|^{-2}$, separating a(n) (anti)caloron center from its periphery where its (anti)selfdual gauge field is that of a dipole [40]. The expression for s contains the modulus $|\phi| = \sqrt{\Lambda^3_{\rm CMB}/(2\pi T)}$ of the emergent, adjoint Higgs field ϕ ($\Lambda_{\rm CMB} \sim 10^{-4}$ eV the Yang–Mills scale of SU(2)$_{\rm CMB}$), associated with densely packed (anti)caloron centers, and, explicitely, temperature T.

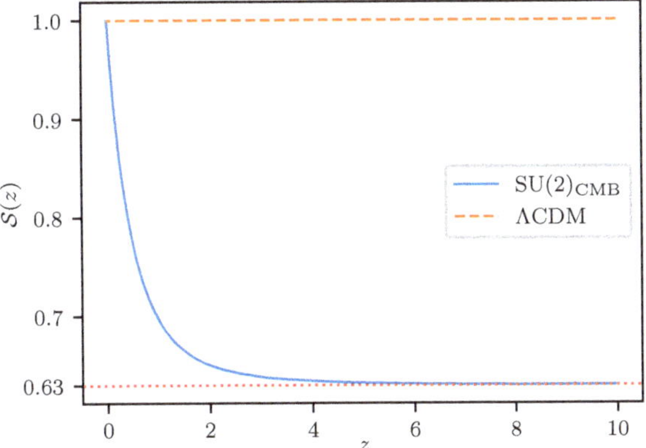

Figure 1. The function $\mathcal{S}(z)$ of Equation (4), indicating the (multiplicative) deviation from the asymptotic T-z relation in Equation (3). Curvature in $\mathcal{S}(z)$ for low z arises from a breaking of scale invariance for $T \sim T_0 = T_c$. There is a rapid approach towards the asymptotics $\left(\frac{1}{4}\right)^{1/3} \approx 0.63$ with increasing z. Figure adopted from [34].

Not yet considering linear and next-to-linear perturbations in $SU(2)_{CMB}$ to shape typical CMB large-angle anomalies in terms of late-time screening/antiscreening effects [37], Equation (5) has implications for the Dark Sector if one wishes to maintain the successes of the standard cosmological model at low z where local cosmography and fits to distance-redshift curves produce a consistent framework within ΛCDM. As it was shown in [34], the assumption of matter domination at recombination, which is not unrealistic even for $SU(2)_{CMB}$ [34], implies that

$$z_{\Lambda\text{CDM},*} \sim \left(\frac{1}{4}\right)^{1/3} z_{\text{SU}(2)_{\text{CMB}},*} \tag{7}$$

and, as a consequence,

$$\Omega_{\Lambda\text{CDM},m,0} \approx 4\,\Omega_{\text{SU}(2)_{\text{CMB}},m,0}\,, \tag{8}$$

where $\Omega_{m,0}$ denotes today's density parameter for nonrelativistic matter (Dark Matter plus baryons), and z_* is the redshift of CMB photon decoupling in either model. Since the matter sector of the $SU(2)_{CMB}$ model, as roughly represented by Equation (8), contradicts ΛCDM at low z one needs to allow for a transition between the two somewhere in the dark ages as z decreases. In [34] a simple model, where the transition is sudden at a redshift z_p and maintains a small dark-energy residual, was introduced: a coherent axion field—a dark-energy like condensate of ultralight axion particles, whose masses derive from $U(1)_A$ anomalies [48–52] invoked by the topological charges of (anti)caloron centers in the thermal ground state of $SU(2)_{CMB}$ and, in a screened way, SU(2)/SU(3) Yang–Mills factors of higher scales—releases solitonic lumps by de-percolation due to the Universe's expansion. Accordingly, we have

$$\Omega_{\text{ds}}(z) = \Omega_\Lambda + \Omega_{\text{pdm},0}(z+1)^3 + \Omega_{\text{edm},0}\begin{cases}(z+1)^3\,, & z < z_p\\ (z_p+1)^3\,, & z \geq z_p\end{cases}\,. \tag{9}$$

Here Ω_Λ and $\Omega_{\text{pdm},0} + \Omega_{\text{edm},0} \equiv \Omega_{\text{cdm},0}$ represent today's density parameters for Dark Energy and Dark Matter, respectively, $\Omega_{\text{pdm},0}$ refers to primordial Dark Matter (that is, dark matter that existed *before* the initial redshift of $z_i \sim 10^4$ used in the CMB Boltzmann code) for all z and $\Omega_{\text{edm},0}$ to emergent Dark Matter for $z < z_p$. In [34] the initial conditions for the evolution of density and velocity perturbations of the emergent Dark-Matter portion at z_p are, up to a rescaling of order unity, assumed to follow those of the primordial Dark Matter.

3. $SU(2)_{CMB}$ Fit of Cosmological Parameters to Planck Data

In [34] a simulation of the new cosmological model subject to $SU(2)_{CMB}$ and the Dark Sector of Equation (9) was performed using a modified version of the Cosmic Linear Anisotropy Solving System (CLASS) Boltzmann code [53]. Best fits to the 2015 Planck data [8] on the angular power spectra of the two-point correlation functions temperature–temperature (TT), electric-mode polarisation–electric-mode polarisation (EE), and temperature–electric-mode polarisation (TE), subject to typical likelihood functions used by the Planck collaboration, were performed. Because temperature perturbations can only be coherently propagated by (low-frequency) massless modes in $SU(2)_{CMB}$ [40] the propagation of the (massive) vector modes was excluded in one version of the code (physically favoured). Furthermore, entropy conservation in e^+e^- annihilation invokes a slightly different counting of relativistic degrees of freedom in $SU(2)_{CMB}$ for this epoch [54]. As a consequence, we have a z-dependent density parameter for (massless) neutrinos given as [34]

$$\Omega_\nu(z) = \frac{7}{8} N_{\text{eff}} \left(\frac{16}{23}\right)^{\frac{4}{3}} \Omega_{\text{SU}(2)_{\text{CMB}},\gamma}(z) \,, \tag{10}$$

where N_{eff} refers to the effective number of neutrino flavours (or any other extra relativistic, free-streaming, fermionic species), and $\Omega_{\text{SU}(2)_{\text{CMB}},\gamma}$ is the density parameter associated with the massless mode only in the expression for $\rho_{\text{SU}(2)_{\text{CMB}}}(z)$ of Equation (6). The value $N_{\text{eff}} = 3.046$ of the Planck result was used as a fixed input in [34]. The statistical goodness of the best fit was found to be comparable to the one obtained by the Planck collaboration [8], see Figure 2 as well as the lower part of Table 1. There is an excess of power in TT for $7 \leq l \leq 30$, however.

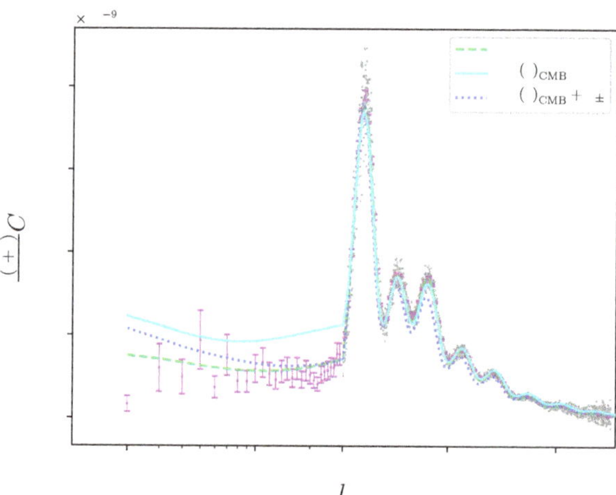

Figure 2. Normalised power spectra of TT correlator for best-fit parameter values quoted in Table 1: Dashed, dotted, and solid lines represent ΛCDM, $SU(2)_{CMB}$ + $V\pm$ (not considered in in the present work), and $SU(2)_{CMB}$, respectively. For $l \leq 29$ the 2015 Planck data points are unbinned and carry error bars, for $l \geq 30$ grey points represent unbinned spectral power. Figure adopted from [34].

This excess could be attributable to the omission of radiative effects in the low-z propagation of the massless mode. The consideration of the modified dispersion law into the CMB Boltzmann code presently is under way. The following table was obtained in [34]:

Table 1. Best-fit cosmological parameters of the $SU(2)_{CMB}$ and the ΛCDM model as obtained in [34]. The best-fit parameters of ΛCDM together with their 68% confidence intervals are taken from [8], employing the TT,TE,EE+lowP+lensing likelihoods. For $SU(2)_{CMB}$ the HiLLiPOP+lowTEB+lensing likelihood (lowP and lowTEB are pixel-based likelihoods) was used, see [55]. The central values associate with $\chi^2 = \chi^2_{ll} + \chi^2_{hl}$ (best fit) as quoted in the lower part of the Table.

Parameter	$SU(2)_{CMB}$	ΛCDM
$\omega_{b,0}$	0.0173 ± 0.0002	0.0225 ± 0.00016
$\omega_{pdm,0}$	0.113 ± 0.002	–
$\omega_{edm,0}$	0.0771 ± 0.0012	–
$100\,\theta_*$	1.0418 ± 0.0022	1.0408 ± 0.00032
τ_{re}	0.02632 ± 0.00218	0.079 ± 0.017
$\ln(10^{10} A_s)$	2.858 ± 0.009	3.094 ± 0.034
n_s	0.7261 ± 0.0058	0.9645 ± 0.0049
z_p	52.88 ± 4.06	–
β	0.811 ± 0.058	–
$H_0/_{\mathrm{km\,s^{-1}Mpc^{-1}}}$	74.24 ± 1.46	67.27 ± 0.66
z_{re}	$6.23^{+0.41}_{-0.42}$	$10^{+1.7}_{-1.5}$
z_*	1715.19 ± 0.19	1090.06 ± 0.30
z_d	1640.87 ± 0.27	1059.65 ± 0.31
$\omega_{cdm,0}$	0.1901 ± 0.0023	0.1198 ± 0.0015
Ω_Λ	0.616 ± 0.006	0.6844 ± 0.0091
$\Omega_{m,0}$	0.384 ± 0.006	0.3156 ± 0.0091
σ_8	0.709 ± 0.020	0.8150 ± 0.0087
Age/Gyr	11.91 ± 0.10	13.799 ± 0.021
χ^2_{ll}	10,640	10,495
$n_{dof,ll}$	9207	9210
$\frac{\chi^2_{ll}}{n_{dof,ll}}$	1.156	1.140
χ^2_{hl}	10,552.6	9951.47
$n_{dof,hl}$	9547	9550
$\frac{\chi^2_{hl}}{n_{dof,hl}}$	1.105	1.042

We would like to discuss the following parameters (the value of z_p will be discussed in Section 4): (i) H_0, (ii) n_s, (iii) σ_8, (iv) z_{re}, and (v) $\omega_{b,0}$. (i): The fitted value of H_0 is in agreement with local-cosmology observation [19,27] but discrepant at the $4.5\,\sigma$ level with the value $H_0 \sim (67.4 \pm 0.5)$ km Mpc^{-1} s^{-1} in ΛCDM of global-cosmology fits as extracted by the Planck collaboration [8], a galaxy-custering survey—the Baryonic Oscillation Spectroscopy Survey (BOSS) [3]—and a galaxy-custering-weak-lensing survey—the Dark Energy Survey (DES) [2]—whose distances are derived from inverse distance ladders using the sound horizons at CMB decoupling or baryon drag as references. Such anchors assume the validity of ΛCDM (or a variant thereof with variable Dark Energy) at high z, and the data analysis employs fiducial cosmologies that are close to those of CMB fits. As a rule of thumb, the inclusion of extra relativistic degrees of freedom within reasonable bounds [2,3] but also sample variance, local matter-density fluctuations, or a directional bias in SNa Ia observations [12–15] cannot explain the above-quoted tension in H_0. Note that the DES Y1 data on three two-point functions (cosmic shear auto correlation, galaxy angular auto correlation, and galaxy-shear cross correlation in up to five redshift bins with $z \leq 1.3$), which are most sensitive to Ω_m and σ_8 (or $S_8 = \sigma_8/\Omega_m$), cannot usefully constrain H_0 by itself. However, the combination of DES 1Y data with those of Planck (no lensing) yields an increases in H_0 on the 1-2 σ level compared to the Planck (no lensing) data alone, and a similar tendency is seen when BAO data [3] are enriched with those of DES Y1, see Table II in [2]. Loosely speaking, one thus may state a mild increase of H_0 with an increasing portion of late-time (local-cosmology) information in the data. (ii): The index of the initial spectrum of adiabatic, scalar curvature perturbations n_s is unusually low, expressing an enhancement of infrared modes as compared to the ultraviolet ones (violation of scale invariance). As discussed in [34], such a tilted spectrum is not consistent with single-field slow-roll inflation and implies that the Hubble parameter during the inflationary epoch has changed appreciably. (iii): There is a low value of σ_8 (initial amplitude of matter-density power spectrum at comoving wavelength $8\,\mathrm{h}^{-1}\,\mathrm{Mpc}$ with $H_0 \equiv h\,100\,\mathrm{km\,s}^{-1}\mathrm{Mpc}^{-1}$) compared to CMB fits [8] and BAO [3] although the DES Y1 fit to cosmic shear data alone would allow for a value $\sigma_8 \sim 0.71$ within the 1-σ margin [2]. This goes also for our high value of today's matter density parameter $\Omega_{\mathrm{m},0}$ (the ratio of physical matter density to the critical density). (iv): The low value of the redshift to (instantaneous) re-ionisation z_{re} (and the correspondingly low optical depth τ_{re}) compared to values obtained in CMB fits [8] is consistent with the one extracted from the observation of the Gunn-Peterson trough in the spectra of high-redshift quasars [35]. (v): The low value of $\omega_{\mathrm{b},0}$—today's physical baryon density—could provide a theoretical solution of the missing baryon problem [34] (missing compared to CMB fits to ΛCDM (CMB fits) and the primordial D/H (Deuterium-to-Hydrogen) ratio in Big-Bang Nucleosynthesis (BBN), the latter yielding $\omega_{\mathrm{b},0} = 0.02156 \pm 0.00020$ [56] which is by $2.3\,\sigma$ lower than the Planck value of 0.0225 ± 0.00016 in Table 1). If this low value of $\omega_{\mathrm{b},0}$ could be consolidated observationally than either the idea that BBN is determined by isolated nuclear reaction cross sections only or the observation of a truly primordial D/H in metal-poor environments or both will have to be questioned in future ivestigations. There is a recent claim, however, that the missing 30–40% of baryons were observed in highly ionized oxygen absorbers representing the warm-hot intergalactic medium when illuminated by the X-ray spectrum of a quasar with $z > 0.4$ [57]. This results is disputed in [58]. In [59] the missing baryon problem is addressed by the measurement of electron column density within the intergalactic medium using the dispersion of localised radio bursts with $z \leq 0.522$. In projecting electron density the measurement appeals to a flat ΛCDM CMB-fitted model (Planck collaboration [8]). Presently, $\omega_{\mathrm{b},0}$ is extracted from the ASKAP data in [59] to be consistent with CMB fits and BBN albeit subject to a 50% error which is expected to decrease with the advent of more powerful radio observatories such as SKA.

4. Axionic Dark Sector and Galactic Dark-Matter Densities

The model for the Dark Sector in Equation (9) is motivated by the possibility that a coherent condensate of ultra-light particles—an axion field [52]—forms selfgravitating lumps or vortices in the course of nonthermal (Hagedorn) phase transitions due to SU(2)/SU(3) Yang–Mills factors, governing the radiation and matter content of the very early Universe, going confining. A portion of the thus created, large abundance of such solitons percolates into a dark-energy like state: the contributions Ω_Λ and $\Omega_{\mathrm{edm},0}(z_p+1)^3$ in Equation (9) of which the latter may de-percolate at $z = z_p$ into a dark-matter like component $\Omega_{\mathrm{edm},0}(z+1)^3$ for $z < z_p$, the expansion of the Universe increasing the average distance between the centers of neighbouring solitons. By virtue of the $\mathrm{U}(1)_A$ anomaly, mediated by topological charge density [48–51], residing in turn within the ground state of an SU(2)/SU(3) Yang–Mills theory [41], the mass m_a of an axion particle due to a single such theory of Yang–Mills scale Λ is given as [52]

$$m_a = \frac{\Lambda^2}{M_P}, \tag{11}$$

where we have assumed that the so-called Peccei–Quinn scale, which associates with a dynamical chiral-symmetry breaking, was set equal to the Planck mass $M_P = 1.221 \times 10^{28}\,\mathrm{eV}$, see [60,61] for motivations. Let (natural units: $c = \hbar = 1$)

$$r_c \equiv 1/m_a \tag{12}$$

denote the Compton wavelength,

$$r_B \equiv \frac{M_P^2}{M} m_a^{-2} \tag{13}$$

the gravitational Bohr radius, where $M \sim 10^{12}\,M_\odot$ is the total dark mass of a typical (spiral) galaxy like the Milky Way, and

$$d_a \equiv \left(\frac{m_a}{\rho_{\rm dm}}\right)^{1/3} \tag{14}$$

the interparticle distance where $\rho_{\rm dm}$ indicates the typical mean energy density in Dark Matter of a spiral. Following [62], we assume

$$\rho_{\rm dm} = (0.2 \cdots 0.4)\,\text{GeV}\,\text{cm}^{-3}\,. \tag{15}$$

For the concept of a gravitational Bohr radius in a selfconsistent, non-relativistic potential model to apply, the axion-particle velocity v_a in the condensate should be much small than unity. Appealing to the virial theorem at a distance to the gravitational center of r_B, one has $v_a \sim \left(\frac{Mm_a}{M_P^2}\right)^2$ [63]. In [63], where a selfgravitating axion condensate was treated non-relativistically by means of a non-linear and non-local Schrödinger equation to represent a typical galactic dark-matter halo, the following citeria on the validity of such an approach were put forward (natural units: $c = \hbar = 1$): (i) $v_a \ll 1$. (ii) $d_a \ll r_c$ is required for the description of axion particles in terms of a coherent Bose condensate to be realistic. (iii) r_B should be the typical extent of a galactic Dark-Matter halo: $r_B \sim 100$–$300\,$kpc. With $\Lambda_{\rm CMB} \sim 10^{-4}\,$eV one obtains $m_a = 8.2 \times 10^{-37}\,$eV, $d_a = (4.1 \cdots 5.2) \times 10^{-34}\,$pc, $r_c = 3.2 \times 10^{25}\,$pc, and $r_B \sim 8 \times 10^{24}\,$kpc. While (i) and (ii) are extremely well satified with $v_a \sim 10^{-28}$ and $\frac{d_a}{r_c} \sim 10^{-59}$ point (iii) is badly violated. (r_B is about 2×10^{18} the size of the visible Universe).

Therefore, the Yang–Mills scale responsible for the axion mass that associates with dark-matter halos of galaxies must be dramatically larger. Indeed, setting $\Lambda = 10^{-2}\Lambda_e$ where $\Lambda_e = \frac{m_e}{118.6}$ is the Yang–Mills scale of an SU(2) theory that could associate with the emergence of the electron of mass $m_e = 511\,$keV [64,65], one obtains $m_a = 1.5 \times 10^{-25}\,$eV, $d_a = (2.3 \cdots 2.9) \times 10^{-30}\,$pc, $r_c = 1.7 \times 10^{14}\,$pc, and $r_B \sim 232\,$kpc. In addition to (i) and (ii) with $v_a \sim 10^{-4}$ and $\frac{d_a}{r_c} \sim 10^{-44}$ also point (iii) is now well satisfied. If the explicit Yang–Mills scale of an SU(2) theory, which is directly imprinted in the spectra of the excitations in the pre - and deconfining phases, acts only in a screened way in the confining phase as far as the axial anomaly is concerned—reducing its value by a factor of one hundred or so—then the above axionic Dark-Sector scenario would link the theory responsible for the emergence of the electron with galactic dark-matter halos! In addition, the axions of $SU(2)_{\rm CMB}$ would provide the Dark-Energy density Ω_Λ of such a scenario.

Finally, we wish to point out that the de-percolation mechanism of axionic solitons (lumps forming out of former Dark-Energy density) in the cosmological model based on $SU(2)_{\rm CMB}$, which may be considered to underly the transition in the Dark Sector at $z_p = 53$ described by Equation (9), is consistent with the Dark-Matter density in the Milky Way. Namely, working with $H_0 = 74\,\text{km}\,\text{s}^{-1}\,\text{Mpc}^{-1}$, see Table 1 and [19], the total (critical) energy density $\rho_{c,0}$ of our spatially flat Universe is at present

$$\rho_{c,0} = \frac{3}{8\pi}\,M_P^2 H_0^2 = 1.75 \times 10^{-9}\,\text{eV}^4\,. \tag{16}$$

The portion of cosmological Dark Matter $\rho_{\rm cdm,0}$ then is, see Table 1,

$$\rho_{\rm cdm,0} \sim 0.35\,\rho_{c,0} \tag{17}$$

which yields a cosmological energy scale $E_{\rm cdm,0}$ in association with Dark Matter of

$$E_{\rm cdm,0} \equiv \rho_{\rm DM,0}^{1/4} \sim 0.00497\,\text{eV}\,. \tag{18}$$

On the other hand, we may imagine the percolate of axionic field profiles, which dissolves at z_p, to be associated with densely packed dark-matter halos typical of today's galaxies. Namely, scaling the typical Dark-Matter energy density of the Milky Way $\rho_{\rm dm}$ of Equation (15) from z_p (at $z_p + 0$ $\rho_{\rm dm}$ yet behaves like a cosmological constant) down to $z = 0$ and allowing for a factor of $(\omega_{\rm pdm,0} + \omega_{\rm edm,0})/\omega_{\rm edm,0} = 2.47$, see Table 1, one extracts the energy scale for cosmological Dark Matter $E_{\rm G,0}$ in association with galactic Dark-Matter halos and primordial dark matter as

$$E_{\rm G,0} \equiv \left(\frac{2.47\rho_{\rm dm}}{(z_p+1)^3}\right)^{1/4} = (0.00879\cdots 0.0105)\,{\rm eV}\,. \tag{19}$$

A comparison of Equations (18) and (19) reveals that $E_{\rm cdm,0}$ is smaller but comparable to $E_{\rm G,0}$. This could be due to the neglect of galactic-halo compactification through the missing pull by neighbouring profiles in the axionic percolate and because of the omission of selfgravitation and baryonic-matter accretion/gravitation during the evolution from $z_p = 53$ to present (that is, the use of the values of Equation (15) in Equation (19) overestimates the homogeneous energy density in the percolate). The de-percolation of axionic solitons at z_p, whose mean, selfgravitating energy density $\rho_{\rm dm}$ in Dark Matter is nearly independent of cosmological expansion but subject to local gravitation, could therefore be linked to cosmological Dark Matter today within the Dark-Sector model of Equation (9).

5. Conclusions

The present article's goal was to address some tensions between local and global cosmology on the basis of the ΛCDM standard model. Cracks in this model could be identified during the last few years thanks to independent tests resting on precise observational data and their sophisticated analysis. To reconcile these results, a change of ΛCDM likely is required before the onset of the formation of non-linear, large-scale structure. Here, we have reviewed a proposal made in [34], which assumes thermal photons to be governed by an SU(2) rather than a U(1) gauge principle, and we have discussed the $SU(2)_{\rm CMB}$-implied changes in cosmological parameters and the structure of the Dark Sector. Noticeably, the tensions in H_0, the baryonic density, and the redshift for re-ionisation are addressed in favour of local measurements. High-z inputs to CMB and BAO simulations, such as n_s and σ_8, are sizeably reduced as compared to their fitted values in ΛCDM. The Dark Sector now invokes a de-percolation of axionic field profiles at a redshift of $z_p \sim 53$. This idea is roughly consistent with typical galactic Dark-Matter halos today, such as the one of the Milky Way, being released from the percolate. Axionic field profiles, in turn, appear to be compatible with Dark-Matter halos in typical galaxies if (confining) Yang–Mills dynamics subject to a much higher mass scales than that of $SU(2)_{\rm CMB}$ is considered to produce the axion mass.

To consolidate such a scenario two immediate fields of investigation suggest themselves: (i) A deep understanding of possible selfgravitating profiles needs to be gained towards their role in actively assisted large-scale structure formation as well as in quasar emergence, strong lensing, cosmic shear, galaxy clustering, and galaxy phenomenology (Tully–Fisher, rotation curves, etc.), distinguishing spirals from ellipticals and satellites from hosts. (ii) More directly, the CMB large-angle anomalies require an addressation in terms of radiative effects in $SU(2)_{\rm CMB}$, playing out at low redshifts, which includes a re-investigation of the CMB dipole.

Funding: This research received no external funding.

Conflicts of Interest: The author declares no conflict of interest.

References

1. Eisenstein, D.J.; Weinberg, D.H.; Agol, E.; Aihara, H.; Prieto, C.A.; Anderson, S.F.; Arns, J.A.; Aubourg, É.; Bailey, S.; Balbinot, E.; et al. SDSS-III Massive Spectroscopic Surveys of the Distant Universe, the Milky Way, and Extra-Solar Planetary Systems. *Astron. J.* **2011**, *142*, 24. [CrossRef]
2. Abbott, T.M.C.; Abdalla, F.B.; Alarcon, A.; Aleksić, J.; Allam, S.; Allen, S.; Amara, A.; Annis, J.; Asorey, J.; Avila, S.; et al. [Dark Energy Survey Collaboration] Dark Energy Survey year 1 results: Cosmological constraints from galaxy clustering and weak lensing. *Phys. Rev. D* **2018**, *98*, 043526. [CrossRef]

3. Alam, S.; Ata, M.; Bailey, S.; Beutler, F.; Bizyaev, D.; Blazek, J.A.; Bolton, A.S.; Brownstein, J.R.; Burden, A.; Chuang, C.-H.; et al. The clustering of galaxies in the completed SDSS-III Baryon Oscillation Spectroscopic Survey: Cosmological analysis of the DR12 galaxy sample. *Month. Not. R. Astron. Soc.* **2017**, *470*, 2617–2652. [CrossRef]
4. Ivezić, Ž.; Kahn, S.M.; Tyson, J.A.; Abel, B.; Acosta, E.; Allsman, R.; Alonso, D.; AlSayyad, Y.; Anderson, S.F.; Andrew, J.; et al. LSST: From Science Drivers to Reference Design and Anticipated Data Products. *Astrophys. J.* **2019**, *873*, 111. [CrossRef]
5. Vargas-Magaña, M.; Brooks, D.D.; Levi, M.M.; Tarle, G.G. Unravelling the Universe with DESI. *arXiv* **2019**, arXiv:1901.01581.
6. Mather, J.C.; Cheng, E.S.; Cottingham, D.A.; Eplee, R.E., Jr.; Fixsen, D.J.; Hewagama, T.; Isaacman, R.B.; Jensen, K.A.; Meyer, S.S.; Noerdlinger, P.D.; et al. Measurement of the cosmic microwave background spectrum by the COBE FIRAS instrument. *Astrophys. J.* **1994**, *420*, 439. [CrossRef]
7. Hinshaw, G.; Larson, D.; Komatsu, E.; Spergel, D.N.; Bennett, C.L.; Dunkley, J.; Nolta, M.R.; Halpern, M.; Hill, R.S.; Odegard, N.; et al. Nine-Year Wilkinson Microwave Anisotropy Probe (WMAP) Observations: Cosmological Parameter Results. *Astrophys. J. Suppl. Ser.* **2013**, *208*, 19. [CrossRef]
8. Aghanim, N.; Akrami, Y.; Ashdown, M.; Aumont, J.; Baccigalupi, C.; Ballardini, M.; Banday, A.J.; Barreiro, R.B.; Bartolo, N.; Basaket, S.; et al. Planck 2018 results. VI Cosmological parameters. *arXiv* **2018**, arXiv:1807.06209.
9. Riess, A.G.; Filippenko, A.V.; Challis, P.; Clocchiattia, A.; Diercks, A.; Garnavich, P.M.; Gilliland, R.L.; Hogan, C.J.; Jha, S.; Kirshner, R.P.; et al. Observational Evidence from Supernovae for an Accelerating Universe and a Cosmological Constant. *Astron. J.* **1998**, *116*, 1009. [CrossRef]
10. Perlmutter, S.; Aldering, G.; Goldhaber, G.; Knop, R.A.; Nugent, P.; Castro, P.G.; Deustua, S.; Fabbro, S.; Goobar, A.; Groom, D.E.; et al. Measurement of Ω and Λ from 42 high-redshift supernovae. *Astrophys. J.* **1998**, *517*, 565. [CrossRef]
11. Dhawan, S.; Brout, D.; Scolnik, D.; Goobar, A.; Riess, A.G.; Miranda, V. Cosmological model insensitivity of local H_0 from the Cepheid distance ladder. *arXiv* **2020**, arXiv:2001.09260.
12. Marra, V.; Amendola, L.; Sawicki, I.; Valkenburg, W. Cosmic Variance and the Measurement of the Local Hubble Parameter. *Phys. Rev. Lett.* **2013**, *110*, 241305. [CrossRef] [PubMed]
13. Oddersov, I.; Hannestad, S.; Haugbølle, T. On the local variation of the Hubble constant. *J. Cosmol. Astropart. Phys.* **2014**, *10*, 028. [CrossRef]
14. Oddersov, I.; Hannestad, S.; Brandbyge, J. The variance of the locally measured Hubble parameter explained with different estimators. *J. Cosmol. Astropart. Phys.* **2017**, *03*, 022. [CrossRef]
15. Wu, H.-Y.; Huterer, D. Sample variance in the local measurement of the Hubble constant. *Month. Not. R. Astron. Soc.* **2017**, *471*, 4946. [CrossRef]
16. Riess, A.G.; Macri, L.M.; Hoffmann, S.L.; Scolnic, D.; Casertano, S.; Filippenko, A.V.; Tucker, B.E.; Reid, M.J.; Jones, D.O.; Silverman, J.M.; et al. A 2.4% Determination of the Local Value of the Hubble Constant. *Astrophys. J.* **2016**, *826*, 56. [CrossRef]
17. De Jaeger, T.; Stahl, B.E.; Zheng, W.; Filippenko, A.V.; Riess, A.G.; Galbany, L. A measurment of the Hubble constant from Type II supernovae. *Month. Not. R. Astron. Soc.* **2020**, *496*, 3402. [CrossRef]
18. Riess, A.G.; Casertano, S.; Yuan, W.; Macri, L.; Anderson, J.; Mackenty, J.W.; Bowers, J.B.; Clubb, K.I.; Filippenko, A.V.; Jones, D.O.; et al. New Parallaxes of Galacic Cepheids from Spatially Scanning the Hubble Space Telescope: Implications for the Hubble Constant. *Astrophys. J.* **2018**, *855*, 136. [CrossRef]
19. Riess, A.G.; Casertano, S.; Yuan, W.; Macri, L.M.; Scolnic, D. Large Magellanic Cloud Cepheid Standards Provide a 1% Foundation for the Determination of the Hubble Constant and Stronger Evidence for Physics beyond ΛCDM. *Astrophys. J.* **2019**, *876*, 85. [CrossRef]
20. Jang, I.S.; Lee, M.G. The Tip of the Red Giant Branch Distances to Type Ia Supernova Host Galaxies. IV. Color Dependence and Zero-Point Calibration. *Astrophys. J.* **2017**, *835*, 28. [CrossRef]
21. Freedman, W.L.; Madore, B.F.; Hatt, D.; Hoyt, T.J.; Jang, I.-S.; Beaton, R.L.; Burns, C.R.; Lee, M.G.; Monson, A.J.; Neeley, J.R.; et al. The Carnegie-Chicago Hubble Program. VIII. An Independent Determination of the Hubble Constant Based on the Tip of the Red Giant Branch. *Astrophys. J.* **2019**, *882*, 34. [CrossRef]
22. Huang, C.D.; Riess, A.G.; Hoffmann, S.L.; Klein, C.; Bloom, J.; Yuan, W.; Macri, L.M.; Jones, D.O.; Whitelock, P.A.; Casertano, S.; et al. A Near-infrared Period-Luminosity Relation for Miras in NGC 4258, an Anchor for a New Distance Ladder. *Astrophys. J.* **2018**, *857*, 67. [CrossRef]

23. Ma, C.-P.; Bertschinger, E. Cosmological Perturbation Theory in the Synchronous and Conformal Newtonian Gauges. *Astrophys. J.* **1995**, *455*, 7. [CrossRef]
24. Raghunathan, S.; Patil, S.; Baxter, E.; Benson, B.A.; Bleem, L.E.; Crawford, T.M.; Holder, G.P.; McClintock, T.; Reichardt, C.L.; Varga, T.N.; et al. Detection of CMB-Cluster Lensing using Polarization Data from SPTpol. *Phys. Rev. Lett.* **2019**, *123*, 181301. [CrossRef] [PubMed]
25. Birrer, S.; Amara, A.; Refregier, A. The mass-sheet degeneracy and time-delay cosmography analysis of the strong lens RXJ1131-1231. *J. Cosmol. Astropart. Phys.* **2016**, *8*, 20. [CrossRef]
26. Bonvin, V.; Courbin, F.; Suyu, S.H.; Marshall, P.J.; Rusu, C.E.; Sluse, D.; Tewes, M.; Wong, K.C.; Collett, T.; Fassnacht, C.D.; et al. HoLiCOW—V. New COSOMGRAIL time delays of HE 0435-1223: H_0 to 3.8 per cent precision from strong lensing in a flat ΛCDM model. *Month. Not. R. Astron. Soc.* **2017**, *465*, 4914. [CrossRef]
27. Wong, K.C.; Suyu, S.H.; Chen, G.C.-F.; Rusu, C.E.; Millon, M.; Sluse, D.; Bonvin, V.; Fassnacht, C.D.; Taubenberger, S.; Auger, M.W.; et al. HoLiCOW-XIII. A 2.4% measurement of H_0 from lensed quasars: 5.3 σ tension between early and late-Universe probes. *Month. Not. R. Astron. Soc.* **2020**. [CrossRef]
28. Schutz, B.F. Determining he Hubble constant from gravitational wave observations. *Nature* **1986**, *323*, 310. [CrossRef]
29. Chen, H.-Y.; Fishbach, M.; Holz, D.E. A two percent Hubble constant measurement from standard sirens within five years. *Nature* **2018**, *562*, 545. [CrossRef]
30. Handley, W.; Lemos, P. Quantifying tensions in cosmological parameters: Interpreting the DES evidence ratio. *Phys. Rev. D* **2019**, *100*, 043504-1. [CrossRef]
31. Schwarz, D.J.; Copi, C.J.; Huterer, D.; Starkman, G.D. CMB anomalies after Planck. *Class. Quantum. Grav.* **2016**, *33*, 184001. [CrossRef]
32. Hofmann, R. Low-frequency line temperatures of the CMB (Cosmic Microwave Background). *Annalen Phys.* **2009**, *18*, 634. [CrossRef]
33. Fixsen, D.J.; Kogut, A.; Levin, S.; Limon, M.; Lubin, P.; Mirel, P.; Seiffert, M.; Singal, J.; Wollack, E.; Villela, T.; et al. ARCADE 2 measurement of the absolute sky brightness at 3–90 GHz. *Astrophys. J.* **2011**, *734*, 5. [CrossRef]
34. Hahn, S.; Hofmann, R.; Kramer, D. $SU(2)_{CMB}$ and the cosmological model: Angular power spectra. *Mon. Not. R. Astron. Soc.* **2019**, *482*, 4290. [CrossRef]
35. Becker, R.H.; Fan, X.; White, R.L.; Strauss, M.A.; Narayanan, V.K.; Lupton, R.H.; Gunn, J.E.; Annis, J.; Bahcall, N.A.; Brinkmann, J.; et al. Evidence for Reionization at $z \sim 6$: Detection of a Gunn-Peterson Trough in a $z = 6.28$ Quasar. *Astrophys. J.* **2001**, *122*, 2850.
36. Ludescher, J.; Hofmann, R. Thermal photon dispersion law and modified black-body spectra. *Annalen Phys.* **2009**, *18*, 271. [CrossRef]
37. Hofmann, R. The fate of statistical isotropy. *Nat. Phys.* **2013**, *9*, 686. [CrossRef]
38. Harrington, B.J.; Shepard, H.K. Periodic Euclidean solutions and the finite-temperature Yang–Mills gas. *Phys. Rev. D* **1978**, *17*, 2122. [CrossRef]
39. Herbst, U.; Hofmann, R. Emergent Inert Adjoint Scalar Field in SU(2) Yang–Mills Thermodynamics due to Coarse-Grained Topological Fluctuations. *ISRN High Energy Phys.* **2012**, *2012*, 373121. [CrossRef]
40. Grandou, T.; Hofmann, R. Thermal ground state and nonthermal probes. *Adv. Math. Phys.* **2015**, *2015*, 197197. [CrossRef]
41. Hofmann, R. *The Thermodynamics of Quantum Yang–Mills Theory: Theory and Application*, 2nd ed.; World Scientific: Singapore, 2016.
42. Schwarz, M.; Hofmann, R.; Giacosa, F. Radiative corrections to the pressure and the one-loop polarization tensor of massless modes in SU(2) Yang–Mills thermodynamics. *Int. J. Mod. Phys. A* **2007**, *22*, 1213. [CrossRef]
43. Falquez, C.; Hofmann, R.; Baumbach, T. Modification of black-body radiance at low temperatures and frequencies. *Ann. Phys.* **2010**, *522*, 904. [CrossRef]
44. Diakonov, D.; Gromov, N.; Petrov, V.; Slizovskiy, S. Quantum weights of dyons and of instantons with nontrivial holonomy. *Phys. Rev. D* **2004**, *70*, 036003. [CrossRef]
45. Szopa, M.; Hofmann, R. A Model for CMB anisotropies on large angular scales. *J. Cosmol. Astropart. Phys.* **2008**, *3*, 001. [CrossRef]
46. Ludescher, J.; Hofmann, R. CMB dipole revisited. *arXiv* **2009** arXiv:0902.3898.
47. Hahn, S.; Hofmann, R. Exact determination of asymptotic CMB temperature-redshift relation. *Mod. Phys. Lett. A* **2018**, *33*, 1850029. [CrossRef]

48. Adler, S.L. Axial-Vector Vertex in Spinor Electrodynamics. *Phys. Rev.* **1969**, *177*, 2426. [CrossRef]
49. Adler, S.L.; Bardeen, W.A. Absence of Higer-Order Corrections in the Anomalous Axial-Vector Divergence Equation. *Phys. Rev.* **1969**, *182*, 1517. [CrossRef]
50. Bell, J.S.; Jackiw, R. A PCAC Puzzle: $\pi^0 \to \gamma\gamma$ in the σ model. *Nuovo Cim. A* **1969**, *60*, 47. [CrossRef]
51. Fujikawa, K. Path-Integral Measure for Gauge-Invariant Fermion Theories. *Phys. Rev. Lett.* **1979**, *42*, 1195. [CrossRef]
52. Peccei, R.D.; Quinn, H.R. Constraints imposed by CP conservation in the presence of pseudoparticles. *Phys. Rev. D* **1977**, *16*, 1791. [CrossRef]
53. Blas, D.; Lesgourges, J.; Tram, T. The Cosmic Linear Anisotropy Solving System (CLASS). Part II: Approximation schemes. *J. Cosmol. Astropart. Phys.* **2011**, *07*, 034. [CrossRef]
54. Hofmann, R. Relic photon temperature versus redshift and the cosmic neutrino background. *Annalen Phys.* **2015**, *527*, 254. [CrossRef]
55. Aghanim, N.; Arnaud, M.; Ashdown, M.; Aumont, J.; Baccigalupi, C.; Banday, A.J.; Barreiro, R.B.; Bartlett, J.G.; Bartolo, N.; Battaner, E.; et al. Planck 2015 results. XI. CMB power spectra, likelyhoods, and robustness of parameters. *Astron. Astrophys.* **2016**, *594*, A11.
56. Cooke, R.J.; Pettini, M.; Nollett, K.M.; Jorgenson, R. The primordial deuterium abundance of the most metal-poor damped Lyα system. *Astrophys. J.* **2016**, *830*, 148. [CrossRef]
57. Nicastro F.; Kaastra, J.; Krongold, Y.; Borgani, S.; Branchini, E.; Cen, R.; Dadina, M.; Danforth, C.W.; Elvis, M.; Fiore, F.; et al. Observations of the missing baryons in the warm hot intergalactic medium. *Nature* **2018**, *558*, 406.
58. Johnson, S.D.; Mulchaey, J.S.; Chen, H.-W.; Wijers, N.A.; Connor, T.; Muzahid, S.; Schaye, J.; Cen, R.; Carlsten, S.G.; Charlton, J.; et al. The Physical Origins of the Identified and Still Missing Components of the Warm–Hot Intergalactic Medium: Insights from Deep Surveys in the Field of Blazar 1ES1553+113. *Astrophys. J. Lett.* **2019**, *884*, L31. [CrossRef]
59. Macquart, J.-P.; Prochaska, J.X.; McQuinn, M.; Bannister, K.W.; Bhandari, S.; Day, C.K.; Deller, A.T.; Ekers, R.D.; James, C.W.; Marnoch, L.; et al. A census of baryons in the Universe from localized fast radio bursts. *Nature* **2020**, *581*, 391. [CrossRef]
60. Frieman, J.A.; Hill, C.T.; Stebbins, A.; Waga, I. Cosmology with Ultralight Pseudo Nambu-Goldstone Bosons. *Phys. Rev. Lett.* **1995**, *75*, 2077. [CrossRef]
61. Giacosa, F.; Hofmann, R. A Planck-scale axion and SU(2) Yang–Mills dynamics: Present acceleration and the fate of the photon. *Eur. Phys. J. C* **2007**, *50*, 635. [CrossRef]
62. Weber, M.; de Boer, W. Determination of the local dark matter density in our Galaxy. *Astron. Astrophys.* **2010**, *509*, A25. [CrossRef]
63. Sin, S.-J. Late-time phase transition and the galactic halo as a Bose liquid. *Phys. Rev. D* **1994**, *50*, 3650. [CrossRef]
64. Hofmann, R. The isolated electron: De Broglie's "hidden" thermodynamics, SU(2) Quantum Yang–Mills theory, and a strongly perturbed BPS monopole. *Entropy* **2017**, *19*, 575. [CrossRef]
65. Grandou, T.; Hofmann, R. On emergent particles and stable neutral plasma balls in SU(2) Yang–Mills thermodynamics. *arXiv* **2020**, arXiv:2007.08460.

© 2020 by the authors. Licensee MDPI, Basel, Switzerland. This article is an open access article distributed under the terms and conditions of the Creative Commons Attribution (CC BY) license (http://creativecommons.org/licenses/by/4.0/).

Review

The Equation of State of Nuclear Matter: From Finite Nuclei to Neutron Stars

G. Fiorella Burgio *,† and Isaac Vidaña †

INFN Sezione di Catania, Via S. Sofia 64, I-95123 Catania, Italy
* Correspondence: fiorella.burgio@ct.infn.it
† These authors contributed equally to this work.

Received: 30 June 2020; Accepted: 5 August 2020; Published: 10 August 2020

Abstract: *Background.* We investigate possible correlations between neutron star observables and properties of atomic nuclei. In particular, we explore how the tidal deformability of a 1.4 solar mass neutron star, $M_{1.4}$, and the neutron-skin thickness of ^{48}Ca and ^{208}Pb are related to the stellar radius and the stiffness of the symmetry energy. *Methods.* We examine a large set of nuclear equations of state based on phenomenological models (Skyrme, NLWM, DDM) and *ab initio* theoretical methods (BBG, Dirac–Brueckner, Variational, Quantum Monte Carlo). *Results:* We find strong correlations between tidal deformability and NS radius, whereas a weaker correlation does exist with the stiffness of the symmetry energy. Regarding the neutron-skin thickness, weak correlations appear both with the stiffness of the symmetry energy, and the radius of a $M_{1.4}$. Our results show that whereas the considered EoS are compatible with the largest masses observed up to now, only five microscopic models and four Skyrme forces are simultaneously compatible with the present constraints on L and the PREX experimental data on the ^{208}Pb neutron-skin thickness. We find that all the NLWM and DDM models and the majority of the Skyrme forces are excluded by these two experimental constraints, and that the analysis of the data collected by the NICER mission excludes most of the NLWM considered. *Conclusion.* The tidal deformability of a $M_{1.4}$ and the neutron-skin thickness of atomic nuclei show some degree of correlation with nuclear and astrophysical observables, which however depends on the ensemble of adopted EoS.

Keywords: neutron star; equation of state; many-body methods of nuclear matter; neutron-skin thickness; GW170817

---

1. Introduction

The equation of state (EoS) of isospin asymmetric nuclear matter plays a major role in many different realms of modern physics, being the fundamental ingredient for the description of heavy-ion collision dynamics, nuclear structure, static and dynamical properties of neutron stars (NS), core-collapse supernova and binary compact-star mergers [1,2]. In principle, it can be expected that in heavy-ion collisions at large enough energy nuclear matter is compressed at density a few times larger than the nuclear saturation density, and that at the same time, the two collision partners produce flows of matter, which should be connected with the nuclear EoS. In the physics of compact objects, the central density likely reached in the inner core of a NS may reach values up to one order of magnitude larger than the saturation density, and this poses several theoretical problems because a complete theory of nuclear interactions at arbitrarily large values of density, temperature and asymmetry, should in principle be derived from the quantum chromodynamics (QCD), and this is a very difficult task which presently cannot be realized. Therefore, theoretical models and methods of the nuclear many-body theory are required to build the EoS, which has to be applied and tested in terrestrial laboratories for the description of ordinary nuclear structure, and in astrophysical observations for the study of compact objects.

Among possible observables regarding NS, the mass and radius are the most promising since they encode unique information on the EoS at supranuclear densities. Currently the masses of several NSs are known with good precision [3–7], but the information on their radii is less accurate [8,9]. The recent observations of NICER [10,11] have reached a larger accuracy for the radius, but future planned missions like eXTP [12] will allow us to statistically infer NS-mass and radius to within a few percent.

A big step forward is represented by the recent detection by the Advanced LIGO and VIRGO collaborations of gravitational waves emitted during the GW170817 NS merger event [13–15]. This has provided important new insights on the mass and radii of these objects by means of the measurement of the tidal deformability [16,17], and allowed to deduce upper and lower limits on it [14,18].

In this paper, we analyze the constraints on the nuclear EoS obtained from the analysis of the NS merger event GW170817, and try to select the most compatible EoS chosen among those derived from both phenomenological and ab initio theoretical models. We also examine possible correlations among properties of nuclear matter close to saturation with the observational quantities deduced from GW170817 and nuclear physics experiments. In particular, we concentrate on the tidal deformability of NS, and the neutron-skin thickness in finite nuclei, thus connecting astrophysical observables with laboratory nuclear physics. The present work is complementary to the recent analysis of Horowitz made in Ref. [19].

The paper is organized as follows. In Section 2 we give a schematic overview of NS phenomenology, whereas in Section 3 we explain the role of the equation of state in determining the main properties of NS, and illustrate the ones we adopt in the present study. The experimental constraints on the nuclear EoS are presented in Section 3.1 whereas the astrophysical ones are discussed in Section 3.2. A brief overview of different EoS of β-stable matter is given in Section 4, along with numerical results. In Section 5 we briefly discuss the NS tidal deformability, and its connection to the neutron-skin thickness in Section 6. Conclusions are drawn in Section 7.

2. Neutron Stars in a Nutshell

Neutron stars are a type of stellar compact remnant that can result from the gravitational collapse of an ordinary star with a mass in the range 8–25$M_\odot$ (with $M_\odot \approx 2 \times 10^{33}$g the mass of the Sun) during a Type II, Ib or Ic supernova event. A supernova explosion will occur when the star has exhausted its possibilities for energy production by nuclear fusion. Then, the pressure gradient provided by the radiation is not sufficient to balance the gravitational attraction, becoming the star unstable and, eventually, collapsing. The inner regions of the star collapse first and the gravitational energy is released and transferred to the outer layers of the star blowing them away.

NS are supported against gravitational collapse mainly by the neutron degeneracy pressure and may have masses in the range M $\sim$ 1–2$M_\odot$ and radii of about 10–12 km. A schematic cross section of the predicted "onion"-like structure of the NS interior is shown in Figure 1. At the *surface*, densities are typically $\rho < 10^6$ g/cm^3. The *outer crust*, with densities ranging from 10^6 g/cm^3 to 10^{11} g/cm^3 is a solid region where heavy nuclei, mainly around the iron mass number, in a Coulomb lattice coexist in β-equilibrium (i.e., in equilibrium with respect to weak interaction processes) with an electron gas. Moving towards the center the density increases and the electron chemical potentials increases, and the electron capture processes on nuclei

$$e^- +^A Z \rightarrow^A (Z-1) + \nu_e \,, \tag{1}$$

opens and the nuclei become more and more neutron-rich. At densities of $\sim$4.3 $\times$ 10^{11} g/cm^3 the only available levels for the neutrons are in the continuum and they start to "drip out" of the nuclei. We have then reached the *inner crust* region, where matter consist of a Coulomb lattice of very neutron-rich nuclei together with a superfluid neutron gas and an electron gas. In addition, due to the competition between the nuclear and Coulomb forces, nuclei in this region lose their spherical shapes and present more exotic topologies (droplets, rods, cross-rods, slabs, tubes, bubbles) giving rise to what has been called "nuclear pasta" phase [20]. At densities of $\sim 10^{14}$ g/cm^3 nuclei start to dissolve, and one enters the *outer core*. In this region matter is mainly composed of superfluid neutrons with a smaller concentration of superconducting protons and normal electrons and muons. In the deepest region of the star, the *inner core*, the density can reach values of $\sim 10^{15}$ g/cm^3. The composition of this region, however, is not known, and it is still subject of speculation. Suggestions range from hyperonic matter, meson condensates, or deconfined quark matter.

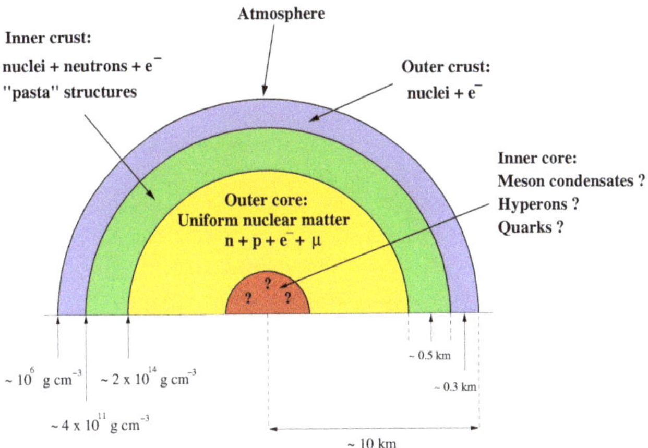

Figure 1. A schematic cross section of a neutron star illustrating the various regions discussed in the text. The different regions shown are not drawn on scale.

The observation of NS requires different types of ground-based and on-board telescopes covering all bands of the electromagnetic spectrum. Radio observations are carried out with ground-based antennas located in different places of the Earth. Three examples of these radio telescopes are the *Arecibo radio telescope* in Puerto Rico, the *Green Bank Observatory* in West Virginia, and the *Nançay decimetric radio telescope* in France. Observations in the near infrared and the optical bands can be performed with the use of large ground-based telescopes such as the *Very Large Telescope* (VLT) in the Atacama Desert in Chile. The *Hubble-Space Telescope* (HST) can be used to cover the optical and ultraviolet regions. Observations in the extreme ultraviolet, X-ray and γ-ray require the use of space observatories such as the *Chandra X-ray Observatory* (CXO), the *X-ray Multi Mirror* (XMM-Newton) and the *Rossi X-ray Timing Explorer* (RXTE) in the case of X-ray observations; and the *High Energy Transient Explorer* (HETE-2), the *International Gamma-Ray Astrophysics Laboratory* (INTEGRAL) and the *Fermi Gamma ray Space Telescope* (FGST), in the case of γ-ray ones.

Information on the properties of NS additional to that obtained from the observation of their electromagnetic radiation can be provided from the detection of the neutrinos emitted during the supernova explosion that signals the birth of the star. Examples of neutrino observatories are: the under-ice *IceCube* observatory placed in the South Pole; the under-water projects ANTARES (*Astronomy with a Neutrino Telescope and Abyss environmental REsearch*) and the future KM3NET (*Cubic Kilometre Neutrino Telescope*) in the Mediterranean sea; and the underground observatories SNO (*Sudbury Neutrino Observatory*) located 2100 m underground in the Vale's Creighton Mine in Canada, and the *Kamioka* observatory placed at the Mozumi Mine near the city of Hida in Japan.

The detection of gravitational waves, originated during the coalescence of two NS as in the GW170817 event recently detected by the Advanced LIGO and Advanced VIRGO collaborations [13–15] or from the oscillation modes of NS, represents presently a new way of observing these objects and constitutes a very valuable new source of information. In particular, observations of NS mergers can potentially provide stringent constraints on the nuclear EoS by comparing model predictions with the precise shape of the detected gravitational wave signal. The interested reader is referred to Ref. [21] for a recent review of this hot and exciting topic.

3. The Nuclear Equation of State

The theoretical description of nuclear matter under extreme density conditions is a very challenging task. Theoretical predictions in this regime are diverse, ranging from purely nucleonic matter with high neutron-proton asymmetry, to baryonic strange matter or a quark deconfined phase of matter. In this work we adopt a conventional description by assuming that the most relevant degrees of freedom are nucleons. Theoretical approaches to determine the nuclear EoS can be classified in two categories: phenomenological and microscopic (*ab initio*).

Phenomenological approaches, either non-relativistic or relativistic, are based on effective interactions that are frequently built to reproduce the properties of nuclei [22]. Skyrme interactions [23,24] and relativistic mean-field (RMF) models [25] are among the most used ones. Many of such interactions are built to describe finite nuclei in their ground state, i.e., close to the isospin symmetric case and, therefore, predictions at high isospin asymmetries should be taken with care. For instance, most Skyrme forces are, by construction, well behaved close to nuclear saturation density $\rho_0 \approx 0.15$–$0.16\ \mathrm{fm}^{-3}$ and moderate values of the isospin asymmetry, but predict very different EoS for pure neutron matter, and therefore give different predictions for NS observables. In this work we use the 27 Skyrme forces that passed the restrictive tests imposed by Stone et al. in Ref. [22] over almost 90 existing Skyrme parametrizations. These forces are: GS and Rs [26], SGI [27], SLy0-SLy10 [28] and SLy230a [29,30] of the Lyon group, the old SV [31], SkI1-Sk5 [32] and SkI6 [33] of the SkI family, SkMP [34], SkO and SkO′ [35], and SkT4 and SkT5 [36].

Similarly, relativistic mean-field models are based on effective Lagrangian densities where the interaction between baryons is described in terms of meson exchanges. The couplings of nucleons with mesons are usually fixed by fitting masses and radii of nuclei and the bulk properties of nuclear matter, whereas those of other baryons, like hyperons, are fixed by symmetry relations and hypernuclear observables. In this work we consider two types of RMF models: models with constant meson-baryon couplings described by the Lagrangian density of the nonlinear Walecka model (NLWM), and models with density-dependent couplings [hereafter referred to as density-dependent models (DDM)]. In particular, within the first type, we consider the models GM1 and GM3 [37], TM1 [38], NL3 and NL3-II [39], and NL-SH [40]. For the DDM, we consider the models DDME1 and DDME2 [41], TW99 [42], and the models PK1, PK1R and PKDD of the Pekin group [43].

Microscopic approaches, on other hand, are based on realistic two- and three-body forces that describe nucleon scattering data in free space and the properties of the deuteron. These interactions are based on meson-exchange theory [44,45] or, very recently, on chiral perturbation theory [46–49]. Then one must solve the complicated many-body problem [50] to obtain the nuclear EoS. The main difficulty is the treatment of the short-range repulsive core of the nucleon-nucleon interaction. Different many-body approaches have been devised for the construction of the nuclear matter EoS, e.g., the Brueckner–Hartree–Fock (BHF) [51] and the Dirac–Brueckner–Hartree–Fock (DBHF) [52–54] theories, the variational method [55], the correlated basis function formalism [56], the self-consistent Green's function technique [57,58], the $V_{\mathrm{low}\,k}$ approach [59] or Quantum Monte Carlo techniques [60,61].

As far as the microscopic approaches are concerned, in this paper we adopt several BHF EoS based on different nucleon-nucleon potentials, namely the Bonn B (BOB) [44,62], the Nijmegen 93 (N93) [63], and the Argonne V_{18} (V18) [64]. In all those cases, the two-body forces are supplemented by nucleonic three-body forces (TBF), which are needed to correctly reproduce the saturation properties of nuclear matter. Currently a complete ab initio theory of TBF is not available yet, and therefore we adopt either phenomenological or microscopic models [65–68]. The microscopic TBF employed in this paper are described in detail in Refs. [68,69], whereas a phenomenological approach based on the

Urbana model [66,70,71], is also adopted. In this case, the corresponding EoS is labelled UIX in Table 1. Within the same theoretical framework, we also studied an EoS based on a potential model which includes explicitly the quark-gluon degrees of freedom, named FSS2 [72,73]. This correctly reproduces the saturation point of symmetric matter and the binding energy of few-nucleon systems without the need for introducing TBF. In the following we use two different EoS versions labelled respectively as FSS2CC and FSS2GC. Moreover, we compare these BHF EoSs with the often-used results of the Dirac-BHF method (DBHF) [54], which employs the Bonn A potential, the APR EoS [55] based on the variational method and the V_{18} potential, and a parametrization of a recent Auxiliary Field diffusion Monte Carlo (AFDMC) calculation of Gandolfi et al. given in Ref. [74].

The above-mentioned methods provide EoSs for homogeneous nuclear matter, $\rho > \rho_t \approx 0.08\,\mathrm{fm}^{-3}$. For the low-density inhomogeneous part we adopt the well-known Negele–Vautherin EoS [75] for the inner crust in the medium-density regime ($0.001\,\mathrm{fm}^{-3} < \rho < 0.08\,\mathrm{fm}^{-3}$), and the ones by Baym–Pethick–Sutherland [76] and Feynman–Metropolis–Teller [77] for the outer crust ($\rho < 0.001\,\mathrm{fm}^{-3}$).

3.1. Laboratory Constraints on the Nuclear EoS

Around saturation density ρ_0 and isospin asymmetry $\delta \equiv (N-Z)/(N+Z) = 0$ [being $N(Z)$ the number of neutrons (protons)], the nuclear EoS can be characterized by a set of a few isoscalar (E_0, K_0) and isovector (S_0, L, K_{sym}) parameters. These parameters can be constrained by nuclear experiments and are related to the coefficients of a Taylor expansion of the energy per particle of asymmetric nuclear matter as a function of density and isospin asymmetry

$$E(\rho,\delta) = E_{\mathrm{SNM}}(\rho) + E_{\mathrm{sym}}(\rho)\delta^2\,, \tag{2}$$

$$E_{\mathrm{SNM}}(\rho) = E_0 + \frac{K_0}{2}x^2\,, \tag{3}$$

$$E_{\mathrm{sym}}(\rho) = S_0 + Lx + \frac{K_{\mathrm{sym}}}{2}x^2\,, \tag{4}$$

where $x \equiv (\rho-\rho_0)/3\rho_0$, E_0 is the energy per particle of symmetric nuclear matter at ρ_0, K_0 the incompressibility and $S_0 \equiv E_{\mathrm{sym}}(\rho_0)$ is the symmetry energy coefficient at saturation. The parameters L and K_{sym} characterize the density dependence of the symmetry energy around saturation. These parameters are defined as

$$K_0 \equiv 9\rho_0^2\frac{d^2E_{\mathrm{SNM}}}{d\rho^2}(\rho_0)\,, \tag{5}$$

$$S_0 \equiv \frac{1}{2}\frac{\partial^2 E}{\partial\delta^2}(\rho_0,0)\,, \tag{6}$$

$$L \equiv 3\rho_0\frac{dE_{\mathrm{sym}}}{d\rho}(\rho_0)\,, \tag{7}$$

$$K_{\mathrm{sym}} \equiv 9\rho_0^2\frac{d^2E_{\mathrm{sym}}}{d\rho^2}(\rho_0)\,. \tag{8}$$

The incompressibility K_0 gives the curvature of $E(\rho)$ at $\rho = \rho_0$, whereas S_0 determines the increase of the energy per nucleon due to a small asymmetry δ.

Properties of the various considered EoSs are listed in Table 1, namely the value of the saturation density ρ_0, the binding energy per particle E_0, the incompressibility K_0, the symmetry energy S_0, and its derivative L at ρ_0. Measurements of nuclear masses [78] and density distributions [79] yield $E_0 = -16 \pm 1$ MeV and $\rho_0 = 0.14 - 0.17$ fm^{-3}, respectively. The value of K_0 can be extracted from the analysis of isoscalar giant monopole resonances in heavy nuclei, and results of Ref. [80] suggest $K_0 = 240 \pm 10$ MeV, whereas in Ref. [81] a value of $K = 248 \pm 8$ MeV is reported. Even heavy-ion collision experiments point to a "soft" EoS, i.e., a low value of K_0, though the constraints inferred from heavy-ion collisions are model dependent because the analysis of the measured data requires the use of transport models [82]. Experimental information on the symmetry energy at saturation S_0 and its derivative L can be obtained from several sources such as the analysis of giant [83] and pygmy [84,85] dipole resonances, isospin diffusion measurements [86], isobaric analog states [87], measurements of the neutron-skin thickness in heavy nuclei [88–92] and meson production in heavy-ion collisions [93]. However, whereas S_0 is more or less well established ($\approx$30 MeV), the values of L ($30\,\text{MeV} < L < 87\,\text{MeV}$), and especially those of K_{sym} ($-400\,\text{MeV} < K_{\text{sym}} < 100\,\text{MeV}$) are still quite uncertain and poorly constrained [94,95]. The reason why the isospin dependent part of the nuclear EoS is so uncertain is still an open question, very likely related to our limited knowledge of the nuclear forces and, in particular, to its spin and isospin dependence.

From Table 1, we notice that all the adopted EoSs in this work agree fairly well with the empirical values. Marginal cases are the slightly too low E_0 and K_0 for V18, too large S_0 for N93, and too low K_0 for UIX and FSS2GC. We notice that the L parameter does not exclude any of the microscopic EoSs, whereas several phenomenological models predict too large L values.

3.2. *Astrophysical Constraints on the Nuclear EoS*

The main astrophysical constraints on the nuclear EoS are those arising from the observation of NS. An enormous amount of data on different NS observables have been collected after 50 years of NS observations. These observables include masses, radii, rotational periods, surface temperatures, gravitational redshifts, quasi-periodic oscillations, magnetic fields, glitches, timing noise and, very recently, gravitational waves. In the next lines we briefly review how masses and radii are measured. Observational constraints derived from the recent observation of the gravitational wave signal from the merger of two NS detected by the Advanced LIGO and Advanced VIRGO collaborations [13–15] will be discussed in detail in Section 5.

NS masses can be directly measured from observations of binary systems. There are five orbital parameters, also known as Keplerian parameters, which can be precisely measured. They are the projection of the pulsar's semi-major axis (a_1) on the line of sight ($x \equiv a_1 \sin i/c$, where i is inclination of the orbit), the eccentricity of the orbit (e), the orbital period (P_b), and the time (T_0) and longitude (ω_0) of the periastron. With the use of Kepler's Third Law, these parameters can be related to the masses of the NS (M_p) and its companion (M_c) though the so-called mass function

$$f(M_p, M_c, i) = \frac{(M_c \sin i)^3}{(M_p + M_c)^2} = \frac{P_b v_1^3}{2\pi G} \tag{9}$$

where $v_1 = 2\pi a_1 \sin i/P_b$ is the projection of the orbital velocity of the NS along the line of sight. The individual masses of the two components of the system cannot be obtained if only the mass function is determined. Additional information is required. Fortunately, deviations from the Keplerian orbit due to general relativity effects can be detected. The relativistic corrections to the orbit are parametrized in terms of one or more parameters called post-Keplerian. The most significant ones are: the combined effect of variations in the transverse Doppler shift and gravitational redshift around an elliptical orbit (γ), the range (r) and shape (s) parameters that characterize the Shapiro time delay of the pulsar signal as it propagates through the gravitational field of its companion, the advance of the periastron of the orbit ($\dot{\omega}$) and the orbital decay due to the emission of quadrupole gravitational radiation ($\dot{P}_b$). These post-Keplerian parameters can be written in terms of measured quantities and the masses of the star and its companion (see e.g., Ref. [100] for specific expressions). The measurement of any two of these post-Keplerian parameters together with the mass function f is sufficient to determine uniquely the masses of the two components of the system.

As the reader can imagine NS radii are very difficult to measure, the reason being that NS are very small objects and are very far away from us (e.g., the closest NS is the object RX J1856.5-3754 in the constellation Corona Australis which is about 400 light-years from the Earth). That is the reason

Table 1. Saturation properties predicted by the considered EoSs. Experimental nuclear parameters are listed for comparison. See text for details.

Model Class	EoS	$\rho_0[\,\mathrm{fm}^{-3}]$	$-E_0$[MeV]	K_0[MeV]	S_0[MeV]	L[MeV]
Skyrme	Gs	0.158	14.68	239.34	39.55	93.55
	Rs	0.158	14.01	248.33	38.45	86.41
	SGI	0.155	14.67	265.35	34.32	63.85
	SLy0	0.16	15.28	226.42	34.82	45.37
	SLy1	0.161	15.23	233.25	36.21	48.88
	SLy2	0.161	15.16	234.54	36.01	48.84
	SLy3	0.161	15.22	232.85	35.45	45.56
	SLy4	0.16	15.18	232.19	35.26	45.38
	SLy5	0.161	15.25	232.03	36.44	50.34
	SLy6	0.159	15.10	230.09	34.74	45.21
	SLy7	0.159	15.05	233.10	36.18	48.11
	SLy8	0.161	15.22	233.34	34.84	45.36
	SLy9	0.151	14.53	228.95	37.72	55.37
	SLy10	0.156	14.92	231.75	35.32	39.24
	SLy230a	0.16	15.22	229.98	35.26	43.99
	SV	0.155	14.65	304.99	42.42	96.51
	SkI1	0.161	15.59	233.87	51.24	160.46
	SkI2	0.158	14.78	245.14	43.38	105.72
	SkI3	0.158	14.99	259.44	44.32	101.16
	SkI4	0.16	15.42	238.92	34.21	59.34
	SkI5	0.156	14.73	257.41	49.44	129.29
	SkI6	0.158	14.98	243.93	41.62	81.76
	SkMP	0.157	14.66	230.16	35.88	69.7
	SkO	0.161	15.04	228.10	38.52	79.92
	SkO′	0.16	14.99	222.28	37.66	69.68
	SkT4	0.159	15.12	235.48	43.19	93.48
	SkT5	0.164	15.48	201.66	44.88	100.32
NLWM	GM1	0.153	16.34	300.28	32.49	93.92
	GM3	0.153	16.36	240.53	32.54	89.83
	TM1	0.145	16.26	281.16	36.89	110.79
	NL3	0.148	16.24	271.54	37.4	118.53
	NL3-II	0.149	16.26	271.74	37.70	119.71
	NL-Sh	0.146	16.36	355.65	36.13	113.68
DDM	DDME1	0.152	16.2	244.72	33.067	55.46
	DDME2	0.152	16.14	250.9	32.3	51.26
	TW99	0.153	16.25	240.26	32.766	55.31
	PK1	0.148	16.27	282.7	37.64	115.88
	PK1R	0.148	16.27	283.68	37.83	116.5
	PKDD	0.149	16.27	262.19	36.79	90.21
Microscopic	BOB	0.170	15.4	238	33.7	70
	V18	0.178	13.9	207	32.3	67
	N93	0.185	16.1	229	36.5	77
	UIX	0.171	14.9	171	33.5	61
	FSS2CC	0.157	16.3	219	31.8	52
	FSS2GC	0.170	15.6	185	31.0	51
	DBHF	0.181	16.2	218	34.4	69
	APR	0.159	15.9	233	33.4	51
	AFDMC	0.160	16.0	239	31.3	60
	Exp.	∼ 0.14–0.17	∼ 15–17	220–260	28.5–34.9	30–87
	Ref.	[96]	[96]	[97,98]	[1,99]	[1,99]

why direct measurements of NS radii do not exist yet. Nevertheless, it is possible to determine them by using the thermal emission of low-mass X-ray binaries (systems where one of the components is a NS and the companion a less massive object ($M_c < M_\odot$) which can be a main sequence star, a red giant or a white dwarf). The observed X-ray flux (F) and estimated surface temperature (T) together with a determination of the distance (D) of the star, can be used to obtain the radius of the NS through the implicit relation

$$R = \sqrt{\frac{FD^2}{\sigma T^4}\left(1 - \frac{2GM}{c^2 R}\right)} . \tag{10}$$

Here σ is the Stefan–Boltzmann constant and M the mass of the NS. The major uncertainties in the measurement of the radius through Equation (10) come from the determination of the temperature, which requires the assumption of an atmospheric model, and the estimation of the distance of the star. However, the analysis of present observations from quiescent low-mass X-ray binaries is still controversial (see e.g., Refs. [101,102]).

We notice that the simultaneous measurement of both mass and radius of the same NS would provide the most definite observational constraint on the nuclear EoS. Very recently the NICER (Neutron Star Interior Composition Explorer) mission has reported a Bayesian parameter estimation of the mass and equatorial radius of the millisecond pulsar PSR J0030+0451 [10,11]. The values inferred from two Bayesian analysis of the collected data are ($1.34^{+0.15}_{-0.16}\ M_\odot$, $12.71^{+1.14}_{-1.19}$ km) [10] and ($1.44^{+0.15}_{-0.14}\ M_\odot$, $13.02^{+1.24}_{-1.09}$ km) [11].

4. EoS for β-Stable Matter

To study the structure of the NS core, we must calculate the composition and the EoS of cold, neutrino-free, catalyzed matter. As stated before, we consider a NS with a core of nucleonic matter without hyperons or other exotic particles. We require that it contains charge neutral matter consisting of neutrons, protons, and leptons (e^-, μ^-) in β-equilibrium, and compute the EoS for charge neutral and β-stable matter in the following standard way [103]. The output of the many-body calculation is the energy density of lepton/baryon matter as a function of the different densities ρ_i of the species $i = n, p, e, \mu$,

$$\epsilon(\rho_n, \rho_p, \rho_e, \rho_\mu) = (\rho_n m_n + \rho_p m_p) + (\rho_n + \rho_p) E(\rho_n, \rho_p) + \epsilon(\rho_e) + \epsilon(\rho_\mu) , \tag{11}$$

where m_i are the corresponding masses, and $E(\rho_n, \rho_p)$ is the energy per particle of asymmetric nuclear matter. We have used ultrarelativistic and relativistic expressions for the energy densities of electrons $\epsilon(\rho_e)$ and muons $\epsilon(\rho_\mu)$, respectively [103]. Since microscopic calculations are very time consuming in the case of these models we have used the parabolic approximation [104–108] of the energy per particle of asymmetric nuclear matter given in Equation (2) with the symmetry energy calculated simply as the difference between the energy per particle of pure neutron matter $E(\rho_n = \rho, \rho_p = 0)$ and symmetric nuclear matter $E(\rho_n = \frac{\rho}{2}, \rho_p = \frac{\rho}{2})$

$$E_{\text{sym}}(\rho) \approx E(\rho_n = \rho, \rho_p = 0) - E(\rho_n = \frac{\rho}{2}, \rho_p = \frac{\rho}{2}) . \tag{12}$$

Once the energy density (Equation (11)) is known the various chemical potentials can be computed straightforwardly,

$$\mu_i = \frac{\partial \epsilon}{\partial \rho_i} , \tag{13}$$

and solving the equations for β-equilibrium,

$$\mu_i = b_i \mu_n - q_i \mu_e , \tag{14}$$

(b_i and q_i denoting baryon number and charge of species i) along with the charge neutrality condition,

$$\sum_i \rho_i q_i = 0\,, \tag{15}$$

allows one to find the equilibrium composition ρ_i at fixed baryon density ρ, and finally the EoS,

$$P(\epsilon) = \rho^2 \frac{d}{d\rho} \frac{\epsilon(\rho_i(\rho))}{d\rho} = \rho \frac{d\epsilon}{d\rho} - \epsilon = \rho\mu_n - \epsilon\,. \tag{16}$$

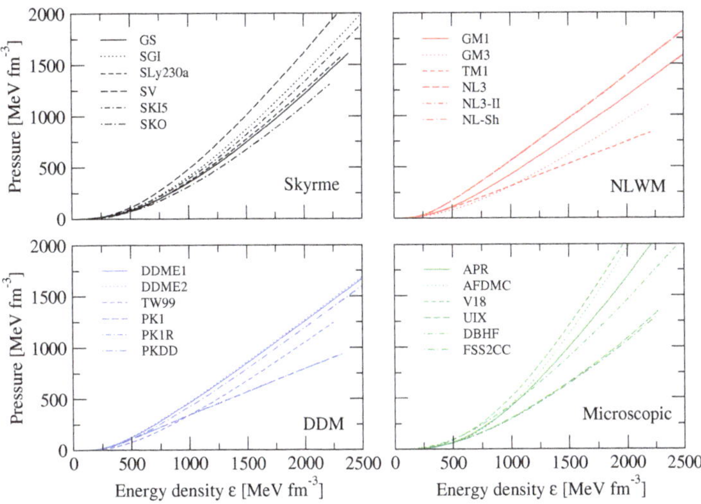

Figure 2. Equation of state of β-stable matter for the four model classes reported in Table 1.

Once the EoS of β-stable matter is known, one can determine the hydrostatical equilibrium configurations just solving the Tolman–Oppenheimer–Volkoff (TOV) [103] equations which describe the structure of a non-rotating spherically symmetric star in general relativity:

$$\begin{aligned} \frac{dP}{dr} &= -G\frac{\epsilon m}{r^2}\left(1+\frac{P}{\epsilon}\right)\left(1+\frac{4\pi P r^3}{m}\right)\left(1-\frac{2Gm}{r}\right)^{-1} \\ \frac{dm}{dr} &= 4\pi r^2 \epsilon\,, \end{aligned} \tag{17}$$

where G is the gravitational constant, P the pressure, ϵ the energy density, m the mass enclosed within a sphere of radius r. The TOV equations have an easy interpretation. Consider a spherical shell of matter of radius r and thickness dr. The second equation gives the mass in this shell whereas the left-hand side of the first one is the net force acting on the surface of the shell by the pressure difference $dP(r)$. The first factor of the right-hand side of this equation is the attractive Newtonian force of gravity acting on the shell by the mass interior to it. The remaining three factors result from the correction of general relativity. So the TOV equations express the balance at each r between the internal pressure as it supports the overlying material against the gravitational attraction of the mass interior to r. The integration of the TOV equations gives the mass and radius of the star for a given central density. It turns out that the mass of the NS has a maximum value as a function of radius (or central density), above which the star is unstable against collapse to a black hole. The value of the maximum mass depends on the nuclear EoS, so that the observation of a mass higher than the maximum mass allowed by a given EoS simply rules out that EoS.

We now turn to the discussion of some results. We display in Figure 2 the β-stable matter EoS obtained for some of the models illustrated in Table 1, a limited sample of each class being plotted in one single panel. We see that the pressure is a monotonically increasing function of the energy density for all EoS. Each EoS is characterized by a given stiffness, which determines the maximum mass value of a NS: the stiffer the EoS the larger the maximum mass predicted.

The corresponding mass-radius relation is displayed in Figure 3. The observed trend is consistent with the EoS displayed in Figure 2. As expected, when the EoS stiffness increases the NS maximum mass increases as well. The considered EoS are compatible with the largest masses observed up to now, $M_{\max} > 2.14^{+0.10}_{-0.09}$ [7] for the object PSR J0740+6620 (cyan hatched area), and PSR J0348+0432 [6], $M_G = 2.01 \pm 0.04\, M_\odot$ (red hatched area). We notice that recent analysis of the GW170817 event also indicate an upper limit of the maximum mass of about 2.2–2.3 $M_\odot$ [109–112], with which most of the models shown in the figure are compatible. The symbols with the error bars show the estimation of the mass and equatorial radius of the millisecond pulsar PSR J0030+0451 inferred from two Bayesian analysis of the data collected by the NICER mission: ($1.34^{+0.15}_{-0.16}\, M_\odot$, $12.71^{+1.14}_{-1.19}$ km) [10] and ($1.44^{+0.15}_{-0.14}\, M_\odot$, $13.02^{+1.24}_{-1.09}$ km) [11]. Please note that this constraint excludes most of the NLWM EoS considered in this work.

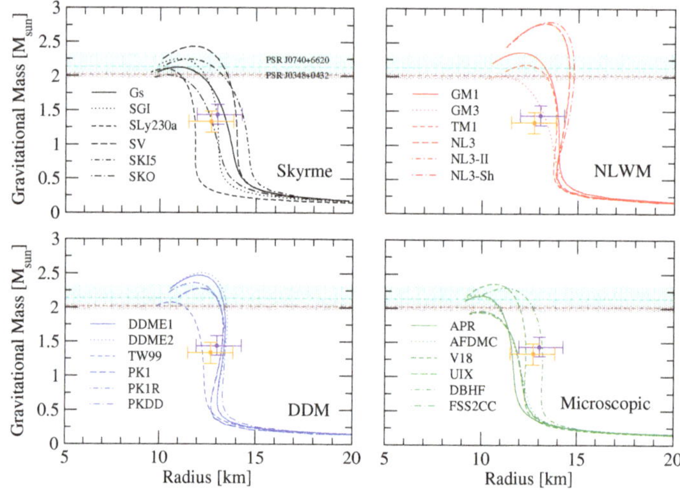

Figure 3. Mass-radius relation predicted by the different EoS displayed in Figure 2. The observed masses of the millisecond pulsars PSR J0740+6620 [4] and PSR J0348+0432 [6] are also shown. The symbols with the error bars show the values of the mass and equatorial radius of the millisecond pulsar PSR J0030+0451inferred from two analysis of the observations reported by the NICER mission: ($1.34^{+0.15}_{-0.16}\, M_\odot$, $12.71^{+1.14}_{-1.19}$ km) [10] and ($1.44^{+0.15}_{-0.14}\, M_\odot$, $13.02^{+1.24}_{-1.09}$ km) [11]. See text for details.

5. The Neutron Star Tidal Deformability

Recently the tidal deformability λ, or equivalently the tidal Love number k_2 of a NS [113–115], has been recognized to provide valuable information and constraints on the related EoS, because it strongly depends on the compactness of the object, i.e., $\beta \equiv M/R$. More specifically, the Love number

$$
\begin{aligned}
k_2 \;&=\; \frac{3}{2}\frac{\lambda}{R^5} = \frac{3}{2}\beta^5\Lambda = \frac{8}{5}\frac{\beta^5 z}{F}\,, \qquad (18)\\
&z \equiv (1-2\beta)^2[2-y_R+2\beta(y_R-1)]\,,\\
&F \equiv 6\beta(2-y_R) + 6\beta^2(5y_R-8) + 4\beta^3(13-11y_R)\\
&\quad + 4\beta^4(3y_R-2) + 8\beta^5(1+y_R) + 3z\ln(1-2\beta)
\end{aligned}
$$

with $\Lambda \equiv \lambda/M^5$, can be obtained by solving the TOV equations (17), along with the following first-order differential equation [116],

$$\frac{dy}{dr} = -\frac{y^2}{r} - \frac{y-6}{r-2m} - rQ\,,$$
$$Q \equiv 4\pi \frac{(5-y)\varepsilon + (9+y)P + (\varepsilon+P)/c_s^2}{1-2m/r} - \left[\frac{2(m+4\pi r^3 P)}{r(r-2m)}\right]^2, \tag{19}$$

with the EoS $P(\varepsilon)$ as input, $c_s^2 = dP/d\varepsilon$ the speed of sound, and boundary conditions given by

$$[P, m, y](r=0) = [P_c, 0, 2]\,, \tag{20}$$

being $y_R \equiv y(R)$, and the mass-radius relation $M(R)$ provided by the condition $P(R) = 0$ for varying central pressure P_c.

For an asymmetric binary NS system, $(M,R)_1 + (M,R)_2$, with mass asymmetry $q = M_2/M_1$, and known chirp mass M_c, which characterizes the GW signal waveform,

$$M_c = \frac{(M_1 M_2)^{3/5}}{(M_1+M_2)^{1/5}}\,, \tag{21}$$

the average tidal deformability is defined by

$$\tilde{\Lambda} = \frac{16}{13}\frac{(1+12q)\Lambda_1 + (q+12)\Lambda_2}{(1+q)^5} \tag{22}$$

with

$$\frac{[M_1, M_2]}{M_c} = \frac{297}{250}(1+q)^{1/5}[q^{-3/5}, q^{2/5}]\,. \tag{23}$$

From the analysis of the GW170817 event [13–15], a value of $M_c = 1.186\,^{+0.001}_{-0.001}\,M_\odot$ was obtained, corresponding to $M_1 = M_2 = 1.36\,M_\odot$ for a symmetric binary system, $q = 0.73$–1 and $\tilde{\Lambda} < 730$ from the phase-shift analysis of the observed signal. It turns out that requiring both NSs to have the same EoS, leads to constraints $70 < \Lambda_{1.4} < 580$ and $10.5 < R_{1.4} < 13.3$ km [14] for a 1.4 solar mass NS.

However, the high luminosity of the kilonova AT2017gfo following the NS merger event, imposes a lower limit on the average tidal deformability, Equation (22), $\tilde{\Lambda} > 400$, which was deduced to justify the amount of ejected material heavier than $0.05\,M_\odot$. This constraint could indicate that $R_{1.4} \gtrsim 12$ km, which was used in Refs. [117–120] to constrain the EoS. This lower limit must be taken with great care and, in fact, it has been recently revised to $\tilde{\Lambda} \gtrsim 300$ [121], but considered of limited significance in Ref. [122]. We notice that the determination of the average tidal deformability of the binary neutron star system GW170817 has imposed constraints for the neutron star radii, thus finding compatible radii between 12 and 13 km [123]. This is complementary to the M-R measurement by NICER, and contributes to selecting the suitable equations of state.

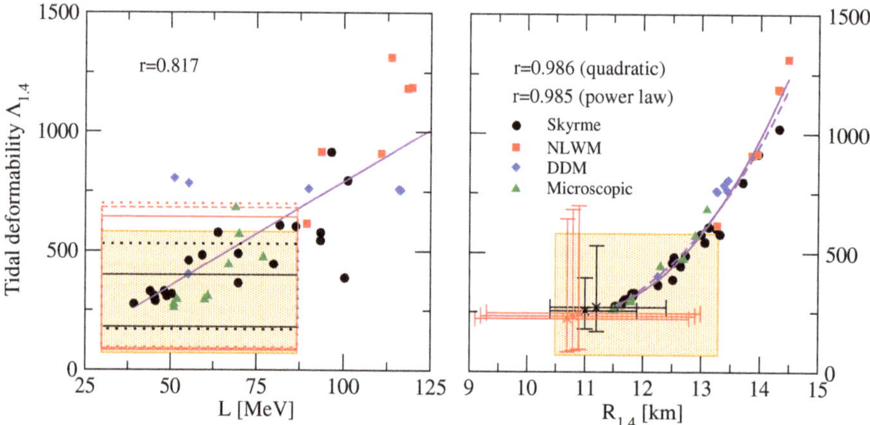

Figure 4. In the left panel the tidal deformability of a 1.4 solar mass NS $\Lambda_{1.4}$ is plotted vs. the symmetry energy derivative at saturation density L, whereas in the right panel it is displayed as a function of the radius of a 1.4 solar mass NS, $R_{1.4}$, for the different EoS shown in Table 1. The orange box indicate the experimental and observational constraints on L (see Table 1) and $\Lambda_{1.4}$ and $R_{1.4}$ from Ref. [14]. Updated values of $\Lambda_{1.4}$ and $R_{1.4}$ from the works of Capano et al. [124] and De et al. [125] are shown by the black and red open boxes (left panel) and by the black and red stars with error bars (right panel). The continuous violet line in the left (right) panel shows a linear (quadratic) fit of the EoS data. The dashed one in the right panel shows an alternative fit of the EoS data of the type $\Lambda_{1.4} \sim R_{1.4}^{6.47}$. The values of the corresponding correlation factors are also given. See text for details.

One of the main theoretical issues, following the detection of gravitational waves from NS mergers, regards the possibility of finding correlations between properties of nuclear matter and NS observables [126]. Along this same line, we further explore this issue, and using the set of microscopic EoS and the several Skyrme forces and relativistic models listed in Table 1, in the left panel of Figure 4 we show the tidal deformability of a 1.4 solar mass NS as a function the symmetry energy parameter L at saturation density. The orange box shows the constraint on $\Lambda_{1.4}$ inferred from the observational data of the GW170817 event [14] together with the experimental limits of L reported in Table 1. We observe some degree of correlation between the tidal deformability and L , for which we can estimate the so-called correlation factor r, defined as

$$r(L, \Lambda_{1.4}) = \frac{1}{n-1} \frac{\sum_L \sum_{\Lambda_{1.4}} (L - \bar{L})(\Lambda_{1.4} - \bar{\Lambda}_{1.4})}{\sigma_L \sigma_{\Lambda_{1.4}}} , \tag{24}$$

being n the number of data pairs, $\bar{L}$ and $\bar{\Lambda}_{1.4}$ the mean values of L and $\Lambda_{1.4}$ over the data set; and σ_L and $\sigma_{\Lambda_{1.4}}$ their standard deviations. We get a value $r = 0.817$, which indicates a weak correlation. We note that several EoS lie outside the orange observational band. In particular, we notice that all DDM EoS (blue diamonds), except TW99, are not compatible with the data, as well as all the NLWM EoS (red squares). On the other hand, most of the Skyrme interactions lie within the orange band, with a few cases incompatible with observations because the predicted L values lie outside the experimental range, and some other are marginally compatible. As far as microscopic calculations are concerned, they are all in agreement with GW observations, except the DBHF EoS.

In the right panel of Figure 4 we report the tidal deformability as a function of the radius for a 1.4 solar mass NS, $R_{1.4}$, for the same set of EoS. The observational constraints on $\Lambda_{1.4}$ and $R_{1.4}$ from GW170817 of Ref. [14] are shown by the orange box. Updated values on these quantities from the works of Capano et al. [124] and De et al. [125] are shown by the black and red open boxes (left panel) and by the black and red stars with error bars (right panel). We note that the stronger constraint on

$R_{1.4}$ from Capano et al. [124] excludes all the NLWM models and all, but one (TW99), of the DDM ones, whereas 6 microscopic models (V18, UIX, FSS2CC, FSS2GC,APR,AFDMC) and 11 Skyrme forces (SLy0-SLy8,SLy10,SLy230a) are compatible with it. Contrary to the weak $\Lambda_{1.4} - L$ correlation found, we observe a strong correlation between $\Lambda_{1.4}$ and $R_{1.4}$. This strong correlation was already noticed by Zhao and Lattimer in Ref. [127] who found it to be of the form $\Lambda_{1.4} \sim R_{1.4}^6$, and later by Tsang et al. in Ref. [128] using a different set of EoS based again on Skyrme and relativistic mean-field models. In this work we find that the EoS data can be fitted assuming a power law (dashed line) of the form $\Lambda_{1.4} = 3.65 \times 10^{-5} R_{1.4}^{6.47}$, in agreement with the work of Refs. [127,128]. We note, however, that a quadratic function (continuous) $\Lambda_{1.4} = 9922.02 - 1757.55 R_{1.4} + 79.91 R_{1.4}^2$ fits equally well the data. In both cases we find a very similar correlation factor: $r = 0.985$ for the power law fit and $r = 0.986$ for the quadratic one. The behavior of the microscopic and phenomenological EoS look very similar.

We notice that since $\Lambda_{1.4}$ is not sensitive to the EoS details at very high density regions, a possible alternative in the analysis of correlations is offered by the use of parameterized EoS above a few times ρ_0 [127–130]. Those can explore large variations of the matter properties above ρ_0, and determine systematic uncertainties related to it. However, this approach does not improve our knowledge of the nuclear interaction in the medium under extreme conditions, simply because those simple parameterizations are not based on any particular model, and therefore they are unable to explain the observational data in terms of the microphysics.

6. The Neutron-Skin Thickness

As stated in the previous Section, correlations between astrophysical observations and microscopic constraints from nuclear measurements, could help to better understand the properties of nuclear matter. For this purpose, the limits derived for the tidal deformability in GW170817 could be very valuable and exploited for studying the neutron-skin thickness, defined as the difference between the neutron (R_n) and proton (R_p) root-mean-square radii: $\delta R = \sqrt{\langle r_n^2 \rangle} - \sqrt{\langle r_p^2 \rangle}$. It has been shown that this is strongly correlated with both L and to the radius of low-mass NS, since the size of a NS and the neutron-skin thickness originate both from the pressure of neutron-rich matter, and are sensitive to the same EoS. As shown by Brown and Typel [88,89], and confirmed later by other authors [90,131–134], the neutron-skin thickness calculated in mean-field models with either non-relativistic or relativistic effective interactions, is very sensitive to the density dependence of the nuclear symmetry energy, and, in particular, to the slope parameter L at normal nuclear saturation density. Using the Brueckner approach and the several Skyrme forces and relativistic models considered here, the authors of Ref. [135] made an estimation of the neutron-skin thickness of ^{208}Pb and ^{132}Sn, adopting the suggestion of Steiner *et al.* in Ref. [131], where δR is calculated to lowest order in the diffuseness corrections as $\delta R \sim \sqrt{\frac{3}{5}} t$, being t the thickness of semi-infinite asymmetric nuclear matter

$$t = \frac{\delta_c}{\rho_0(\delta_c)(1-\delta_c^2)} \frac{E_s}{4\pi r_0^2} \frac{\int_0^{\rho_0(\delta_c)} \rho^{1/2} [S_0/E_{sym}(\rho) - 1][E_{SNM}(\rho) - E_0]^{-1/2} d\rho}{\int_0^{\rho_0(\delta_c)} \rho^{1/2} [E_{SNM}(\rho) - E_0]^{1/2} d\rho} . \tag{25}$$

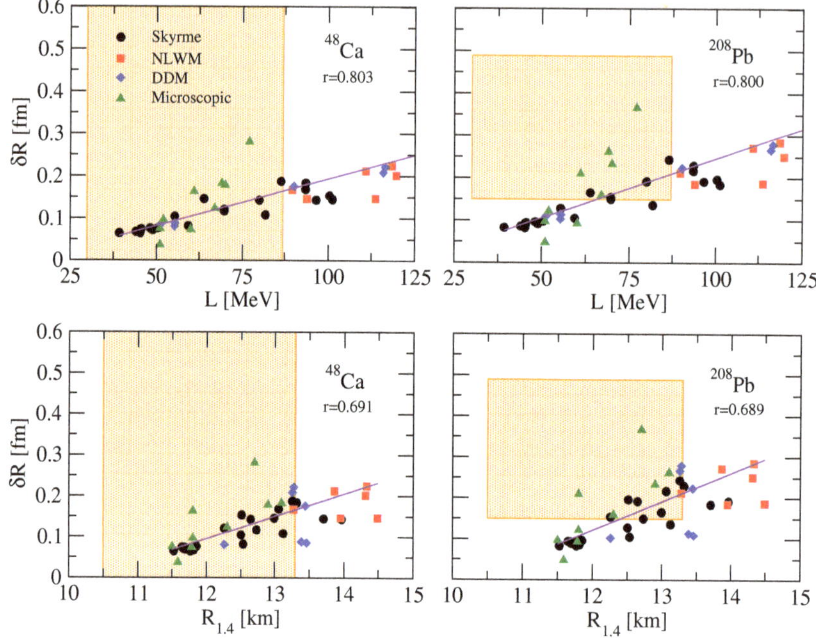

Figure 5. The neutron-skin thickness is displayed as a function of the symmetry energy derivative at saturation density L (upper panels) and of the radius of a $1.4M_{\odot}$ NS (lower panels) for the different EoS displayed in Table 1. In the left (right) panel calculations are shown for ^{48}Ca (^{208}Pb). The band on the left panel shows the experimental constraint on L, whereas the box on the right one shows in addition the constraint from the PREX experiment [136]. The violet line indicates a linear fit of the EoS data, Equation (25). The values of the corresponding correlation factors are also given.

In the above expression E_s is the surface energy taken from the semi-empirical mass formula equal to 17.23 MeV, r_0 is obtained from the normalization condition $(4\pi r_0^3/3)(0.16) = 1$, and δ_c is the isospin asymmetry in the center of the nucleus taken as $\delta_c = \delta/2$ according to Thomas-Fermi calculations. In this paper, we use the same prescription for the calculation of the neutron-skin thickness of ^{208}Pb and ^{48}Ca, and we show the results in Figure 5. The orange bands represent the predicted data for ^{48}Ca (left panels) for which the Calcium Radius Experiment (CREX) has not been run yet [137], whereas in the right panels experimental data obtained in the Lead Radius Experiment (PREX) [136] for ^{208}Pb, $\delta R = 0.33^{+0.16}_{-0.18}$ fm, are plotted. In the upper panels, results are shown for the neutron-skin thickness as a function of the derivative of the symmetry energy L. We notice that all the theoretical predictions from phenomenological models and some of the microscopic ones show some correlation between the neutron-skin thickness and L, as indicated by the linear fits (violet curve) and by the value of the correlation coefficient, $r = 0.803$ for ^{48}Ca ($r = 0.800$ for ^{208}Pb). Almost all the microscopic EoS turn out to be compatible with the PREX experimental data, whereas some phenomenological models, e.g., those of the NLWM class, give predictions out of the experimental range. The linear increase of δR with L is not surprising since the thickness of the neutron skin in heavy nuclei is determined by the pressure difference between neutrons and protons, which is proportional to the parameter L, i.e., $P(\rho_0, \delta) \approx L\rho_0\delta^2/3$. On the other hand, in the lower panels, the neutron-skin thickness is displayed as a function of $R_{1.4}$, and in both cases the correlation is very scarce, $r = 0.691$ for ^{48}Ca ($r = 0.689$ for ^{208}Pb), with a few Skyrme, DDM and all the NLWM EoS incompatible with the observational data. The experimental data from PREX [136], and the upcoming campaigns: PREX-II at Jefferson Lab and the Mainz Radius Experiment (MREX) [138] at the future Mainz Energy-Recovering Superconductor

Accelerator, can put further strong constraints on the nuclear matter properties, thus selecting the most compatible EoS.

7. Conclusions

In this work we have analyzed the existence of possible correlations between NS observables and properties of atomic nuclei. In particular, we have examined correlations of the tidal deformability $\Lambda_{1.4}$ of a 1.4 $M_\odot$ NS and the neutron-skin thickness δR of ^{48}Ca and ^{208}Pb with the stellar radius $R_{1.4}$ and the symmetry energy derivative L. To such end we have used a large set of different models for the nuclear equation of state that include microscopic calculations based on the Brueckner–Hartree–Fock and Dirac–Brueckner–Hartree–Fock theories, the variational method and Quantum Monte Carlo techniques, and several phenomenological Skyrme and relativistic mean-field models. We have found a strong quadratic correlation between $\Lambda_{1.4}$ and $R_{1.4}$ in agreement with the results of the recent work by Tsang et al. [128]. On the contrary, we have observed a weaker linear correlation between $\Lambda_{1.4}$ and L. Our results have confirmed the existence of a quite linear correlation between the neutron-skin thickness of ^{48}Ca and ^{208}Pb with L, already pointed out by several authors using non-relativistic and relativistic phenomenological models. A much weaker correlation has been found between δR and $R_{1.4}$. The existence of these correlations, predicted by models based on approaches of different nature, suggest that their origin goes beyond the mean-field character of the models employed.

To select the most compatible EoS among the ones predicted by the different models considered in this work, we have employed the experimental constraints on L and δR together with the observational ones on the mass, radius and tidal deformability imposed by the mass measurement of the millisecond pulsars PSR J1614-2230 [4] and PSR J0348+0432 [6], the GW170817 NS merger event [13–15] and the data of the NICER mission [10,11]. Our results have shown that only five microscopic models (BOB, V18, N93, UIX and DBHF) and four Skyrme forces (SGI, SkMP, SkO and SkO') are simultaneously compatible with the present constraints on L (30 MeV $< L <$ 87 MeV) and the PREX experimental data on the ^{208}Pb neutron-skin thickness. All the NLWM and DDM models and the majority of the Skyrme forces are excluded by these two experimental constraints. We have also found that almost all the models considered are compatible with the largest masses observed up to now, $M_{\max} > 2.14^{+0.10}_{-0.09}$ [7] for the object PSR J0740+6620, and PSR J0348+0432 [6], $M_G = 2.01 \pm 0.04\, M_\odot$, and with the upper limit of the maximum mass of about 2.2–2.3 $M_\odot$ [109–112] deduced from the analysis of the GW170817 event. Finally, we have seen that the estimation of the mass ($1.34^{+0.15}_{-0.16}\, M_\odot$) and equatorial radius ($12.71^{+1.14}_{-1.19}$ km) of the millisecond pulsar PSR J0030+0451 inferred from the Bayesian analysis of the data collected by the NICER mission [10,11] excludes most of the NLWM EoS considered in this work.

We would like to stress that the current study on correlations does not allow us to select the best EoS, but only limit the number of EoS models.

The major experimental, observational and theoretical advances on understanding the nuclear EoS done in recent decades have led to constrain rather well its isoscalar part. Nevertheless, the isovector part of the nuclear EoS is less well constraint due mainly to our still limited knowledge of the nuclear force and, in particular, of its in medium modifications and its spin and isospin dependence. Future laboratory experiments being planned in existing or next-generation radioactive ion beam facilities together with further NS observations, particularly a precise simultaneous measurement of the mass and radius of a single object, are fundamental to provide more stringent constraints on the nuclear EoS, and are very much awaited.

Author Contributions: All authors have read and agreed to the published version of the manuscript.

Funding: This research received no external funding.

Acknowledgments: This work has been supported by the COST Action CA16214 "PHAROS: The multimessenger physics and astrophysics of neutron stars".

Conflicts of Interest: The authors declare no conflict of interest.

References

1. Oertel, M.; Hempel, M.; Klähn, T.; Typel, S. Equations of state for supernovae and compact stars. *Rev. Mod. Phys.* **2017**, *89*, 015007. [CrossRef]
2. Burgio, G.F.; Fantina, A.F. Nuclear Equation of state for Compact Stars and Supernovae. *Astrophys. Space Sci. Libr.* **2018**, *457*, 255.
3. Lattimer, J.M. The Nuclear Equation of State and Neutron Star Masses. *Ann. Rev. Nucl. Sci.* **2012**, *62*, 485–515. [CrossRef]
4. Demorest, P.B.; Pennucci, T.; Ransom, S.M.; Roberts, M.S.; Hessels, J.W. A two-solar-mass neutron star measured using Shapiro delay. *Nature* **2010**, *467*, 1081–1083. [CrossRef] [PubMed]
5. Fonseca, E. The NANOGrav Nine-year Data Set: Mass and Geometric Measurements of Binary Millisecond Pulsars. *Astrophys. J.* **2016**, *832*, 167. [CrossRef]
6. Antoniadis, J. A Massive Pulsar in a Compact Relativistic Binary. *Science* **2013**, *340*, 6131. [CrossRef]
7. Cromartie, H.T. Relativistic Shapiro delay measurements of an extremely massive millisecond pulsar. *Nat. Astron.* **2019**. [CrossRef]
8. Özel, F.; Freire, P. Masses, Radii, and the Equation of State of Neutron Stars. *Ann. Rev. Astron. Astrophys.* **2016**, *54*, 401–440. [CrossRef]
9. Guillot, S.; Servillat, M.; Webb, N.A.; Rutledge, R.E. Measurement of the Radius of Neutron Stars with High Signal-to-noise Quiescent Low-mass X-Ray Binaries in Globular Clusters. *Astrophys. J.* **2013**, *772*, 7. [CrossRef]
10. Riley, T.E.; Watts, A.L.; Bogdanov, S.; Ray, P.S.; Ludlam, R.M.; Guillot, S.; Arzoumanian, Z.; Baker, C.L.; Bilous, A.V.; Chakrabarty, D.; et al. A NICER View of PSR J0030+0451: Millisecond Pulsar Parameter Estimation. *Astrophys. J. Lett.* **2019**, *887*, L21. [CrossRef]
11. Miller, M.C.; Lamb, F.K.; Dittmann, A.J.; Bogdanov, S.; Arzoumanian, Z.; Gendreau, K.C.; Guillot, S.; Harding, A.K.; Ho, W.C.G.; Lattimer. J.M.; et al. PSR J0030+0451 Mass and Radius from NICER Data and Implications for the Properties of Neutron Star Matter. *Astrophys. J. Lett.* **2019**, *887*, L24. [CrossRef]
12. Watts, A.L. Dense matter with eXTP. *Sci. China Phys. Mech. Astron.* **2019**, *62*, 29503. [CrossRef]
13. Abbott, B. GW170817: Observation of Gravitational Waves from a Binary Neutron Star Inspiral. *Phys. Rev. Lett.* **2017**, *119*, 161101. [CrossRef] [PubMed]
14. Abbott, B.P. GW170817: Measurements of neutron star radii and equation of state. *Phys. Rev. Lett.* **2018**, *121*, 161101. [CrossRef] [PubMed]
15. Abbott, B.P. Properties of the Binary Neutron Star Merger GW170817. *Phys. Rev. X* **2019**, *9*, 011001. [CrossRef]
16. Hartle, J.B. Slowly rotating relativistic stars. 1. Equations of structure. *Astrophys. J.* **1967**, *150*, 1005–1029. [CrossRef]
17. Flanagan, E.E.; Hinderer, T. Constraining neutron star tidal Love numbers with gravitational wave detectors. *Phys. Rev. D* **2008**, *77*, 021502. [CrossRef]
18. Radice, D.; Perego, A.; Zappa, F.; Bernuzzi, S. GW170817: Joint Constraint on the Neutron Star Equation of State from Multimessenger Observations. *Astrophys. J.* **2018**, *852*, L29. [CrossRef]
19. Horowitz, C.J. Neutron rich matter in the laboratory and in the heavens after GW170817. *Ann. Phys.* **2019**, *411*, 167992. [CrossRef]
20. Ravenhall, D.G.; Pethick, C.J.; Wilson, J.R. Structure of Matter below Nuclear Saturation Density. *Phys. Rev. Lett.* **1983**, *50*, 2066. [CrossRef]
21. Blaschke, D.; Colpi, M.Y.; Horowitz, C.J.; Radice, D.E. First joint gravitational wave and electromagnetic observations: Implications for nuclear and particle physics. *Eur. Phys. J. A* **2019**, *55*. Available online: https://epja.epj.org/component/toc/?task=topic&id=1005 (accessed on 9 August 2020).
22. Stone, J.R.; Reinhard, P.G. The Skyrme interaction in finite nuclei and nuclear matter. *Prog. Part. Nucl. Phys.* **2007**, *58*, 587–657. [CrossRef]
23. Vautherin, D.; Brink, D.M. Hartree-Fock Calculations with Skyrme's Interaction. I. Spherical Nuclei. *Phys. Rev. C* **1972**, *5*, 626–647. [CrossRef]
24. Quentin, P.; Flocard, H. Self-Consistent Calculations of Nuclear Properties with Phenomenological Effective Forces. *Annu. Rev. Nucl. Part. Sci.* **1978**, *28*, 523–594. [CrossRef]
25. Boguta, J.; Bodmer, A.R. Relativistic calculation of nuclear matter and the nuclear surface. *Nucl. Phys. A* **1977**, *292*, 413–428. [CrossRef]

26. Friedrich, J.; Reinhard, P.G. Skyrme-force parametrization: Least-squares fit to nuclear ground-state properties. *Phys. Rev. C* **1986**, *33*, 335–351. [CrossRef]
27. Giai, N.V.; Sagawa, H. Spin-isospin and pairing properties of modified Skyrme interactions. *Phys. Lett. B* **1981**, *106*, 379–382. [CrossRef]
28. Chabanat, E. Report No. LYCENT 9501 (Unpublished). Ph.D. Thesis, Universitè Claude Bernard Lyon-1: Lyon, France, 1995.
29. Chabanat, E.; Bonche, P.; Haensel, P.; Meyer, J.; Schaeffer, R. A Skyrme parametrization from subnuclear to neutron star densities. *Nucl. Phys. A* **1997**, *627*, 710–746. [CrossRef]
30. Chabanat, E.; Bonche, P.; Haensel, P.; Meyer, J.; Schaeffer, R. A Skyrme parametrization from subnuclear to neutron star densitiesPart II. Nuclei far from stabilities. *Nucl. Phys. A* **1998**, *635*, 231–256. [CrossRef]
31. Beiner, M.; Flocard, H.; Van Giai, N.; Quentin, P. Nuclear ground-state properties and self-consistent calculations with the skyrme interaction (I). Spherical description. *Nucl. Phys. A* **1975**, *238*, 29–69. [CrossRef]
32. Reinhard, P.G.; Flocard, H. Nuclear effective forces and isotope shifts. *Nucl. Phys. A* **1995**, *584*, 467–488. [CrossRef]
33. Nazarewicz, W.; Dobaczewski, J.; Werner, T.R.; Maruhn, J.A.; Reinhard, P.G.; Rutz, K.; Chinn, C.R.; Umar, A.S.; Strayer, M.R. Structure of proton drip-line nuclei around doubly magic ^{48}Ni. *Phys. Rev. C* **1996**, *53*, 740–751. [CrossRef] [PubMed]
34. Bennour, L.; Heenen, P.H.; Bonche, P.; Dobaczewski, J.; Flocard, H. Charge distributions of ^{208}Pb, ^{206}Pb, and ^{205}Tl and the mean-field approximation. *Phys. Rev. C* **1989**, *40*, 2834–2839. [CrossRef] [PubMed]
35. Reinhard, P.G.; Dean, D.J.; Nazarewicz, W.; Dobaczewski, J.; Maruhn, J.A.; Strayer, M.R. Shape coexistence and the effective nucleon-nucleon interaction. *Phys. Rev. C* **1999**, *60*, 014316. [CrossRef]
36. Tondeur, F.; Brack, M.; Farine, M.; Pearson, J.M. Static nuclear properties and the parametrisation of Skyrme forces. *Nucl. Phys. A* **1984**, *420*, 297–319. [CrossRef]
37. Glendenning, N.K.; Moszkowski, S.A. Reconciliation of neutron-star masses and binding of the Lambda in hypernuclei. *Phys. Rev. Lett.* **1991**, *67*, 2414–1417. doi:10.1103/PhysRevLett.67.2414. [CrossRef]
38. Sumiyoshi, K.; Kuwabara, H.; Toki, H. Relativistic mean-field theory with non-linear σ and ω terms for neutron stars and supernovae. *Nucl. Phys. A* **1995**, *581*, 725–746. [CrossRef]
39. Lalazissis, G.A.; König, J.; Ring, P. New parametrization for the Lagrangian density of relativistic mean field theory. *Phys. Rev. C* **1997**, *55*, 540–543. [CrossRef]
40. Sharma, M.M.; Nagarajan, M.A.; Ring, P. Rho meson coupling in the relativistic mean field theory and description of exotic nuclei. *Phys. Lett. B* **1993**, *312*, 377–381. [CrossRef]
41. Nikšić, T.; Vretenar, D.; Finelli, P.; Ring, P. Relativistic Hartree-Bogoliubov model with density-dependent meson-nucleon couplings. *Phys. Rev. C* **2002**, *66*, 024306. [CrossRef]
42. Typel, S.; Wolter, H.H. Relativistic mean field calculations with density-dependent meson-nucleon coupling. *Nucl. Phys. A* **1999**, *656*, 331–364. [CrossRef]
43. Long, W.; Meng, J.; Giai, N.V.; Zhou, S.G. New effective interactions in relativistic mean field theory with nonlinear terms and density-dependent meson-nucleon coupling. *Phys. Rev. C* **2004**, *69*, 034319. [CrossRef]
44. Machleidt, R.; Holinde, K.; Elster, C. The Bonn Meson Exchange Model for the Nucleon Nucleon Interaction. *Phys. Rep.* **1987**, *149*, 1–89. [CrossRef]
45. Nagels, M.M.; Rijken, T.A.; de Swart, J.J. Low-energy nucleon-nucleon potential from Regge-pole theory. *Phys. Rev. D* **1978**, *17*, 768–776. [CrossRef]
46. Weinberg, S. Effective chiral lagrangians for nucleon-pion interactions and nuclear forces. *Nucl. Phys. B* **1991**, *363*, 3–18. [CrossRef]
47. Weinberg, S. Nuclear forces from chiral lagrangians. *Phys. Lett. B* **1990**, *251*, 288–292. [CrossRef]
48. Entem, D.R.; Machleidt, R. Accurate charge-dependent nucleon-nucleon potential at fourth order of chiral perturbation theory. *Phys. Rev. C* **2003**, *68*, 041001. [CrossRef]
49. Epelbaum, E.; Hammer, H.W.; Meißner, U.G. Modern theory of nuclear forces. *Rev. Mod. Phys.* **2009**, *81*, 1773–1825. [CrossRef]
50. Baldo, M. Nuclear Methods And The Nuclear Equation Of State. *Int. Rev. Nucl. Phys.* **1999**, *8*. [CrossRef]
51. Day, B.D. Elements of the Brueckner-Goldstone Theory of Nuclear Matter. *Rev. Mod. Phys.* **1967**, *39*, 719–744. [CrossRef]
52. Brockmann, R.; Machleidt, R. Relativistic nuclear structure. I. Nuclear matter. *Phys. Rev. C* **1990**, *42*, 1965–1980. [CrossRef] [PubMed]

53. Li, G.Q.; Machleidt, R.; Brockmann, R. Properties of dense nuclear and neutron matter with relativistic nucleon-nucleon interactions. *Phys. Rev. C* **1992**, *45*, 2782–2794. [CrossRef] [PubMed]
54. Gross-Boelting, T.; Fuchs, C.; Faessler, A. Covariant representations of the relativistic Brueckner T-matrix and the nuclear matter problem. *Nucl. Phys. A* **1999**, *648*, 105–137. [CrossRef]
55. Akmal, A.; Pandharipande, V.R.; Ravenhall, D.G. Equation of state of nucleon matter and neutron star structure. *Phys. Rev. C* **1998**, *58*, 1804–1828. [CrossRef]
56. Fabrocini, A.; Fantoni, S. Correlated basis function results for the Argonne models of nuclear matter. *Phys. Lett. B* **1993**, *298*, 263–266. [CrossRef]
57. Kadanoff, L.; Baym, G. *Quantum Statistical Mechanics*; W.A. Benjamin Inc.: New York, NY, USA, 1962.
58. Dickhoff, W.; Van Neck, D. *Many-Body Theory Exposed!* World Scientific Publishing, Singapore: 2005; 752p.
59. Bogner, S.K.; Furnstahl, R.J.; Schwenk, A. From low-momentum interactions to nuclear structure. *Prog. Part. Nucl. Phys.* **2010**, *65*, 94–147. [CrossRef]
60. Wiringa, R.B.; Pieper, S.C.; Carlson, J.; Pand haripande, V.R. Quantum Monte Carlo calculations of A=8 nuclei. *Phys. Rev. C* **2000**, *62*, 014001. [CrossRef]
61. Gandolfi, S.; Illarionov, A.Y.; Schmidt, K.E.; Pederiva, F.; Fantoni, S. Quantum Monte Carlo calculation of the equation of state of neutron matter. *Phys. Rev. C* **2009**, *79*, 054005. [CrossRef]
62. Machleidt, R. The Meson theory of nuclear forces and nuclear structure. *Adv. Nucl. Phys.* **1989**, *19*, 189–376. [CrossRef]
63. Stoks, V.G.J.; Klomp, R.A.M.; Terheggen, C.P.F.; de Swart, J.J. Construction of high quality N N potential models. *Phys. Rev. C* **1994**, *49*, 2950–2962. [CrossRef]
64. Wiringa, R.B.; Stoks, V.G.J.; Schiavilla, R. An Accurate nucleon-nucleon potential with charge independence breaking. *Phys. Rev. C* **1995**, *51*, 38–51. [CrossRef] [PubMed]
65. Grangé, P.; Lejeune, A.; Martzolff, M.; Mathiot, J.F. Consistent three-nucleon forces in the nuclear many-body problem. *Phys. Rev. C* **1989**, *40*, 1040–1060. [CrossRef] [PubMed]
66. Baldo, M.; Bombaci, I.; Burgio, G.F. Microscopic nuclear equation of state with three-body forces and neutron star structure. *Astron. Astrophys.* **1997**, *328*, 274–282.
67. Zuo, W.; Lejeune, A.; Lombardo, U.; Mathiot, J.F. Microscopic three-body force for asymmetric nuclear matter. *Eur. Phys. J. A* **2002**, *14*, 469–475. [CrossRef]
68. Li, Z.H.; Lombardo, U.; Schulze, H.J.; Zuo, W. Consistent nucleon-nucleon potentials and three-body forces. *Phys. Rev. C* **2008**, *77*, 034316. [CrossRef]
69. Li, Z.H.; Schulze, H.J. Neutron star structure with modern nucleonic three-body forces. *Phys. Rev. C* **2008**, *78*, 028801. [CrossRef]
70. Pudliner, B.S.; Pandharipande, V.R.; Carlson, J.; Wiringa, R.B. Quantum Monte Carlo calculations of A <= 6 nuclei. *Phys. Rev. Lett.* **1995**, *74*, 4396–4399. [CrossRef]
71. Pudliner, B.S.; Pandharipande, V.R.; Carlson, J.; Pieper, S.C.; Wiringa, R.B. Quantum Monte Carlo calculations of nuclei with A <= 7. *Phys. Rev. C* **1997**, *56*, 1720–1750. [CrossRef]
72. Baldo, M.; Fukukawa, K. Nuclear Matter from Effective Quark-Quark Interaction. *Phys. Rev. Lett.* **2014**, *113*, 242501. [CrossRef]
73. Fukukawa, K.; Baldo, M.; Burgio, G.F.; Lo Monaco, L.; Schulze, H.J. Nuclear matter equation of state from a quark-model nucleon-nucleon interaction. *Phys. Rev. C* **2015**, *92*, 065802. [CrossRef]
74. Gandolfi, S.; Illarionov, A.Y.; Fantoni, S.; Miller, J.C.; Pederiva, F.; Schmidt, K.E. Microscopic calculation of the equation of state of nuclear matter and neutron star structure. *Mon. Not. Roy. Astron. Soc.* **2010**, *404*, L35–L39. [CrossRef]
75. Negele, J.W.; Vautherin, D. Neutron star matter at subnuclear densities. *Nucl. Phys. A* **1973**, *207*, 298–320. [CrossRef]
76. Baym, G.; Pethick, C.; Sutherland, P. The Ground state of matter at high densities: Equation of state and stellar models. *Astrophys. J.* **1971**, *170*, 299–317. [CrossRef]
77. Feynman, R.P.; Metropolis, N.; Teller, E. Equations of State of Elements Based on the Generalized Fermi-Thomas Theory. *Phys. Rev.* **1949**, *75*, 1561–1573. [CrossRef]
78. Audi, G.; Wapstra, A.H.; Thibault, C. The AME2003 atomic mass evaluation. (II). Tables, graphs and references. *Nucl. Phys. A* **2003**, *729*, 337–676. [CrossRef]
79. de Vries, H.; de Jager, C.W.; de Vries, C. Nuclear Charge-Density-Distribution Parameters from Electron Scattering. *At. Data Nucl. Data Tables* **1987**, *36*, 495. [CrossRef]

80. Colò, G.; van Giai, N.; Meyer, J.; Bennaceur, K.; Bonche, P. Microscopic determination of the nuclear incompressibility within the nonrelativistic framework. *Phys. Rev. C* **2004**, *70*, 024307. [CrossRef]
81. Piekarewicz, J. Unmasking the nuclear matter equation of state. *Phys. Rev. C* **2004**, *69*, 041301. [CrossRef]
82. Fuchs, C.; Faessler, A.; Zabrodin, E.; Zheng, Y.M. Probing the Nuclear Equation of State by K^+ Production in Heavy-Ion Collisions. *Phys. Rev. Lett.* **2001**, *86*, 1974–1977. [CrossRef]
83. Garg, U.; Li, T.; Okumura, S.; Akimune, H.; Fujiwara, M.; Harakeh, M.N.; Hashimoto, H.; Itoh, M.; Iwao, Y.; Kawabata, T.; et al. The Giant Monopole Resonance in the Sn Isotopes: Why is Tin so "Fluffy"? *Nucl. Phys. A* **2007**, *788*, 36–43. [CrossRef]
84. Klimkiewicz, A.; Paar, N.; Adrich, P.; Fallot, M.; Boretzky, K.; Aumann, T.; Cortina-Gil, D.; Pramanik, U.D.; Elze, T.W.; Emling, H.; et al. Nuclear symmetry energy and neutron skins derived from pygmy dipole resonances. *Phys. Rev. C* **2007**, *76*, 051603. [CrossRef]
85. Carbone, A.; Colò, G.; Bracco, A.; Cao, L.G.; Bortignon, P.F.; Camera, F.; Wieland, O. Constraints on the symmetry energy and neutron skins from pygmy resonances in Ni68 and Sn132. *Phys. Rev. C* **2010**, *81*, 041301. [CrossRef]
86. Chen, L.W.; Ko, C.M.; Li, B.A. Determination of the Stiffness of the Nuclear Symmetry Energy from Isospin Diffusion. *Phys. Rev. Lett.* **2005**, *94*, 032701. [CrossRef] [PubMed]
87. Danielewicz, P.; Lee, J. Symmetry energy I: Semi-infinite matter. *Nucl. Phys. A* **2009**, *818*, 36–96. [CrossRef]
88. Brown, B.A. Neutron Radii in Nuclei and the Neutron Equation of State. *Phys. Rev. Lett.* **2000**, *85*, 5296–5299. [CrossRef] [PubMed]
89. Typel, S.; Brown, B.A. Neutron radii and the neutron equation of state in relativistic models. *Phys. Rev. C* **2001**, *64*, 027302. [CrossRef]
90. Horowitz, C.J.; Pollock, S.J.; Souder, P.A.; Michaels, R. Parity violating measurements of neutron densities. *Phys. Rev. C* **2001**, *63*, 025501. [CrossRef]
91. Roca-Maza, X.; Centelles, M.; Viñas, X.; Warda, M. Neutron Skin of Pb208, Nuclear Symmetry Energy, and the Parity Radius Experiment. *Phys. Rev. Lett.* **2011**, *106*, 252501. [CrossRef]
92. Centelles, M.; Roca-Maza, X.; Viñas, X.; Warda, M. Origin of the neutron skin thickness of Pb208 in nuclear mean-field models. *Phys. Rev. C* **2010**, *82*, 054314. [CrossRef]
93. Fuchs, C. pp i Kaon production in heavy ion reactions at intermediate energies [review article]. *Prog. Part. Nucl. Phys.* **2006**, *56*, 1–103. [CrossRef]
94. Tews, I.; Lattimer, J.M.; Ohnishi, A.; Kolomeitsev, E.E. Symmetry Parameter Constraints from a Lower Bound on Neutron-matter Energy. *Astrophys. J.* **2017**, *848*, 105. [CrossRef]
95. Zhang, N.B.; Cai, B.J.; Li, B.A.; Newton, W.G.; Xu, J. How tightly is nuclear symmetry energy constrained by unitary Fermi gas? *Nucl. Sci. Tech.* **2017**, *28*, 181. [CrossRef]
96. Margueron, J.; Hoffmann Casali, R.; Gulminelli, F. Equation of state for dense nucleonic matter from metamodeling. I. Foundational aspects. *Phys. Rev. C* **2018**, *97*, 025805. [CrossRef]
97. Shlomo, S.; Kolomietz, V.M.; Colò, G. Deducing the nuclear-matter incompressibility coefficient from data on isoscalar compression modes. *EPJA* **2006**, *30*, 23 30. [CrossRef]
98. Piekarewicz, J. Do we understand the incompressibility of neutron-rich matter? *J. Phys. Nucl. Phys.* **2010**, *37*, 064038. [CrossRef]
99. Li, B.A.; Han, X. Constraining the neutron-proton effective mass splitting using empirical constraints on the density dependence of nuclear symmetry energy around normal density. *Phys. Lett. B* **2013**, *727*, 276–281. [CrossRef]
100. Taylor, J.H. Pulsar Timing and Relativistic Gravity. *Philos. Trans. R. Soc. Lond. Ser. A* **1992**, *341*, 117–134. [CrossRef]
101. Lattimer, J.M.; Steiner, A.W. Neutron Star Masses and Radii from Quiescent Low-Mass X-ray Binaries. *Astrophys. J.* **2014**, *784*, 123. [CrossRef]
102. Guillot, S.; Rutledge, R.E. Rejecting Proposed Dense Matter Equations of State with Quiescent Low-mass X-Ray Binaries. *Astroph. J. Lett.* **2014**, *796*, L3. [CrossRef]
103. Shapiro, S.L.; Teukolsky, S.A. *Black Holes, White Dwarfs and Neutron Stars: The Physics of Compact Objects*; John Wiley & Sons, New York: NY, USA, 2008.
104. Baldo, M.; Burgio, G.F.; Schulze, H.J. Onset of hyperon formation in neutron star matter from Brueckner theory. *Phys. Rev. C* **1998**, *58*, 3688–3695. [CrossRef]

105. Baldo, M.; Burgio, G.F.; Schulze, H.J. Hyperon stars in the Brueckner-Bethe-Goldstone theory. *Phys. Rev. C* **2000**, *61*, 055801. [CrossRef]
106. Lejeune, A.; Grange, P.; Martzolff, M.; Cugnon, J. Hot nuclear matter in an extended Brueckner approach. *Nucl. Phys. A* **1986**, *453*, 189–219. [CrossRef]
107. Zuo, W.; Bombaci, I.; Lombardo, U. Asymmetric nuclear matter from an extended Brueckner-Hartree-Fock approach. *Phys. Rev. C* **1999**, *60*, 024605. [CrossRef]
108. Bombaci, I.; Lombardo, U. Asymmetric nuclear matter equation of state. *Phys. Rev. C* **1991**, *44*, 1892–1900. [CrossRef]
109. Shibata, M.; Fujibayashi, S.; Hotokezaka, K.; Kiuchi, K.; Kyutoku, K.; Sekiguchi, Y.; Tanaka, M. Modeling GW170817 based on numerical relativity and its implications. *Phys. Rev. D* **2017**, *96*, 123012. [CrossRef]
110. Margalit, B.; Metzger, B.D. Constraining the Maximum Mass of Neutron Stars From Multi-Messenger Observations of GW170817. *Astrophys. J.* **2017**, *850*, L19. [CrossRef]
111. Rezzolla, L.; Most, E.R.; Weih, L.R. Using gravitational-wave observations and quasi-universal relations to constrain the maximum mass of neutron stars. *Astrophys. J.* **2018**, *852*, L25. [CrossRef]
112. Shibata, M.; Zhou, E.; Kiuchi, K.; Fujibayashi, S. Constraint on the maximum mass of neutron stars using GW170817 event. *Phys. Rev. D* **2019**, *100*, 023015. [CrossRef]
113. Hinderer, T. Tidal Love numbers of neutron stars. *Astrophys. J.* **2008**, *677*, 1216–1220. [CrossRef]
114. Hinderer, T. Erratum: "Tidal Love Numbers of Neutron Stars" (2008, ApJ, 677, 1216). *Astrophys. J.* **2009**, *697*, 964. [CrossRef]
115. Hinderer, T.; Lackey, B.D.; Lang, R.N.; Read, J.S. Tidal deformability of neutron stars with realistic equations of state and their gravitational wave signatures in binary inspiral. *Phys. Rev. D* **2010**, *81*, 123016. [CrossRef]
116. Lattimer, J.M.; Prakash, M. Neutron star observations: Prognosis for equation of state constraints. *Phys. Rep.* **2007**, *442*, 109–165. [CrossRef]
117. Most, E.R.; Weih, L.R.; Rezzolla, L.; Schaffner-Bielich, J. New constraints on radii and tidal deformabilities of neutron stars from GW170817. *Phys. Rev. Lett.* **2018**, *120*, 261103. [CrossRef] [PubMed]
118. Lim, Y.; Holt, J.W. Neutron star tidal deformabilities constrained by nuclear theory and experiment. *Phys. Rev. Lett.* **2018**, *121*, 062701. [CrossRef] [PubMed]
119. Malik, T.; Alam, N.; Fortin, M.; Providência, C.; Agrawal, B.K.; Jha, T.K.; Kumar, B.; Patra, S.K. GW170817: Constraining the nuclear matter equation of state from the neutron star tidal deformability. *Phys. Rev. C* **2018**, *98*, 035804. [CrossRef]
120. Burgio, G.F.; Drago, A.; Pagliara, G.; Schulze, H.J.; Wei, J.B. Are Small Radii of Compact Stars Ruled out by GW170817/AT2017gfo? *Astrophys. J.* **2018**, *860*, 139. [CrossRef]
121. Radice, D.; Dai, L. Multimessenger parameter estimation of GW170817. *EPJA* **2019**, *55*, 50. [CrossRef]
122. Kiuchi, K.; Kyutoku, K.; Shibata, M.; Taniguchi, K. Revisiting the Lower Bound on Tidal Deformability Derived by AT 2017gfo. *Astrophys. J. Lett.* **2019**, *876*, L31. [CrossRef]
123. Wei, J.B.; A., F.; Burgio, G.F.; Chen, H.; Schulze, H.J. Neutron star universal relations with microscopic equations of state. *J. Phys. G Nucl. Part. Phys.* **2019**, *46*, 034001. [CrossRef]
124. Capano, C.D.; Tews, I.; Brown, S.M.; Margalit, B.; De S.; Kumar, S.; Brown, D.A.; Krishnan, B.; Reddy, S. Stringent constraints on neutron-star radii from multimessenger observations and nuclear theory. *Nat. Astron.* **2020**, *4*, 625. [CrossRef]
125. De S.; Finstad, D.; Lattimer, J.M.; Brown, D.A.; Berger, E.; Biwer, C.M. Tidal Deformabilities and Radii of Neutron Stars from the Observation of GW170817. *Phys. Rev. Lett.* **2018**, *121*, 091102; Erratum in **2018**, *121*, 259902. [CrossRef] [PubMed]
126. Wei, J.B.; Lu, J.J.; Burgio, G.F.; Li, Z.H.; Schulze, H.J. Are nuclear matter properties correlated to neutron star observables? *Eur. Phys. J. A* **2020**, *56*, 63. [CrossRef]
127. Zhao, T.; Lattimer, J. Tidal deformabilities and neutron star mergers. *Phys. Rev. D* **2018**, *98*, 063020. [CrossRef]
128. Tsang, C.Y.; Tsang, M.B.; Danielewicz, P.; Fattoyev, F.J.; Lynch, W.G. Insights on Skyrme parameters from GW170817. *Phys. Lett. B* **2019**, *796*, 1–5. [CrossRef]
129. Hebeler, K.; Lattimer, J.M.; Pethick, C.J.; Schwenk, A. Constraints on Neutron Star Radii Based on Chiral Effective Field Theory Interactions. *Phys. Rev. Lett.* **2010**, *105*, 161102. [CrossRef] [PubMed]
130. Hebeler, K.; Lattimer, J.M.; Pethick, C.J.; Schwenk, A. Equation of State and Neutron Star Properties Constrained by Nuclear Physics and Observation. *Astrophys. J.* **2013**, *773*, 11. [CrossRef]

131. Steiner, A.W.; Prakash, M.; Lattimer, J.M.; Ellis, P.J. Isospin asymmetry in nuclei and neutron stars [review article]. *Phys. Rep.* **2005**, *411*, 325–375. [CrossRef]
132. Centelles, M.; Roca-Maza, X.; Viñas, X.; Warda, M. Nuclear Symmetry Energy Probed by Neutron Skin Thickness of Nuclei. *Phys. Rev. Lett.* **2009**, *102*, 122502. [CrossRef]
133. Horowitz, C.J.; Piekarewicz, J. Neutron Star Structure and the Neutron Radius of ^{208}Pb. *Phys. Rev. Lett.* **2001**, *86*, 5647–5650. [CrossRef]
134. Furnstahl, R.J. Neutron radii in mean-field models. *Nucl. Phys. A* **2002**, *706*, 85–110. [CrossRef]
135. Vidaña, I.; Providência, C.; Polls, A.; Rios, A. Density dependence of the nuclear symmetry energy: A microscopic perspective. *Phys. Rev. C* **2009**, *80*, 045806. [CrossRef]
136. Abrahamyan, S.; others. Measurement of the Neutron Radius of 208Pb Through Parity-Violation in Electron Scattering. *Phys. Rev. Lett.* **2012**, *108*, 112502. [CrossRef] [PubMed]
137. PREX/CREX Collaboration. CREX Run Plan. Available online: https://hallaweb.jlab.org/parity/prex/CREXrunPlan_2019Oct6_mcnulty.pdf (accessed on 7 August 2020).
138. MREX Collaboration. Available online: https://indico.mitp.uni-mainz.de/event/47/contributions/1634/attachments/1355/1426/MREX.pdf (accessed on 7 August 2020).

© 2020 by the authors. Licensee MDPI, Basel, Switzerland. This article is an open access article distributed under the terms and conditions of the Creative Commons Attribution (CC BY) license (http://creativecommons.org/licenses/by/4.0/).

Article

Evolution of Quasiperiodic Structures in a Non-Ideal Hydrodynamic Description of Phase Transitions

D. N. Voskresensky [1,2]

[1] Joint Institute for Nuclear Research, RU-141980 Dubna, Moscow Region, Russia; d.voskresen@gmail.com
[2] National Research Nuclear University (MEPhI), Kashirskoe shosse 31, 115409 Moscow, Russia

Received: 28 January 2020; Accepted: 3 March 2020; Published: 7 March 2020

Abstract: Various phase transitions could have taken place in the early universe, and may occur in the course of heavy-ion collisions and supernova explosions, in proto-neutron stars, in cold compact stars, and in the condensed matter at terrestrial conditions. Most generally, the dynamics of the density and temperature at first- and second-order phase transitions can be described with the help of the equations of non-ideal hydrodynamics. In the given work, some novel solutions are found describing the evolution of quasiperiodic structures that are formed in the course of the phase transitions. Although this consideration is very general, particular examples of quark-hadron and nuclear liquid-gas first-order phase transitions to the uniform $k_0 = 0$ state and of a pion-condensate second-order phase transition to a non-uniform $k_0 \neq 0$ state in dense baryon matter are considered.

Keywords: dynamics of phase transitions; spinodal instability; heavy-ion collisions; neutron stars

---

1. Introduction

Cosmological observations of the two last decades [1] have supplied us with some extraordinary results and puzzles. Particularly important is the fact that the universe undergoes an accelerated expansion and the fact that only 5% of its mass is contained in baryons, 26% is in dark matter, and the remaining part is in dark energy. It is commonly believed that at least two cosmic phase transitions occurred in the early universe, the electro-weak and the QCD phase transitions [2,3]. The standard model of particle physics predicts that, after the inflation, the hot expanding universe was filled with deconfined quarks in the state of quark-gluon plasma [4]. The quark-gluon plasma in baryon-poor matter persists down to a temperature of $T \simeq 160$ MeV. Whether the quark-hadron transition is a first-order phase transition, a second-order transition, or a crossover is still not completely settled. This view on the early universe is supported by simulations done in various cosmological and relativistic heavy-ion collision models [5,6] and by the lattice calculations. The latter calculations support the QCD crossover transition obtained by the HotQCD Collaboration [7,8]. Nucleosynthesis is affected by remnant inhomogeneities in the baryon-to-entropy ratio and in isospin [9]. These problems can be considered within the standard model. However, the standard model does not account for the presence of the dark matter with which additional cosmic phase transitions may be associated during the cooling of the expanding universe to its present temperature $T \simeq 2.7$ K, cf. [10].

Another piece of important information about strongly interacting matter can be extracted from neutrino and photon radiation of compact stars formed in supernova events [11,12] and from analysis of gravitational waves in gamma ray bursts. A strong phase transition may result in a second neutrino burst occurring during a supernova explosion and a hot neutron star formation with a typical minute time delay. It might be also associated with a larger time delay related to the slow heat transport to the neutron-star surface, if the system is close to the pion-condensate phase transition [12,13]. Recently, new arguments have been expressed showing that indeed two neutrino bursts were measured during the 1987A explosion, and that one delayed respectively the other by 4.7 h, cf. [14]. The second burst and a blowing of some amount of matter could then be related to the phase transition of the neutron star to the pion-condensate state. A first-order deconfinement transition can serve as an explosion mechanism for massive blue supergiant stars of $M \sim 50 M_{\odot}$, for which so far no explosion mechanism is known [15]. In old neutron stars, the first-order phase transition, if it occurs, could result in a blowing of matter or in a strong star-quake [12,16]. The detection of merging compact stars in the gravitational wave spectra [17] and the detection of massive compact stars [18–21] provide constraints on the equation of state of strongly interacting dense matter and strong phase transitions in it. The dominant postmerger gravitational-wave frequency $f_{\rm peak}$ may exhibit a significant deviation from an empirical relation between $f_{\rm peak}$ and the tidal deformability, if a strong first-order phase transition leads to the formation of a gravitationally stable extended quark matter core in the postmerger remnant [22]. Thus, one may feasibly identify observable imprints of a first-order hadron-quark phase transition at supranuclear densities on the gravitational-wave emission of neutron star mergers.

An experimental study of the ultrarelativistic heavy-ion collisions helps to simulate at the terrestrial conditions the processes that have occurred in the very early universe, in supernova explosions and in gamma ray bursts. Experimental data and lattice calculations [7,8] indicate that the hadron-quark transition in heavy-ion collisions at RHIC and LHC collision energies is the crossover transition, cf. [23–25]. For lower collision energies relevant for NICA and FAIR facilities, one expects to find signatures of the strong first-order quark-hadron phase transition [26]. There are experimental evidences that, in the very low-energetic heavy-ion collisions of approximately isospin-symmetrical nuclei, there is a first-order nuclear liquid-gas phase transition (for temperatures $T \lesssim 20$ MeV and baryon densities $n \lesssim 0.7 n_0$, where n_0 is the nuclear saturation density) [27–29].

In a many-component system, a mechanical instability is accompanied by a chemical instability, see [30,31]. The inclusion of the Coulomb interaction, see [32,33], leads to a possibility of the pasta phase in the neutron star crusts for densities $0.3 n_0 \lesssim n \lesssim 0.7 n_0$. For higher densities in dense neutron star interiors, there may be phase transitions to the pion [12,34], kaon [35,36], and charged rho [37] condensate states and to the quark matter [35,38–41]. The quark-hadron, pion, kaon, and charged rho-meson condensate phase transitions may occur during the iso-entropical falling of the baryon-rich matter in supernova explosions [11], in proto-neutron stars, and in cold compact stars, cf. [12]. In some models, these phase transitions are considered first-order phase transitions leading to mixed phases in dense matter. The formation of the pasta non-uniform phases is one of the possibilities [36,40,41]. We can add here the possibilities of the phase transitions between various superfluid [42,43] and ferromagnetic-superfluid [44] phases in the cold neutron stars and in the color-superconducting hybrid compact stars [45], as well as numerous possibilities of the phase transitions in the condensed matter physics at terrestrial conditions, such as liquid-gas, liquid-glass, and glass-metal transitions.

The liquid-gas phase transition, the transition to the superfluid state in quantum liquids, and many other transitions occur in a uniform state characterized by the wave number $k_0 = 0$. Other phase transitions, such as the transitions in solids and liquid crystals, are transitions to inhomogeneous states characterized by the non-zero wave-vectors, $\vec{k}_{0,i} \neq 0$, cf. [46,47]. In glasses, the order, characterized by $k_0 \neq 0$, appears at rather short distances but disappears at long distances [47]. The phase transition to the pion-condensate state [12,34] possible in the interiors of neutron stars may occur due to a strong p-wave pion-baryon attraction, which increases with the increase of the baryon density. Thereby, the pion-condensation occurs in a non-uniform state, $k_0 \neq 0$. Chiral condensate, which is constant in a vacuum, may also become spatially modulated at high densities, where, in the traditional picture of the QCD phase diagram, a first-order chiral phase transition occurs. Examples of inhomogeneous phases are the chiral density wave, the Skyrme crystal, and crystalline color superconductors, cf. [48]. Perhaps the antikaon condensation in dense baryon matter also occurs in the non-uniform state, $k_0 \neq 0$, cf. [49].

Some of the mentioned phase transitions, such as the transition of the normal matter to superfluid in metals and in ^{4}He, are transitions of the second order [50]. Other phase transitions mentioned above, such as the liquid-gas phase transition, are transitions of the first order. The search for the critical endpoint separating the crossover and first-order quark-hadron transitions is one of the benchmarks for future experiments at NICA and FAIR.

Finite size structures are important in the context of the general relativistic evolution of density perturbations in the early universe [10]. Even the simple case of dust baryon matter gives the right order of magnitude for globular star clusters with the corresponding Jeans mass and wavelength. The presence of a substantial amount of homogeneous scalar field energy density at low redshifts inhibits the growth of perturbations in the baryonic fluid [51]. For example, dark matter may result from the transition of a non-minimally coupled scalar field from radiation to collision-less matter. Dynamical instabilities of the field fluctuations, which are typical for oscillatory scalar field regimes, can be amplified and transmitted by the coupling to dark matter perturbations, cf. [52]. The presence of the dark matter may also trigger strong electro-weak phase transition in the early universe [53].

In the early universe, at the processes of the formation of compact stars in supernova explosions, collisions of compact binary stars, and in heavy-ion collisions, one deals with a rapid thermalization of a strongly interacting quark-gluon matter and then the hadronic matter. These processes can be described within non-ideal hydrodynamics, where viscosity and thermal conductivity effects are of crucial importance. The dynamics of the phase transitions can also be considered within non-ideal hydrodynamics, cf. [47,54–56].

Below, some novel solutions will be found describing the evolution of periodic structures at second-order phase transitions to the non-uniform state with the wave number $k_0 \neq 0$. Quasiperiodic time-dependent structures appear in the course of the spinodal instabilities at the first-order phase transitions to the uniform state with $k_0 = 0$ and in the dynamics of the second-order phase transitions occurring in the uniform state. Although consideration is very general, quark-hadron and nuclear liquid-gas first-order phase transitions and the pion condensation second-order transition will be considered as examples.

The presentation is organized as follows. In Section 2, the main features of the van der Waals-like equation of state are reviewed. In Section 3, a hydrodynamical description of the first- and second-order phase transitions to the uniform, $k_0 = 0$ state and for the second-order phase transition to the nonuniform, $k_0 \neq 0$ state is formulated assuming a small overcriticality. The dynamics of seeds at the first-order phase transition from a metastable to the stable state is considered in Section 4. The dynamics of fluctuations in the unstable region is studied in Section 5. Some novel solutions describing the time evolution of quasiperiodic and periodic structures are found. Section 6 contains concluding remarks.

2. Van Der Waals-Like Equation of State for a Description of First-Order Phase Transitions

The dynamical trajectories of the expanding baryon-rich matter in the heavy-ion collisions and of the falling matter in supernova explosions before phase transition can be characterized by an approximately constant entropy, while the volume V and the temperature T are time-dependent. A description of the first-order phase transition is more involved. In the simplest case of one-component matter, e.g., of the baryon matter, the pressure–baryon number density isotherms $P(n)|_T$ describing the liquid-like (with a higher density) or gas-like (with a smaller density) states demonstrate a monotonous behavior for the values of temperature T above the critical temperature of the first-order phase transition of the liquid-gas type. However, for T values below the critical temperature, $P(n)|_T$ isotherms acquire a convex-concave form [47], see Figure 1. The horizontal dashed line connecting points A and D shows the Maxwell construction (MC) describing thermal equilibrium of phases. At equilibrium, the baryon chemical potentials are $\mu_A = \mu_D$. The interval AB corresponds to a metastable supercooled vapor (SV) and the interval CD relates to a metastable overheated liquid (OL). The interval BC shows an unstable spinodal region.

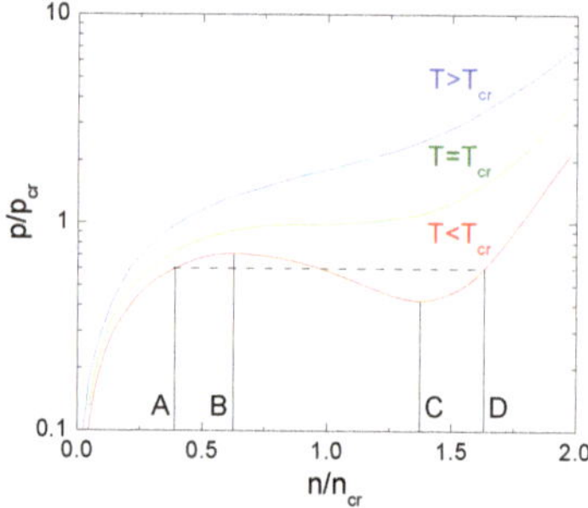

Figure 1. Schematic pressure isotherms as functions of the number density n at a liquid-gas-like phase transition. P_{cr}, n_{cr}, and T_{cr} are the pressure, density, and temperature at the critical point.

Adiabatic trajectories $\tilde{s}_{cr}$ and $\tilde{s}_m$, where $\tilde{s} \equiv s/n \simeq$ const, and s is the entropy density, are shown in Figure 2 on the plot of $T/T_{cr} = f(n/n_{cr})$ by the short dashed lines. The upper convex curve, MC, the bold solid line, demonstrates the boundary of the MC, the bold dashed line, ITS, shows the boundary of the isothermal spinodal region, and the bold dash-dotted curve, AS, indicates the boundary of the adiabatic spinodal region. At the ITS line, $u_T^2 = (\partial P/\partial \rho)_T = 0$; at the AS line, $u_{\tilde{s}}^2 = (\partial P/\partial \rho)_{\tilde{s}} = 0$, where u_T and $u_{\tilde{s}}$ have the meaning of the isothermal and adiabatic sound velocities, respectively, $\rho = m^* n$, and m^* is the baryon quasiparticle mass. The supercooled vapor (SV) and the overheated liquid (OL) regions are situated between the MC and the ITS curves, on the left and on the right, respectively. For $\tilde{s}_{cr} > \tilde{s} > \tilde{s}_{MC2}$, where $\tilde{s}_{cr}$ is the value of the specific entropy $\tilde{s}$ at the critical point and the line with $\tilde{s}_{MC2}$ in the example shown in Figure 2 passes through the point $n/n_{cr} = 3$ at $T = 0$, the system traverses the OL state (the region OL in Figure 2), the ITS region (below the ITS line), and the AS region (below the AS line). For $\tilde{s} > \tilde{s}_{cr}$, the system trajectory passes through the SV state (the region SV in Figure 2) and the ITS region.

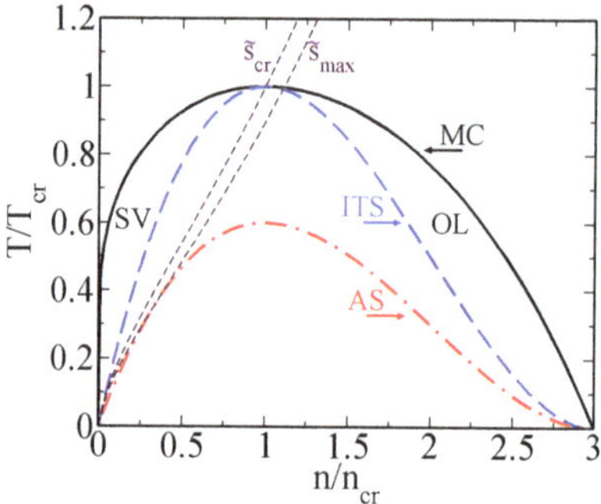

Figure 2. The phase diagram of the van der Waals equation of state on the $T(n)$-plane. The bold solid, dashed, and dash-dotted curves show the boundaries of the MC, the spinodal region at T =const, and $\tilde{s}$ =const, respectively. The short dashed lines show two adiabatic trajectories of the system evolution: the curve labeled $\tilde{s}_{cr}$ passes through the critical point; $\tilde{s}_{max}$ passes through the maximum pressure point $P(n_{P,max})$ on the $P(n)$ plane, cf. [56].

Note that, in reality, for the quark-hadron first-order phase transition, the phase diagram looks slightly different, since then T_{cr} increases with a decrease of the baryon density [57,58]. However, this peculiarity does not change a general analysis given here.

When the adiabatic trajectory $\tilde{s} = const$ enters the region of the first-order phase transition (the region below the solid curve in Figure 2), the approximation of the constant entropy fails, and a further description of the dynamics of the system requires a solution of non-ideal hydrodynamical equations [54–56]. Similarly, the description of the dynamics of the second-order phase transition requires a solution of non-ideal hydrodynamical equations in the case where the density and the temperature (or entropy) can be considered appropriate order parameters.

3. Hydrodynamical Description of First- and Second-Order Phase Transitions at Small Overcriticality

Assume that the dynamics of a second-order phase transition and of a first-order phase transition can be described by the variables n and s (or T), cf. [54–56]. Moreover, assume that the system is rather close to the critical point of the phase transition. Since all the processes in the vicinity of the critical point are slowed down, the velocity of a seed of a new phase prepared in the old phase, $\vec{u}$, is much less than the mean thermal velocity, and thereby we may use equations of non-relativistic non-ideal hydrodynamics: the Navier-Stokes equation, the continuity equation, and the equation for the heat transport, even if we deal with violent heavy-ion collisions:

$$
\begin{aligned}
m^* n\left[\partial_t u_i + (\vec{u}\nabla) u_i\right] &= -\nabla_i P + \nabla_k\left[\eta\left(\nabla_k u_i + \nabla_i u_k - \frac{2}{\nu}\delta_{ik}\mathrm{div}\vec{u}\right) + \zeta\delta_{ik}\mathrm{div}\vec{u}\right], \quad (1)\\
\partial_t n + \mathrm{div}(n\vec{u}) &= 0, \quad (2)\\
T\left[\frac{\partial s}{\partial t} + \mathrm{div}(s\vec{u})\right] &= \mathrm{div}(\kappa\nabla T) + \eta\left(\nabla_k u_i + \nabla_i u_k - \frac{2}{\nu}\delta_{ik}\mathrm{div}\vec{u}\right)^2 + \zeta(\mathrm{div}\vec{u})^2. \quad (3)
\end{aligned}
$$

Here, as above, n is the number density of the conserving charge, to be specific, the baryon density, m^* is the baryon quasiparticle mass, and P is the pressure. The quantities η and ζ are the shear and bulk viscosities, $\nu = 3, 2, 1$ shows the geometry of the seed under consideration (droplets, rods, and slabs), and κ is the thermal conductivity. The treatment of the evolution of seeds within relativistic hydrodynamics is more involved, but this is beyond our scope, cf. [55,59].

All thermodynamical quantities can be expanded near a reference point (n_r, T_r), which we assume to be close to the critical point but still outside the fluctuation region, which we assume to be narrow. This circumstance is important for the determination of the specific heat density $c_{V,r}$ and, m.b., transport coefficients, which may diverge in the critical point, whereas other quantities are smooth functions of n, T, and, by calculating them, one can have that $n_r = n_{cr}, T_r = T_{cr}$.

The Landau free energy, δF_L, counted from the value at $n_r \simeq n_{cr}, T_r \simeq T_{cr}$ in the variables $\delta n = n - n_{cr}, \delta T = T - T_{cr}$, and $\delta(\delta F_L)/\delta(\delta n) = P - P_f + P_{MC}$, can be presented as [54–56]

$$
\delta F_L = \int \frac{d^3x}{n_{cr}}\left\{\frac{cm^*[\nabla(\delta n)]^2}{2} + \frac{\lambda m^{*3}(\delta n)^4}{4} - \frac{\lambda v^2 m^*(\delta n)^2}{2} - \epsilon\delta n\right\} + \delta F_L(k_0), \quad (4)
$$

where $\epsilon = P_f - P_{MC} \sim n_{cr}(\mu_i - \mu_f)$ is expressed through the (final) value of the pressure after the first-order phase transition has occurred, and the pressures at the MC, μ_i and μ_f, are the chemical potentials of the initial and final configurations (at fixed P and T). The quantity $\epsilon \neq 0$ if one deals with a first-order phase transition, and $\epsilon = 0$ if a transition is of the second order. The maximum of the quantity ϵ is $\epsilon_m = 4\lambda v^3/(3\sqrt{3})$. For the description of phase transitions to the uniform state, $k = 0$, one may retain only the term $\propto c[\nabla(\delta n)]^2$ in the expansion of the free energy in the density gradients using $c > 0$. For the description of phase transitions to the non-uniform state, $k_0 \neq 0$, one should perform expansion retaining terms at least up to $\propto d[\Delta(\delta n)]^2$ assuming $c < 0$ and $d > 0$. Therefore, the last term in Equation (4) appears only if $k = k_0 \neq 0$ [47], as is the case for the phase transition to the solid state, liquid crystal state, or pion-condensate state in dense nuclear matter. Thus, for $k_0 \neq 0$ and $c < 0, d > 0$, we have

$$
\delta F_L(k_0) = \int \frac{d^3x}{n_{cr}}\left\{\frac{dm^*}{2}(\Delta\delta n)^2 - \left(\frac{cm^*k_0^2}{2} + \frac{dm^*k_0^4}{2}\right)(\delta n)^2\right\}, \quad (5)
$$

where $k_0^2 = -\frac{c}{2d} > 0$ follows from the minimization of $\delta F_L(k_0)$. In the case of the phase transition to the uniform state, one should have $k_0 = 0, d = 0$ (then $\delta F_L(k_0) = 0$), and $c > 0$. Thus, the first term $\propto c$ in Equation (4) is associated with the positive surface tension, $\delta F_L^{\rm surf} = \sigma S$, where S is the surface of the seed.

The Landau free energy density and pressure as functions of the order parameter $\delta\rho/v$ for the equation of state determined by Equation (4) are shown in Figure 3. For $\epsilon = 0$, two minima of the Landau free energy coincide and correspond to the MC on the curve $\delta P(1/\rho)$ (shown by horizontal lines in the plot $\delta P(\delta\rho)$ in the right panel). If, in the initial state, $(\delta\rho)_i = \rho_i - \rho_{cr} = 0$, we deal with the spontaneous symmetry breaking and the second-order phase transition. For $(\delta\rho)_i = \rho_i - \rho_{cr} \neq 0$, $\epsilon > 0$, or $\epsilon < 0$, we deal either with the first-order phase transition from the metastable to the stable state, if ρ_i corresponds to the metastable state, or with the second-order phase transition either to the metastable state or to the stable state. For $\epsilon > 0$ (solid lines), the liquid state is stable and the gas state is metastable (SV); for $\epsilon < 0$ (dash-dotted lines), the liquid state is metastable (OL), whereas the gas state is stable. The dynamics of the transition starting from a point within the spinodal region for $\epsilon \neq 0$ (but small) is described similarly to that for the second-order phase transition for $\epsilon = 0$.

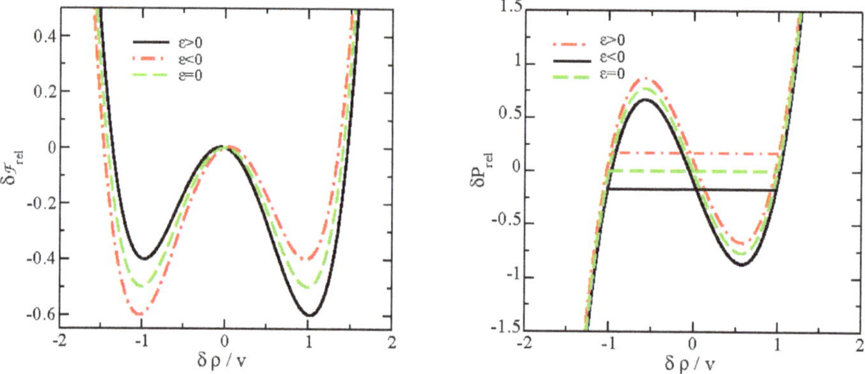

Figure 3. The Landau free energy density $\delta\mathcal{F}_{\rm rel} = \delta\mathcal{F}_L/\mathcal{F}_L(T_{cr},\rho_{cr})$ and the value $\delta P_{\rm rel} = \rho_{cr}\frac{\delta[F_L(T,\delta\rho)]}{\delta(\delta\rho)}|_T/P(T_{cr},\rho_{cr})$, as functions of the order parameter $\delta\rho = m^*\delta n$ for the EoS determined by Equation (4), at $T < T_{cr}$. The dashed horizontal line ($\epsilon = 0$) in the right panel shows MC, cf. [55].

For the purely van der Waals equation of state (in this case, $k_0 = 0$), one obtains [55]:

$$v^2(T) = -4\frac{\delta T n_{cr}^2 m^{*2}}{T_{cr}}, \quad \sigma = \sigma_0\frac{|\delta T|^{3/2}}{T_{cr}^{3/2}}, \quad \sigma_0^2 = 32m^* n_{cr}^2 T_{cr} c. \tag{6}$$

Applying operator div to Equation (1) and replacing div$\vec{u}$ from Equation (2) for small $\delta\rho$ and u, keeping only linear terms in u, which is legitimate, since near the critical point processes develop slowly ($v^2 \propto -\delta T$), we rewrite Equation (1) as

$$\begin{aligned} -\frac{\partial^2\delta n}{\partial t^2} &= \Delta\left[c\Delta\delta n + \lambda v^2\delta n - \lambda m^{*2}(\delta n)^3 + \epsilon/m^* - (m^* n_{cr})^{-1}\left(\tilde{\nu}\eta_{cr} + \zeta_{cr}\right)\frac{\partial\delta n}{\partial t}\right] \\ &\quad -\Delta\left[d\Delta^2\delta n + (ck_0^2 + dk_0^4)\delta n\right], \end{aligned} \tag{7}$$

$\tilde{\nu} = 2(\nu - 1)/\nu$, cf. [12,47,55]. The second line in Equation (7) yields a non-zero term only for the description of the condensation to the inhomogeneous state.

Consider $T < T_{cr}$. In the dimensionless variables $m^*\delta n = v\psi$, $\tau = t/t_0$, $\xi_i = x_i/l$, $i = 1, \cdots, \nu$, $\nu = 3$ for seeds of spherical geometry, Equation (7) is presented as

$$-\beta\frac{\partial^2\psi}{\partial\tau^2} = \Delta_\xi\left(\Delta_\xi\psi + 2\psi(1-\psi^2) + \widetilde{\epsilon} - \frac{\partial\psi}{\partial\tau} - \frac{\lambda v^2 d}{2c^2}\Delta_\xi^2\psi + \frac{2(ck_0^2 + dk_0^4)}{\lambda v^2}\psi\right), \tag{8}$$

$$l = \left(\frac{2c}{\lambda v^2}\right)^{1/2}, \quad t_0 = \frac{2\tilde{\eta}_{\rm r}}{\lambda v^2}, \quad \widetilde{\epsilon} = \frac{2\epsilon}{\lambda v^3}, \quad \beta = \frac{c}{\tilde{\eta}_{\rm r}^2}, \quad \tilde{\eta}_{\rm r} = \frac{(\tilde{v}\eta_{\rm r} + \zeta_{\rm r})}{m^* n_{cr}}.$$

It is important to notice that, even for $k_0 = 0$, Equation (8) differs in form from the standard Ginzburg-Landau equation broadly exploited in the condensed matter physics, since Equation (8) is of the second order in time derivatives, whereas the standard Ginzburg-Landau equation is of the first order in time derivatives. The difference disappears if one sets the bracketed-term in the r.h.s. of Equation (8) to zero. Such a procedure is, however, not legitimate at least for a description of the order parameter on an initial time-stage, since two initial conditions, such as $\delta n(t=0,\vec{r}) = 0$ and $\partial_t\delta n(t,\vec{r})|_{t=0} \simeq 0$, should be fulfilled to describe the evolution of an initially formed fluctuation (seed). Thereby, there exists at least an initial stage of the dynamics of seeds (for $t \lesssim t_{\rm init}$), which is not described by the standard Ginzburg-Landau equation [54,55]. The bracketed term in the r.h.s. of Equation (8) can indeed be set to zero, see below, if one considers an effectively very viscous medium at $\tau \gg 1$. Note also that Equation (8), with the bracketed term in the r.h.s. equal to zero, can be derived from the first-gradient order kinetic equation of Kadanoff-Baym [60].

Equation (8) should be supplemented by Equation (3) for the heat transport, which, owing to Equation (2) after its linearization, reads as

$$T_{cr}\left[\partial_t\delta s - s_{cr}(n_{cr})^{-1}\partial_t\delta n\right] = \kappa_{\rm r}\Delta\delta T\,. \tag{9}$$

The variation of the temperature is related to the variation of the entropy density $s[n,T]$ by

$$\delta T \simeq T_{cr}(c_{V,{\rm r}})^{-1}\left(\delta s - (\partial s/\partial n)_{T,cr}\delta n\right), \tag{10}$$

where c_V is the density of the heat capacity.

3.1. Typical Time Scales

Let us perform some rough dimensional estimates of typical time scales in the problem. The evolution of a seed of one phase in another phase is governed by the slowest mode ($\delta\rho$ or δs, respectively). The time scale for the relaxation of the density following Equation (8) is $t_0 \propto \tilde{\eta}$. Thus, the non-zero viscosity plays the role of the driving force managing the time evolution of the density mode. Moreover, $t_0 \propto 1/(T_{cr} - T)$. Thereby, the processes are slowed down near the critical point of the phase transition. The time scale for the relaxation of the entropy/temperature mode, following Equation (9), is

$$t_T = R_{\rm seed}^2 c_{V,{\rm r}}/\kappa_{\rm r} \propto R_{\rm seed}^2, \tag{11}$$

i.e., the relaxation time of the temperature/entropy is proportional to the surface of the seed. Thus, for $t_T(R_{\rm seed}) < t_0$, i.e for $R_{\rm seed} < R_{\rm fog} = \sqrt{\kappa_{\rm r} t_0/c_{V,{\rm r}}}$, where $R_{\rm fog}$ is the typical size of the seed at $t \sim t_0 = t_T$, the dynamics of the seeds is controlled by Equation (8) for the density mode. For seeds with sizes $R_{\rm seed} > R_{\rm fog}$, the quantity $t_T \propto R_{\rm seed}^2$ exceeds t_0, and the growth of seeds is slowed down. Thereby, the number of seeds with the typical size $R_{\rm seed} \sim R_{\rm fog}$ is increased with the passage of time, and a state of fog is formed. For the quark-hadron phase transition in energetic heavy-ion collisions, one [55] estimates $R_{\rm fog} \sim (0.1 - 1)$ fm and, for the nuclear liquid-gas transition at low energies, $R_{\rm fog} \sim (1-10)$ fm $\lesssim R(t_{\rm f.o.})$, where $R(t_{\rm f.o.})$ is the size of the fireball at the freeze-out, and $t_{\rm f.o.}$ is the fireball evolution time until freeze-out.

There are only two dimensionless parameters in Equation (8): $\widetilde{\epsilon}$ and β. The parameter $\widetilde{\epsilon}$ is responsible for a difference between the Landau free energies of the metastable and stable states. For $t_0 \gg t_T$ (the isothermal stage), $\widetilde{\epsilon} \simeq const$ and the dependence on this quantity disappears because of $\Delta_\xi \widetilde{\epsilon} \simeq 0$. Then, the dynamics is controlled by the parameter β, which characterizes the inertia. It is expressed in terms of the surface tension and the viscosity as

$$\beta = (32T_{cr})^{-1}[\widetilde{\nu}\eta_{\rm r} + \zeta_{\rm r}]^{-2}\sigma_0^2 m^*. \tag{12}$$

The larger viscosity and the smaller surface tension, the effectively more viscous (inertial) is the fluidity of seeds. For $\beta \ll 1$, one deals with the regime of effectively viscous (inertial) fluidity and at $\beta \gg 1$, one deals with the regime of almost perfect fluidity. Estimates [55] show that, for the nuclear liquid-gas phase transition, typically $\beta \sim 0.01$. For the quark-hadron transition, $\beta \sim 0.02 - 0.2$, even for a very low value of the $\eta/s \simeq 1/(4\pi)$ ratio. The latter quantity characterizes the fluidity of the matter at ultra-relativistic heavy-ion collisions [25]. Thus, as we argued, in the case of baryon-rich matter, one effectively deals with a very viscous (inertial) evolution of density fluctuations, in cases of nuclear liquid-gas and quark-hadron phase transitions.

In neutron stars, an overcritical pion-condensate drop reaches a size $R \sim 0.1$ km for $t \sim 10^{-3}$ s by the growth of the density mode. Then, it may reach $R \sim (1 - 10)$ km for typical time t_T varying from ~ 10 s for up to several hours (rather than for a typical collapse time $\sim 10^{-3}$ s). A delay appears owing to neutrino heat transport to the surface (an effect of neutrino thermal conductivity), and this delay strongly depends on the value of the pion softening, which is stronger for most massive neutron stars [12]. One should also take into account that the bulk viscosity is significantly increased in the presence of soft modes [61,62], e.g., near the pion condensation critical point [63]. Notice also that the description of the dynamics of the pion-condensate phase transition is specific, since the transition occurs to the inhomogeneous liquid-crystal-like state characterized by $\vec{k} \neq 0$. The seeds of the liquid-crystal-like state prove to be elongated in the process of their growth [47]. A similar effect is observed in liquid crystals.

Thus, the interplay between viscosity, surface tension, and thermal conductivity effects is responsible for the typical time and size scales of fluctuations.

3.2. Stationary Solutions

Now let us find stationary solutions of Equation (7). For the condensation in the state $k \neq 0$, we find a solution in the form

$$m^*\delta n = a[\sin(kx + \chi) + \frac{c_1\widetilde{\omega}^2(k^2)}{\widetilde{\omega}^2(9k^2)}\sin(3kx + \chi) + ...] + O(\epsilon), \tag{13}$$

where χ is a constant phase,

$$\widetilde{\omega}^2(k^2) = -\lambda v^2 + ck^2 + dk^4 - ck_0^2 - dk_0^4. \tag{14}$$

For the condensation in the uniform state $k_0 \neq 0$, $c < 0$, and $d > 0$, the gap $\widetilde{\omega}^2(k^2)$ has a minimum for $k = k_0$. The phase transition arises for $\widetilde{\omega}^2(k_0^2) < 0$. Setting Equation (13) in Equation (7), we find

$$a^2 = -\frac{4}{3}\widetilde{\omega}^2(k_0^2)/\lambda > 0, \quad c_1 = -1/3. \tag{15}$$

Minimization of the free energy in k yields $k = k_0$. $\widetilde{\omega}^2(k_0^2) = -\lambda v^2$, and $\widetilde{\omega}^2(k_0^2) > 0$ for $T > T_{cr}$. $\widetilde{\omega}^2(k_0^2) < 0$ for $T < T_{cr}$. $\widetilde{\omega}^2(9k_0^2) = -\lambda v^2 + 16c^2/d \gg |\widetilde{\omega}^2(k_0^2)|$. Thereby, with appropriate accuracy, we may use $\delta n \simeq a[\sin(k_0x + \chi)$, which yields $\delta F_L(k_0) \simeq -\lambda v^4 V/(6m^*n) + O(\epsilon^2)$, where V is the volume of the system. Thus, the solution expressed in Equation (13) describes the stationary state at the second-order phase transition.

For the condensation in the uniform state $k_0 = 0$, we have [55]

$$\widetilde{\omega}^2(k^2) = -\lambda v^2 + ck^2, \quad k^2 < \lambda v^2/c, \quad c > 0. \tag{16}$$

Two spatially constant stationary solutions minimizing the free energy for $T < T_{cr}$ correspond to $k = 0$. They describe metastable and stable states:

$$\delta n_{\rm st} \simeq \pm v/m^* + \epsilon/(2\lambda v^2 m^*). \tag{17}$$

The free energy corresponding to these solutions is given by

$$\delta F_{\rm L}(k = 0, k_0 = 0) \simeq -\frac{\lambda v^4 V}{4m^* n_{cr}}\left(1 \pm \frac{4\epsilon}{\lambda v^3}\right). \tag{18}$$

For $k \neq 0$, solutions in the form of Equation (13) are valid for $|\widetilde{\omega}^2(k^2)| \ll \widetilde{\omega}^2(9k^2)$. For $k_0 = 0$, they yield

$$\delta F_{\rm L}(k \neq 0, k_0 = 0) \simeq -\frac{\lambda v^4(1 - ck^2/(\lambda v^2))V}{6m^* n_{cr}}. \tag{19}$$

Although the minimum of the free energy for $k_0 = 0$ is given by Equation (18), corresponding to solutions expressed in Equation (17) obtained for $k = 0$ rather than by solutions expressed in Equation (13) corresponding to the free energy expressed in Equation (19), as we will demonstrate below, solutions expressed in Equation (13) characterized by $k \neq 0$ have a physical meaning.

4. Dynamics of Seeds at a First-Order Phase Transition from a Metastable State to a Stable State

The rate of the formation of seeds in fluctuations has been extensively studied in the literature, e.g., [9,59,64–66]. Let us assume that an initial seed of the new phase has been formed in a fluctuation and consider its subsequent time evolution. Consider the limit of a high thermal conductivity, when, in Equation (7), the temperature can be made constant. The solution expressed in Equation (7) describing the dynamics of the initial density fluctuation developing from the metastable state to the stable state is then presented in the form [56]

$$\delta n(t, r) \simeq \frac{v(T)}{m}\left[\pm \mathrm{th}\frac{r - R_{\rm seed}(t)}{l} + \frac{\epsilon}{2\lambda v^3(T)}\right] + (\delta n)_{\rm cor}, \tag{20}$$

where the upper sign corresponds to the evolution of bubbles of the gas, the lower sign solution describes the evolution of droplets of liquid for $\nu = 3$, and the solution is valid for $|\epsilon/(\lambda v^3(T))| \ll 1$. Compensating correction $(\delta n)_{\rm cor}$ is introduced to fulfill the baryon number conservation. Considering spatial coordinate r in the vicinity of a bubble/droplet boundary, we obtain an equation describing the evolution of the seed size [55,56]:

$$\frac{m^{*2}\beta t_0^2}{2l}\frac{d^2 R_{\rm seed}}{dt^2} = m^{*2}\left[\frac{3\epsilon}{2\lambda v^3(T)} - \frac{2l}{R_{\rm seed}}\right] - \frac{m^{*2}t_0}{l}\frac{dR_{\rm seed}}{dt}. \tag{21}$$

This equation reminds us of Newton's second law for a one-dimensional system, where the quantity $M = \frac{m^{*2}\beta t_0^2}{2l} \propto (T_{cr} - T)^{-3/2}$ has the meaning of the mass, $m^{*2}[\frac{3\epsilon}{2\lambda v^3(T)} - \frac{2l}{R_{\rm seed}}]$ is an external force, and $-\frac{m^{*2}t_0}{l}\frac{dR_{\rm seed}}{dt}$ is the friction force, with a viscous-friction coefficient that is proportional to an effective viscosity and inversely proportional to $\sqrt{T_{cr} - T}$. Following Equation (21), a bubble of an overcritical size $R_{\rm seed} > R_{cr} = 4l\lambda v^3(T)/(3\epsilon)$ of the stable gas phase, or respectively a droplet of the stable liquid phase, is initially prepared in a fluctuation inside a metastable phase and then grows. In an early stage of the evolution, the size of the overcritical bubble/droplet $R_{\rm seed}(t)$ (for $R_{\rm seed} > R_{cr}$) grows with acceleration. Thus, it reaches a steady growth regime with a constant velocity

$u_{\text{as}} = \frac{3\epsilon l}{\lambda v^3(T)t_0} \propto |(T_{cr} - T)/T_{cr}|^{1/2}$. In the interior of the seed $\delta n \simeq \mp v(T)/m^*$. The correction $(\delta n)_{\text{cor}} \simeq v(T)R^3_{\text{seed}}(t)/(m^* R^3)$ is very small for $R_{\text{seed}}(t) \ll R$, where R is the radius of the whole system. In cases of the quark-hadron and nuclear liquid-gas phase transitions in heavy-ion collisions, $R(t)$ is the radius of the expanding fireball. Usage of the isothermal approximation in Equation (20) needs the fulfillment of inequality $t_\rho \sim \frac{R_{\text{seed}}(t_{\text{f.o.}})}{u_{\text{as}}} \gg t_T$. For $R_{\text{seed}} \sim R_{cr}$ and for $\epsilon \sim \epsilon_m$, we obtain $t_\rho \sim t_0$, and the isothermal approximation is valid for $R_{\text{seed}} < R_{\text{fog}}$. For $\epsilon \ll \epsilon_m$, we obtain $t_\rho \gg t_0$, and the isothermal approximation remains correct for seeds of the size $R_{\text{seed}} < R_{\text{fog}}\epsilon_m/\epsilon$.

Substituting Equation (20) in Equation (9) for $T \simeq$ const (which is correct in linear approximation), we obtain

$$\delta s = \left(\frac{\partial s}{\partial n}\right)_T \left\{\frac{v(T)}{m}\left[\pm \text{th}\frac{r - R_{\text{seed}}(t)}{l} + \frac{\epsilon}{2\lambda_{cr} v^3(T)}\right] + (\delta n)_{\text{cor}}\right\}. \tag{22}$$

Note that, for the description of the expanding fireball formed in heavy-ion collisions, the approximation of a quasi-adiabatic expansion can be used even in the presence of a weak first-order phase transition (for $\delta s \ll s$ and $\delta n \ll n$). The evolution of droplets/bubbles in the metastable region can be considered at a fixed size of the fireball, provided an expansion time lasting until freeze-out $t_{\text{f.o.}} \gg (t_\rho, t_T)$.

5. Dynamics of Fluctuations in Unstable Region

5.1. Growth of Fluctuations of Small Amplitude. Linear Regime

In this section, the "r" reference point can be taken as arbitrary, so we suppress the subscript "r." To find solutions of the linearized hydrodynamical equations, we have, cf. [56],

$$\delta n = \delta n_0 \exp[\gamma t + i\vec{k}\vec{r}] - \frac{\epsilon}{m^* \lambda v^2},, \quad \delta s = \delta s_0 \exp[\gamma t + i\vec{k}\vec{r}], \quad T = T_> + \delta T_0 \exp[\gamma t + i\vec{k}\vec{r}], \tag{23}$$

where $T_>$ is the temperature of the uniform matter. For $|\delta n| \gg |\frac{\epsilon}{m^* \lambda v^2}|$, i.e., for $\epsilon \ll \epsilon_m$, the description of the fluctuation in the spinodal region at the first-order phase transition and the description of the second-order phase transition are the same. We may set $\epsilon \to 0$. Thus, from linearized equations of non-ideal hydrodynamics, expressed in Equations (7) and (9), we find the increment $\gamma(k)$, cf. [56],

$$\gamma^2 = -k^2 \left[\tilde{\omega}^2(k^2) + \tilde{\eta}\gamma + \frac{u_{\tilde{s}}^2 - u_T^2}{1 + \kappa k^2/(c_V \gamma)}\right], \tag{24}$$

where $\tilde{\eta} = \frac{(\tilde{v}\eta + \zeta)}{m^* n}$. This equation differs from that derived in [67] by the presence of an extra surface tension term, and it differs from that in [59], which was based on other assumptions. Equation (24) has three solutions corresponding to the growth of the density and thermal modes. For $\kappa k^2/(c_V|\gamma|) \gg 1$, the temperature in the seed can be made constant, and we may deal with only one equation for the density mode expressed in Equation (7), which yields

$$\gamma^2 = -k^2 \left[\tilde{\omega}^2(k^2) + \tilde{\eta}\gamma\right], \tag{25}$$

from which we find two solutions for the density modes,

$$\gamma_{1,2} = -\frac{k^2\tilde{\eta}}{2} \pm \sqrt{\frac{k^4\tilde{\eta}^2}{4} - k^2\tilde{\omega}^2(k^2)}. \tag{26}$$

For $\omega^2\tilde{(}k^2) < 0$, which corresponds to the region of the phase transition, the upper-sign solution, $\gamma_1 > 0$, describes the growing mode, and the lower sign solution, $\gamma_2 < 0$, describes the damping mode. For $k^2\tilde{\eta}^2/|\tilde{\omega}^2(k^2)| \ll 1$, we have

$$\gamma_1 \simeq \sqrt{-k^2\tilde{\omega}^2(k^2)} - \frac{k^2\tilde{\eta}}{2} + O(k^3\tilde{\eta}^2/|\tilde{\omega}(k^2)|) \tag{27}$$

for the growing mode. In the opposite limit $k^2\tilde{\eta}^2/|\tilde{\omega}^2(k^2)| \gg 1$, we obtain

$$\gamma_1 \simeq -\tilde{\omega}^2(k^2)/\tilde{\eta} + O(\tilde{\omega}^4(k^2)/(k^2\tilde{\eta}^3)). \tag{28}$$

Note that, in condensed matter physics, a transition from a liquid to a glass state can be interpreted as a first-order phase transition occurring within a spinodal region at a very high viscosity [47]. Thus, there is an order at a scale of several Å, which transforms in a disorder at larger distances.

For $k_0 = 0, c > 0$, for the most rapidly growing mode (for $\gamma_m = \max\{\gamma_1\}$ corresponding to $k = k_m$), we find

$$\gamma_m \simeq \frac{\lambda v^2}{(2\sqrt{\beta}+1)\tilde{\eta}}, \quad k_m^2 \simeq \frac{\lambda v^2\sqrt{\beta}}{(2\sqrt{\beta}+1)c}.$$

For $k_0 \neq 0, c < 0, d > 0$, the most rapidly growing mode corresponds to $k = k_0$; thus, $\tilde{\omega}^2(k_0^2) < 0$ and $|\tilde{\omega}^2(k_0^2)|$ as a function of k^2 are the largest.

5.2. The Growth of Fluctuations of Arbitrary Amplitude: A Nonlinear Regime

Now we will find the solution to the non-linear Equation (7). We search the solution in the form

$$m^*\delta n = af(t)\left[\sin(kx+\chi) + \frac{c_1\tilde{\omega}^2(k^2)}{\tilde{\omega}^2(9k^2)}\sin(3kx+\chi) + ...\right] + O(\epsilon), \tag{29}$$

as Equation (13) with $a^2 = -\frac{4\tilde{\omega}^2}{3\lambda} > 0$, but now with $f(t)$, satisfying

$$\partial_t^2 f = -k^2\tilde{\omega}^2(k^2)f(1-f^2) - k^2\tilde{\eta}\partial_t f. \tag{30}$$

For $k^2\tilde{\eta}^2/|\tilde{\omega}^2(k^2)| \gg 1$, i.e., for $\beta \ll 1$ or $\tilde{\eta} \gg \sqrt{c}$, the term $\partial_t^2 f$ on the l.h.s. of Equation (30) can be dropped, and the amplitude

$$f(t) = \frac{f_0 e^{\gamma t}}{\sqrt{1+f_0^2 e^{2\gamma t}}} \tag{31}$$

fulfills the resulting Equation (30). $f_0/\sqrt{1+f_0^2}$ shows the amplitude of the fluctuation at $t = 0$, and f_0 is an arbitrary constant. For $k \sim k_m$ at $k_0 = 0$, this solution holds for $k_m^2\tilde{\eta}^2/|\tilde{\omega}^2(k_m^2)| \gg 1$. For $k = k_0 \neq 0$, the criterion of applicability renders as $k_0^2\tilde{\eta}^2/|\tilde{\omega}^2(k_0^2)| \gg 1$. In both cases $k_0 = 0$ and $k_0 \neq 0$, with the density distribution given by Equations (29) and (31), the free energy renders

$$\delta F_L(t) = -\frac{V\tilde{\omega}^4(k^2)}{6\lambda m^* n} f^2(t)\left(2 - f^2(t)\right). \tag{32}$$

For $t \to \infty$, we have $f(t \to \infty) \to 1$, and δF_L reaches the minimum. For $k = k_0$, this value coincides with Equation (19), which is given by the stationary solution.

In the general case, Equation (31) yields an interpolation between two approximate solutions that are valid for the limit cases $\gamma t \ll 1$ and $\gamma t \gg 1$. Replacing Equation (31) in Equation (30), we obtain then the same solutions expressed in Equation (26) as those in the linear case.

Let first $k_0 = 0$. For $t \to \infty$, using Equation (31) at $\gamma = \gamma_m = \gamma(k_m)$, we find

$$\delta F_{\rm L}(t \to \infty) = -\frac{\widetilde{\omega}^4(k_m^2)V}{6\lambda m^* n}. \tag{33}$$

For the case of a large effective viscosity/inertia, $\beta \ll 1$, we obtain $\delta F_{\rm L}(t \to \infty) \simeq -\frac{\lambda v^4 V}{6m^* n}$, which coincides with Equation (19) but is still larger than the value given by Equation (18). For the case of a small effective viscosity/inertia, $\beta \gg 1$, we find $\delta F_L(t \to \infty) \simeq -\frac{\lambda v^4 V}{24m^* n}$, which is much higher than the free energy given by both stationary solutions expressed in Equations (18) and (19). Thus, one may expect that Equation (33) either describes a metastable state or a state that slowly varies on a time scale $t_k \gg t_\gamma \sim 1/\gamma_m$, reaching, for $t \gg t_k$, the stationary state with the free energy given by Equation (18). To show the latter possibility, consider the case $\beta \gg 1$ and assume k in Equation (29) to be a slow function of time, i.e., $k = k(t)$, for a typical time scale $t_k \gg t_\gamma$. One can see that, for $R_{\rm seed} \ll t_k|u_T|$, the quantity $k(t)$ satisfies $(d^2k/dt^2) = -k^2\tilde{\eta}(dk/dt)$ with the solution

$$k(t) = k_{00}[1 + \tilde{\eta}\lambda v^2 t/(3c)]^{-1/2} \tag{34}$$

such as $k(t \to \infty) \to 0$, and the free energy for $t \to \infty$ indeed reaches the limit expressed in Equation (18) provided we set $\sin\chi \simeq \frac{\sqrt{3}}{2} - \frac{\sqrt{3}m^*\tilde{\epsilon}}{8}$. From Equation (34), we easily find that the typical time scale is $t_k \sim \beta t_0$, and we confirm that indeed $t_k \gg t_\gamma$. For $R_{\rm seed} \gtrsim t_k|u_T| \sim l\sqrt{\beta}$, the solution expressed in Equation (29) with Equation (34) does not hold and should be modified.

For $\beta \ll 1$, $k_0 = 0$, Equation (34) with a slowly varying $k(t)$ does not hold. At realistic conditions, convection and sticking processes (at sizes $\sim l$) may be allowed, which destroy periodicity, and owing to these processes the system may finally reach the ground state with the free energy given by Equation (18). Thus, one possibility is that, for the typical time $t \sim t_\gamma \sim t_0$, the quasiperiodic solution expressed in Equation (33) is formed with a typical $k \simeq k_m$, corresponding to a metastable state with the free energy given by Equation (19). Such a distribution is formed most rapidly. Another possibility is that, for the typical time scale $t_{\rm unif} > t_\gamma$ in a system of a large size, an approximately uniform solution, expressed in Equation (35), is developed. In the latter case, to proceed, consider the case $k \sim 1/R \ll k_m$, where R is the typical size of the system ($R = R_{\rm f.o.}$ for the fireball formed in heavy-ion collisions). The spatially uniform solution of

$$\Delta_\xi\psi + 2\psi(1-\psi^2) + \tilde{\epsilon} = \partial_\tau\psi,$$

which follows from Equation (8) in this case (as well as for seeds of a size $R_{\rm seed} \ll R$ at $\beta \ll 1$, as we have argued above), is given by

$$\psi(t) = \pm 1/\sqrt{1 + e^{-\tau}(1-\psi_0^2)/\psi_0^2}, \tag{35}$$

where we, for simplicity, have $\tilde{\epsilon} \to 0$. The typical time needed for the initial amplitude $\psi_0 \ll 1$ to grow to $\psi(t \to \infty) \simeq \pm 1$ is $t_{\rm unif} \sim t_0 \ln(1/\psi_0^2) \gg t_\gamma$.

Thus, we found some novel solutions describing the evolution of fluctuations in the region of instability in addition to the uniform solution expressed in Equation (35). For $k = $const$\neq 0$, we found periodic solutions given by Equations (29) and (31). For $k = k_0 \neq 0$, the solution yields the minimum of the free energy for $t \to \infty$. For $k_0 = 0$, $\beta \gg 1$, we found quasiperiodic solutions expressed in Equations (29) and (31) with $k = k(t)$ from Equation (34), yielding the minimum of the free energy for $t \to \infty$.

6. Conclusions

According to our findings, signatures of QCD spinodal instabilities might in principle be observed in experiments with heavy ions in a collision energy interval that corresponds to the first-order phase transition region of the QCD phase diagram. If the typical times of the growth of a fluctuation in the unstable region t_γ and of that of the fireball expansion $t_{\rm f.o.}$ satisfy the condition $t_\gamma \lesssim t_{\rm f.o.}$, one of the possible experimental signatures of the spinodal region would be a manifestation of a spatially quasiperiodic structure with a typical period $r \simeq 2\pi/k_m$. If the parameter characterizing effective viscosity/inertia β were $\gg 1$, cf. Equation (12), then for $t_\gamma \ll t_{\rm f.o.}$ one of the possible experimental signatures of the spinodal region would be a manifestation of spatially quasiperiodic fluctuations with a typical size $r \sim 2\pi/k(t_{\rm f.o.}) \gg 2\pi/k_m$. However, rough estimates made for the quark-hadron and nuclear gas-liquid first-order phase transitions in heavy-ion collisions [54–56] indicate that $\beta \ll 1$. Future experimental programs at NICA and FAIR will scan the collision energy interval, in which various manifestations of the first-order quark-hadron phase transition are expected, including possible signatures of quasiperiodic structures. It would also be interesting to search for the consequences of the possible formation of quasiperiodic structures during the quark-hadron phase transition in the early universe.

Concluding, we note that viscosity and thermal conductivity are the driving forces of the first-order liquid-gas and quark-hadron phase transitions to the state with $k_0 = 0$, and the spinodal instability occurs for T below the ITS line. The manifestation of a spatially quasiperiodic structure with a typical period of $2\pi/k_m$, cf. Equation (29), in the rapidity spectra of heavy-ion collisions in a collision energy interval could be interpreted as a signature of the occurrence of the spinodal instability at the first-order phase transition. For the second-order phase transition to the state with $k_0 \neq 0$, as for the case of the pion condensation in dense nuclear matter, the periodic solution expressed in Equation (29) holds for $k = k_0 \neq 0$, where k_0 does not depend on time.

Funding: This research received no external funding.

Acknowledgments: I thank E. E. Kolomeitsev for the discussions.

Conflicts of Interest: The author declares no conflict of interest.

References

1. Aghanim, N.; Akrami, Y.; Ashdown, M.; Aumont, J.; Baccigalupi, C.; Ballardini, M.; Banday, A.J.; Barreiro, R.B.; Bartolo, N.; Basak, S.; et al. Planck Collaboration. *arXiv* **2018**, arXiv:1807.06209.
2. Linde, A.D. Phase Transitions in Gauge Theories and Cosmology. *Rept. Prog. Phys.* **1979**, *42*, 389. [CrossRef]
3. Witten, E. Cosmic separation of phases. *Phys. Rev. D* **1984**, *30*, 272. [CrossRef]
4. Rafelski, J. Connecting QGP-Heavy Ion Physics to the Early Universe. *Nucl. Phys. Proc. Suppl.* **2013**, *243*, 155. [CrossRef]
5. Busza, W.; Rajagopal, K.; van der Schee, W. Heavy Ion Collisions: The Big Picture, and the Big Questions. *Ann. Rev. Nucl. Part. Sci.* **2018**, *68*, 339. [CrossRef]
6. Jacak, B.V.; Muller, B. The exploration of hot nuclear matter. *Science.* **2012**, *337*, 310. [CrossRef]
7. Fodor, Z. Selected results in lattice quantum chromodynamics. *PTEP* **2012**, *2012*, 01A108. [CrossRef]
8. Bazavov, A.; Ding, H.T.; Hegde, P.; Kaczmarek, O.; Karsch, F.; Karthik, N.; Laermann, E.; Lahiri, A.; Larsen, R.; Li, S.T.; et al. Chiral crossover in QCD at zero and non-zero chemical potentials. *Phys. Lett. B* **2019**, *795*, 15–21. [CrossRef]
9. Kapusta, J.; Gale, C. *Finite-Temperature Field Theory Principles and Applications*; Cambridge Univ. Press: Cambridge, UK, 2006.
10. Peebles, P.J.E. *Principles of Physical Cosmology*; Princeton Univ. Press: Princeton, NJ, USA, 1993.
11. Shapiro, S.; Teukolsky, S.A. *Black Holes, White Dwarfs and Neutron Stars: The Physics of Compact Objects*; Wiley: New York, NY, USA, 1983.
12. Migdal, A.B.; Saperstein, E.E.; Troitsky, M.A.; Voskresensky, D.N. Pion degrees of freedom in nuclear matter. *Phys. Rept.* **1990**, *192*, 179. [CrossRef]
13. Haubold, H.; Kampfer, B.; Senatorov, A.; Voskresensky, D. A Tentative approach to the second neutrino burst in SN1987A. *Astron. Astrophys.* **1988**, *191*, L22.
14. Galeotti, P.; Pizzella, G. New analysis for the correlation between gravitational wave and neutrino detectors during SN1987A. *Eur. Phys. J. C* **2016**, *76*, 426. [CrossRef]

15. Fischer, T.; Bastian, N.U.; Wu, M.R.; Baklanov, P.; Sorokina, E.; Blinnikov, S.; Typel, S.; Klähn, T.; Blaschke, D.B. Quark deconfinement as a supernova explosion engine for massive blue supergiant stars. *Nat. Astron.* **2018**, *2*, 980–986. [CrossRef]
16. Prakash, M.; Bombaci, I.; Prakash, M.; Ellis, P.; Lattimer, J.; Knorren, R. Composition and structure of protoneutron stars. *Phys. Rept.* **1997**, *280*, 1. [CrossRef]
17. LIGO Scientific and Virgo and 1M2H and Dark Energy Camera GW-E and DES and DLT40 and Las Cumbres Observatory and VINROUGE and MASTER Collaborations. A gravitational-wave standard siren measurement of the Hubble constant. *Nature* **2017**, *551*, 85. [CrossRef]
18. Demorest, P.; Pennucci, T.; Ransom, S.; Roberts, M.; Hessels, J. Shapiro delay measurement of a two solar mass neutron star. *Nature* **2010**, *467*, 1081. [CrossRef]
19. Fonseca, E.; Pennucci, T.T.; Ellis, J.A.; Stairs, I.H.; Nice, D.J.; Ransom, S.M.; Demorest, P.B.; Arzoumanian, Z.; Crowter, K.; Dolch, T.; et al. The NANOGrav nine-year data set: Mass and geometric measurements of binary millisecond pulsars. *Astrophys. J.* **2016**, *832*, 167. [CrossRef]
20. Antoniadis, J.; Freire, P.C.C.; Wex, N.; Tauris, T.M.; Lynch, R.S.; van Kerkwijk, M.H.; Kramer, M.; Bassa, C. A massive pulsar in a compact relativistic binary. *Science* **2013**, *340*, 1233232. [CrossRef]
21. Cromartie, H.T.; Fonseca, E.; Ransom, S.M.; Demorest, P.B.; Arzoumanian, Z.; Blumer, H.; Brook, P.R.; DeCesar, M.E.; Dolch, T.; Ellis, J.A.; et al. Relativistic Shapiro delay measurements of an extremely massive millisecond pulsar. *Nat. Astron.* **2019**, *4*, 72. [CrossRef]
22. Bauswein, A.; Bastian, N.; Blaschke, D.; Chatziioannou, K.; Clark, J.; Fischer, T.; Oertel, M. Identifying a first-order phase transition in neutron star mergers through gravitational waves. *Phys. Rev. Lett.* **2019**, *122*, 061102. [CrossRef]
23. Shuryak, E. What RHIC experiments and theory tell us about properties of quark-gluon plasma? *Nucl. Phys. A* **2005**, *750*, 64. [CrossRef]
24. Teaney, D.; Lauret, J.; Shuryak, E. A Hydrodynamic description of heavy ion collisions at the SPS and RHIC. *arXiv* **2001**, arXiv:nucl-th/0110037.
25. Romatschke, P.; Romatschke, U. Viscosity Information from Relativistic Nuclear Collisions: How Perfect is the Fluid Observed at RHIC? *Phys. Rev. Lett.* **2007**, *99*, 172301. [CrossRef] [PubMed]
26. Shuryak, E. Physics of Strongly coupled Quark-Gluon Plasma. *Prog. Part. Nucl. Phys.* **2009**, *62*, 48. [CrossRef]
27. Röpke, G.; Münchow, L.; Schulz, H. Particle clustering and Mott transitions in nuclear matter at finite temperature. *Nucl. Phys. A* **1982**, *379*, 536. [CrossRef]
28. Schulz, H.; Voskresensky, D.N.; Bondorf, J. Dynamical aspects of the liquid-vapor phase transition in nuclear systems. *Phys. Lett. B* **1983**, *133*, 141. [CrossRef]
29. Chomaz, P.; Colonna, M.; Randrup, J. Nuclear spinodal fragmentation. *Phys. Rept.* **2004**, *389*, 263. [CrossRef]
30. Margueron, J.; Chomaz, P. A Unique spinodal region in asymmetric nuclear matter. *Phys. Rev. C* **2003**, *67*, 041602. [CrossRef]
31. Maslov, K.; Voskresensky, D. RMF models with σ -scaled hadron masses and couplings for the description of heavy-ion collisions below 2 A GeV. *Eur. Phys. J. A* **2019**, *55*, 100. [CrossRef]
32. avenhall, D.G.; Pethick, C.J.; Wilson, J.R. Structure of Matter Below Nuclear Saturation Density. *Phys. Rev. Lett.* **1983**, *50*, 2066. [CrossRef]
33. Maruyama, T.; Tatsumi, T.; Voskresensky, D.N.; Tanigawa, T.; Chiba, S. Nuclear pasta structures and the charge screening effect. *Phys. Rev. C* **2005**, *72*, 015802. [CrossRef]
34. Migdal, A.B. Pion fields in nuclear matter. *Rev. Mod. Phys.* **1978**, *50*, 107. [CrossRef]
35. Glendenning, N. Phase transitions and crystalline structures in neutron star cores. *Phys. Rept.* **2001**, *342*, 393. [CrossRef]
36. Maruyama, T.; Tatsumi, T.; Voskresensky, D.; Tanigawa, T.; Endo, T.; Chiba, S. Finite size effects on kaonic pasta structures. *Phys. Rev. C* **2006**, *73*, 035802. [CrossRef]
37. Voskresensky, D.N. On the possibility of the condensation of the charged rho meson field in dense isospin asymmetric baryon matter. *Phys. Lett. B* **1997**, *392*, 262. [CrossRef]
38. Glendenning, N. First order phase transitions with more than one conserved charge: Consequences for neutron stars. *Phys. Rev. D* **1992**, *46*, 1274. [CrossRef] [PubMed]
39. Heiselberg, H.; Pethick, C.; Staubo, E. Quark matter droplets in neutron stars. *Phys. Rev. Lett.* **1993**, *70*, 1355. [CrossRef]
40. Voskresensky, D.; Yasuhira, M.; Tatsumi, T. Charge screening at first order phase transitions and hadron quark mixed phase. *Nucl. Phys. A* **2003**, *723*, 291. [CrossRef]

41. Maslov, K.; Yasutake, N.; Ayriyan, A.; Blaschke, D.; Grigorian, H.; Maruyama, T.; Tatsumi, T.; Voskresensky, D. Hybrid equation of state with pasta phases and third family of compact stars. *Phys. Rev. C* **2019**, *100*, 025802. [CrossRef]
42. Sedrakian, A.; Clark, J.W. Superfluidity in nuclear systems and neutron stars. *Eur. Phys. J. A* **2019**, *55*, 167. [CrossRef]
43. Kolomeitsev, E.E.; Voskresensky, D.N. Superfluid nucleon matter in and out of equilibrium and weak interactions. *Phys. Atom. Nucl.* **2011**, *74*, 1316. [CrossRef]
44. Voskresensky, D.N. Vector-boson condensates, spin-triplet superfluidity of paired neutral and charged fermions, and $3P_2$ pairing of nucleons. *arXiv* **2019**, arXiv:1911.07502.
45. Alford, M.G.; Schmitt, A.; Rajagopal, K.; Schäfer, T. Color superconductivity in dense quark matter. *Rev. Mod. Phys.* **2008**, *80*, 1455. [CrossRef]
46. Voskresensky, D.N. The phase transition to an inhomogeneous condensate state. *Phys. Scr.* **1984**, *29*, 259. [CrossRef]
47. Voskresensky, D.N. Quasiclassical description of condensed systems by a complex order parameter. *Phys. Scr.* **1993**, *47*, 333. [CrossRef]
48. Buballa, M.; Carignano, S. Inhomogeneous chiral condensates. *Prog. Part. Nucl. Phys.* **2015**, *81*, 39. [CrossRef]
49. Kolomeitsev, E.E.; Voskresensky, D.N. Negative kaons in dense baryonic matter. *Phys. Rev. C* **2003**, *68*, 015803. [CrossRef]
50. Tilley, D.R.; Tilley, J. *Superfluidity and Superconductivity*; Van Nostrand Reinhold Comp.: New York, NY, USA, 1974.
51. Ratra, B.; Peebles, P.J.E. Cosmological Consequences of a Rolling Homogeneous Scalar Field. *Phys. Rev. D* **1988**, *37*, 3406. [CrossRef]
52. Agarwal, S.; Corasaniti, P.S.; Das, S.; Rasera, Y. Small scale clustering of late forming dark matter. *Phys. Rev. D* **2015**, **92**, 063502. [CrossRef]
53. Chowdhury, T.A.; Nemevsek, M.; Senjanovic, G.; Zhang, Y. Dark Matter as the Trigger of Strong Electroweak Phase Transition. *J. Cosmol. Astropart. Phys.* **2012**, *1202*, 029. [CrossRef]
54. Skokov, V.V.; Voskresensky, D.N. Hydrodynamical description of a hadron-quark first-order phase transition. *JETP Lett.* **2009**, *90*, 223. [CrossRef]
55. Skokov, V.V.; Voskresensky, D.N. Hydrodynamical description of first-order phase transitions: Analytical treatment and numerical modeling. *Nucl. Phys. A* **2009**, *828*, 401. [CrossRef]
56. Skokov, V.V.; Voskresensky, D.N. Thermal conductivity in dynamics of first-order phase transition. *Nucl. Phys. A* **2010**, *847*, 253. [CrossRef]
57. Steinheimer, J.; Randrup, J. Spinodal density enhancements in simulations of relativistic nuclear collisions. *Phys. Rev. C* **2013**, *87*, 054903. [CrossRef]
58. Steinheimer, J.; Randrup, J. Spinodal amplification and baryon number fluctuations in nuclear collisions at NICA. *Eur. Phys. J. A* **2016**, *52*, 239. [CrossRef]
59. Ruggeri, F.; Friedman, W. Nucleation rate of hadron bubbles in baryon - free quark - gluon plasma. *Phys. Rev. D* **1996**, *53*, 6543. [CrossRef]
60. Voskresensky, D.N. Hydrodynamics of Resonances. *Nucl. Phys. A* **2011**, *849*, 120. [CrossRef]
61. Mandelstam, L.I.; Leontovich, M.A. To the theory of sound absorption in liquids. *Z. Eksp. Teor. Fiz.* **1937**, *7*, 438.
62. Landau, L.D.; Lifshitz, E.M. *Fluid Mechanics*; Pergamon Press: Oxford, UK, 1987.
63. Kolomeitsev, E.E.; Voskresensky, D.N. Viscosity of neutron star matter and r-modes in rotating pulsars. *Phys. Rev. C* **2015**, *91*, 025805. [CrossRef]
64. Langer, J.S. Statistical theory of the decay of metastable states. *Ann. Phys.* **1969**, *54*, 258. [CrossRef]
65. Langer, J.S.; Turski, L.A. Hydrodynamic model of the condensation of a vapor near its critical point. *Phys. Rev. A* **1973**, *8*, 3230. [CrossRef]
66. Lifshiz, E.M.; Pitaevskii, L.P. *Physical Kinetics*; Pergamon Press: Oxford, UK, 1981.
67. Pethick, C.J.; Ravenhall, D.G. Instabilities in hot nuclear matter and the fragmentation process. *Nucl. Phys. A* **1987**, *471*, 19. [CrossRef]

© 2020 by the author. Licensee MDPI, Basel, Switzerland. This article is an open access article distributed under the terms and conditions of the Creative Commons Attribution (CC BY) license (http://creativecommons.org/licenses/by/4.0/).

Article

Parton Distribution Functions and Tensorgluons

Roland Kirschner [1] and George Savvidy [2,*]

[1] ITP, Universität Leipzig, Augustusplatz 10, D-04109 Leipzig, Germany; kirschne@itp.uni-leipzig.de
[2] INPP, Demokritos National Research Center, Ag. Paraskevi, 15341 Athens, Greece
* Correspondence: savvidy@inp.demokritos.gr; Tel.: +30-693-916-7222

Received: 20 May 2020; Accepted: 10 June 2020; Published: 29 June 2020

Abstract: We derive the regularised evolution equations for the parton distribution functions that include tensorgluons.

Keywords: QCD; DGLAP equations; physics beyond the standard model; tensorgluons; extended DGLAP equations; tensorgluon splitting functions

---

1. Introduction

The Bjorken scaling [1] is broken by a logarithmically dependent function [2,3] of the transverse momentum Q^2 and is due to the interaction of quarks and gluons inside the hadrons [2–16]. In this article, we shall consider a possibility [9] that inside hadrons there are additional partons–tensorgluons, which can carry a part of the proton momentum [17–24]. The extension of the Yang–Mills theory was formulated in terms of the gauge invariant Lagrangian [18–21]. For tensorgluons of rank-2, it has the following form [18–20]:

$$\begin{aligned} \mathcal{L} = &-\frac{1}{4}G^a_{\mu\nu}G^a_{\mu\nu} - \frac{1}{4}G^a_{\mu\nu,\lambda}G^a_{\mu\nu,\lambda} - \frac{1}{4}G^a_{\mu\nu}G^a_{\mu\nu,\lambda\lambda} \\ &+ \frac{1}{4}G^a_{\mu\nu,\lambda}G^a_{\mu\lambda,\nu} + \frac{1}{4}G^a_{\mu\nu,\nu}G^a_{\mu\lambda,\lambda} + \frac{1}{2}G^a_{\mu\nu}G^a_{\mu\lambda,\nu\lambda} + \dots \end{aligned} \quad (1)$$

where the field strength tensors have the form:

$$\begin{aligned} G^a_{\mu\nu} &= \partial_\mu A^a_\nu - \partial_\nu A^a_\mu + g f^{abc}\, A^b_\mu\, A^c_\nu, \\ G^a_{\mu\nu,\lambda} &= \partial_\mu A^a_{\nu\lambda} - \partial_\nu A^a_{\mu\lambda} + g f^{abc}(\, A^b_\mu\, A^c_{\nu\lambda} + A^b_{\mu\lambda}\, A^c_\nu\,), \\ G^a_{\mu\nu,\lambda\rho} &= \partial_\mu A^a_{\nu\lambda\rho} - \partial_\nu A^a_{\mu\lambda\rho} + g f^{abc}(\, A^b_\mu\, A^c_{\nu\lambda\rho} + A^b_{\mu\lambda}\, A^c_{\nu\rho} + A^b_{\mu\rho}\, A^c_{\nu\lambda} + A^b_{\mu\lambda\rho}\, A^c_\nu\,), \\ &\dots\dots\dots\dots \end{aligned} \quad (2)$$

The first term in (1) corresponds to the standard Yang–Mills Lagrangian. The expression for the full Lagrangian can be found in [18–20]. For illustration purposes, we shall present the next term of the Lagrangian that describes the interaction of the rank-3 tensorgluons

$$\begin{aligned} \mathcal{L} = &-\frac{1}{4}G^a_{\mu\nu,\lambda\rho}G^a_{\mu\nu,\lambda\rho} - \frac{1}{8}G^a_{\mu\nu,\lambda\lambda}G^a_{\mu\nu,\rho\rho} - \frac{1}{2}G^a_{\mu\nu,\lambda}G^a_{\mu\nu,\lambda\rho\rho} - \frac{1}{8}G^a_{\mu\nu}G^a_{\mu\nu,\lambda\lambda\rho\rho} + \\ &+ \frac{1}{3}G^a_{\mu\nu,\lambda\rho}G^a_{\mu\lambda,\nu\rho} + \frac{1}{3}G^a_{\mu\nu,\nu\lambda}G^a_{\mu\rho,\rho\lambda} + \frac{1}{3}G^a_{\mu\nu,\nu\lambda}G^a_{\mu\lambda,\rho\rho} + \\ &+ \frac{1}{3}G^a_{\mu\nu,\lambda}G^a_{\mu\lambda,\nu\rho\rho} + \frac{2}{3}G^a_{\mu\nu,\lambda}G^a_{\mu\rho,\nu\lambda\rho} + \frac{1}{3}G^a_{\mu\nu,\nu}G^a_{\mu\lambda,\lambda\rho\rho} + \frac{1}{3}G^a_{\mu\nu}G^a_{\mu\lambda,\nu\lambda\rho\rho}. \end{aligned} \quad (3)$$

A spinor helicity technique [25–42] was used to calculate tensorgluon scattering amplitudes [22] and to extract the splitting amplitudes of gluons and tensorgluons [23]. The tensorgluon splitting amplitudes are singular at the boundary values similar to the case of the standard splitting amplitudes in QCD, though the singularities are of higher order compared to the standard case and require a special regularisation technique to be developed before they can be placed into the equations which are describing the evolution of the generalised parton distribution functions (10). Here, we shall further develop a technique proposed earlier in [9] which will allow for regularising the tensorgluon splitting amplitudes.

The present paper is organised as follows. In Section 2, the basic formulae for scattering amplitude and splitting functions are recalled, definitions and notations are specified, and the details of the regularisation scheme are presented. In Section 3, we derive the regularised evolution equations for the parton distribution functions that take into account the creation of tensorgluons. Section 4 contains concluding remarks and summarises the physical consequences of the tensorgluons' hypothesis.

2. Splitting Functions

It was proposed in [9] that a possible emission of tensorgluons inside a hadron will produce a tensorgluon "cloud" inside a hadron in addition to the quark and gluon "clouds". Our goal here is to specify the regularisation of the generalised evolution equations introduced in [9] such that it will be consistent with the regularisation scheme used to regularised the DGLAP equations [4–8,10–12]. The splitting probabilities for tensorgluons have the form [9]:

$$
\begin{aligned}
P_{TG}(z) &= C_2(G)\left[\frac{z^{2s+1}}{(1-z)_+^{2s-1}} + \frac{(1-z)^{2s+1}}{z^{2s-1}}\right],\\
P_{GT}(z) &= C_2(G)\left[\frac{1}{z(1-z)_+^{2s-1}} + \frac{(1-z)^{2s+1}}{z}\right],\\
P_{TT}(z) &= C_2(G)\left[\frac{z^{2s+1}}{(1-z)_+} + \frac{1}{(1-z)_+ z^{2s-1}}\right].
\end{aligned}
\tag{4}
$$

The invariant operator C_2 for the representation R is defined by the equation $t^a t^a = C_2(R)\,1$ and $tr(t^a t^b) = T(R)\delta^{ab}$. These functions satisfy the relations

$$
P_{TG}(z) = P_{TG}(1-z), \quad P_{GT}(z) = P_{TT}(1-z), \qquad z < 1. \tag{5}
$$

One should define the regularisation procedure for the singular factors $(1-z)^{-2s+1}$ and z^{-2s+1} reinterpreting them as the distributions $(1-z)_+^{-2s+1}$ and z_+^{-2s+1}. The regularisation has been defined in the following way [9]:

$$
\begin{aligned}
\int_0^1 dz \frac{f(z)}{(1-z)_+^{2s-1}} &= \int_0^1 dz \frac{f(z) - \sum_{k=0}^{2s-2} \frac{(-1)^k}{k!} f^{(k)}(1)(1-z)^k}{(1-z)^{2s-1}},\\
\int_0^1 dz \frac{f(z)}{z_+^{2s-1}} &= \int_0^1 dz \frac{f(z) - \sum_{k=0}^{2s-2} \frac{1}{k!} f^{(k)}(0) z^k}{z^{2s-1}},\\
\int_0^1 dz \frac{f(z)}{z_+(1-z)_+} &= \int_0^1 dz \frac{f(z) - (1-z)f(0) - zf(1)}{z(1-z)},
\end{aligned}
\tag{6}
$$

where $f(z)$ is an arbitrary test function that is sufficiently regular at the points $z = 0$ and $z = 1$ and, as one can be convinced, the defined substraction guarantees the convergence of the integrals. Using the same arguments as in the standard case [4], we will add the delta function terms into the definition of the diagonal kernels so that they will completely determine the behaviour of $P_{qq}(z)$, $P_{GG}(z)$ and $P_{TT}(z)$ functions with the coefficients that can be determined by using the momentum sum rule [9]:

$$
\begin{aligned}
P_{qq}(z) &= C_2(R)\left[\frac{1+z^2}{(1-z)_+}+\frac{3}{2}\delta(z-1)\right],\\
P_{GG}(z) &= 2C_2(G)\left[\frac{z}{(1-z)_+}+\frac{1-z}{z}+z(1-z)\right]+\frac{\sum_s(12s^2-1)C_2(G)-4n_f T(R)}{6}\,\delta(z-1),\\
P_{TT}(z) &= C_2(G)\left[\frac{z^{2s+1}}{(1-z)_+}+\frac{1}{(1-z)_+z^{2s-1}}+\sum_{j=1}^{2s+1}\frac{1}{j}\,\delta(z-1)\right].
\end{aligned} \tag{7}
$$

For completeness, we shall present also quark and gluon splitting functions [4]:

$$P_{Gq}(z) = C_2(R)\frac{1+(1-z)^2}{z}, \tag{8}$$

$$P_{qG}(z) = T(R)[z^2+(1-z)^2], \tag{9}$$

where $C_2(G)=N, C_2(R)=\frac{N^2-1}{2N}, T(R)=\frac{1}{2}$ for the SU(N) groups.

3. Regularisation of Generalised DGLAP Equations

The deep inelastic structure functions can be expressed in terms of parton distribution densities [4–8,10–12]. If $q^i(x,Q^2)$ is the density of quarks of type i (summed over colors) inside a nucleon target with fraction x of the proton longitudinal momentum in the infinite momentum frame, then the unpolarised structure functions can be represented in the following form:

$$2F_1(x,Q^2) = F_2(x,Q^2)/x = \sum_i e_i^2[q^i(x,Q^2)+\bar{q}^i(x,Q^2)].$$

The Q^2 dependence of the parton densities is described by the integro-differential equations for quark $q^i(x,t)$ and gluon densities $G(x,t)$, where $t=\ln(Q^2/Q_0^2)$ [4–8,10–12]. If there is an additional emission of tensorgluons in the proton, then one should introduce the corresponding density $T(x,t)$ of tensorgluons and the integro-differential equations that describe the Q^2 dependence of parton densities in this general case has the following form [9]:

$$
\begin{aligned}
\frac{dq^i(x,t)}{dt} &= \frac{\alpha(t)}{2\pi}\int_x^1\frac{dy}{y}[\sum_{j=1}^{2n_f}q^j(y,t)\,P_{q^iq^j}(\frac{x}{y})+G(y,t)\,P_{q^iG}(\frac{x}{y})],\\
\frac{dG(x,t)}{dt} &= \frac{\alpha(t)}{2\pi}\int_x^1\frac{dy}{y}[\sum_{j=1}^{2n_f}q^j(y,t)\,P_{Gq^j}(\frac{x}{y})+G(y,t)\,P_{GG}(\frac{x}{y})+T(y,t)\,P_{GT}(\frac{x}{y})],\\
\frac{dT(x,t)}{dt} &= \frac{\alpha(t)}{2\pi}\int_x^1\frac{dy}{y}[G(y,t)\,P_{TG}(\frac{x}{y})+T(y,t)\,P_{TT}(\frac{x}{y})].
\end{aligned} \tag{10}
$$

In (10), we ignore contribution of the high-spin fermions $\tilde{q}^i$ of spin $s+1/2$, which are the partners of the standard quarks [18–21], supposing that they are even heavier than the top quark. In this article, we shall limit ourselves by considering only emissions that always involve the standard gluons and spin-2 tensorgluons ignoring infinite "stairs" of transitions between tensorgluons of higher spin. In (10), the $\alpha(t)$ is the running coupling ($\alpha=g^2/4\pi$) and has the following form [17]:

$$\frac{\alpha}{\alpha(t)} = 1+b\,\alpha\,t\ , \tag{11}$$

where

$$b = \frac{\sum_s (-1)^{2s}(12s^2-1)C_2(G) - 4n_f T(R)}{12\pi}, \qquad s = 0, 1/2, 1, 3/2, 2, ... \tag{12}$$

is the one-loop Callan–Symanzik coefficient [17]. In particular, the presence of the spin-two tensorgluons in the proton will give

$$b = \frac{58C_2(G) - 4n_f T(R)}{12\pi}. \tag{13}$$

The tensorgluon density $T(x,t)$ changes when a gluon splits into two tensorgluons or when a tensorgluon radiates a gluon. This process is described by the third equation in (10).

The tensorgluon kernels (4) are singular at the boundary values similar to the case of the standard kernels (8), though the singularities are of higher order compared to the standard case. The '+' prescription in

$$P(z) = \begin{pmatrix} P_{qq}(z) & 2n_f P_{qG}(z) & 0 \\ P_{Gq}(z) & P_{GG}(z) & P_{GT}(z) \\ 0 & P_{TG}(z) & P_{TT}(z) \end{pmatrix} \tag{14}$$

is defined as

$$[g(x)]_+ = g(x) - \delta(1-x)\int_0^1 g(z)dz, \tag{15}$$

and so

$$\int_x^1 f(z)[g(z)]_+ dz = \int_x^1 [f(z) - f(1)]g(z)dz - f(1)\int_x^1 g(z)dz. \tag{16}$$

Considering the splitting probabilities for spin two tensorgluons, we have to define '+++' prescription as

$$\begin{aligned}[g(x)]_{+++} =& g(x) - \delta(1-x)\int_0^1 g(z)dz - \delta'(1-x)\int_0^1 g(z)(1-z)dz - \\ & -\frac{1}{2}\delta''(1-x)\int_0^1 g(z)(1-z)^2 dz,\end{aligned} \tag{17}$$

and so

$$\begin{aligned}\int_x^1 f(z)[g(z)]_{+++} dz =& \int_x^1 [f(z) - f(1) + f'(1)(1-z) - \frac{1}{2}f''(1)(1-z)^2]g(z)dz - \\ & - \int_0^x [f(1) - f'(1)(1-z) + \frac{1}{2}f''(1)(1-z)^2]g(z)dz.\end{aligned} \tag{18}$$

For the helicity-2 tensorgluons, the $s = 2$ in (4), we will have

$$\begin{aligned} P_{TG}(z) &= C_2(G)\left[\frac{z^5}{(1-z)^3} + \frac{(1-z)^5}{z^3}\right], \\ P_{GT}(z) &= C_2(G)\left[\frac{1}{z(1-z)^3} + \frac{(1-z)^5}{z}\right], \\ P_{TT}(z) &= C_2(G)\left[\frac{z^5}{(1-z)} + \frac{1}{(1-z)z^3} + \sum_{j=1}^{5}\frac{1}{j}\,\delta(1-z)\right] \end{aligned} \tag{19}$$

and the regularisation of these kernels can be performed using the regularisation prescription (18). The regular splitting functions will take the following form:

$$\begin{aligned}
&\int_x^1 P_{TG}(z) f(\frac{x}{z}) dz = \int_x^1 \left[\frac{z^5}{(1-z)^3}_{+++} + \frac{(1-z)^5}{z^3} \right] f(\frac{x}{z}) dz = I_1 + I_2 + I_3, \\
&I_1 = \int_x^1 \frac{dz}{(1-z)^3} \Big(z^5 f(\frac{x}{z}) - f(x) + (5f(x) - xf'(x))(1-z) - \\
&\qquad - (10f(x) - 4xf'(x) + \frac{1}{2}x^2 f''(x))(1-z)^2 \Big), \\
&I_2 = -\int_0^x \frac{dz}{(1-z)^3} \Big(f(x) - (5f(x) - xf'(x))(1-z) + \\
&\qquad + (10f(x) - 4xf'(x) + \frac{1}{2}x^2 f''(x))(1-z)^2 \Big), \\
&I_3 = \int_x^1 \frac{(1-z)^5}{z^3} f(\frac{x}{z}) dz.
\end{aligned} \tag{20}$$

For the gluon–tensor splitting function, we will get:

$$\begin{aligned}
&\int_x^1 P_{GT}(z) f(\frac{x}{z}) dz = \int_x^1 \left[\frac{3-3z+z^2}{(1-z)^3}_{+++} + \frac{1+(1-z)^5}{z} \right] f(\frac{x}{z}) dz = J_1 + J_2 + J_3, \\
&J_1 = \int_x^1 \frac{1}{(1-z)^3} \Big((3-3z+z^2) f(\frac{x}{z}) - f(x) - (f(x) + xf'(x))(1-z) - \\
&\qquad - (f(x) + 2xf'(x) + \frac{1}{2}x^2 f''(x))(1-z)^2 \Big) dz \\
&J_2 = -\int_0^x \frac{dz}{(1-z)^3} [f(x) + (f(x) + xf'(x)(1-z) + (f(x) + 2xf'(x) + \frac{1}{2}x^2 f''(x)(1-z)^2] \\
&J_3 = \int_x^1 \frac{1+(1-z)^5}{z} f(\frac{x}{z}) dz.
\end{aligned} \tag{21}$$

Using the regularisation (16) for the tensor-tensor splitting function, we will get the following expression:

$$\begin{aligned}
\int_x^1 P_{TT}(z) f(\frac{x}{z}) dz &= \int_x^1 \left[\frac{1+z^5}{(1-z)}_{+} + \frac{1+z+z^2}{z^3} + \sum_{j=1}^{5} \frac{1}{j}\, \delta(1-z) \right] f(\frac{x}{z}) dz \\
&= \int_x^1 \frac{dz}{(1-z)} \left[(1+z^5) f(\frac{x}{z}) - 2f(x) \right] + f(x) \left[2\ln(1-x) + \frac{137}{60} \right] \\
&+ \int_x^1 \frac{1+z+z^2}{z^3} f(\frac{x}{z}) dz.
\end{aligned} \tag{22}$$

All new splitting functions which have been added to the standard evolution equation are now well defined and can be calculated using the above Equations (20)–(22).

4. Discussion

The gluon density $G(x,t)$ inside the hadrons is one of the least constrained functions since it does not couple directly to the photon in deep-inelastic scattering measurements. The process of gluon splitting leads to the emission of tensorgluons and therefore a part of the proton momentum that is carried by the neutral constituents can be shared between gluons and tensorgluons and the density of neutral partons is the sum of two density functions: $G(x,t) + T(x,t)$. Because tensorgluons have a larger spin, they can influence the spin structure of the nucleon. The details can be found in [9,43]. To disentangle the contributions of gluons and tensorgluons to the partons densities of a nucleon and to decide which piece of the neutral partons is generated by gluons and which one by tensorgluons, one should measure the helicities of the neutral components.

In supersymmetric extensions of the Standard Model [44,45], the gluons and quarks have natural partners–gluinos of spin $s = 1/2$ and squarks of spin $s = 0$. If the gluinos appear as elementary constituents of the hadrons, then the theory predicts the existence of new hadronic states, the R-hadrons [46,47]. The experimental data provide the evidence that most probably they have to be very heavy [48,49].

The existence of tensorgluon partons inside the proton does not predict a new hadronic state, a proton remains a proton. The tensorgluons will alternate the parton distribution functions of a proton. The question is to which extent the tensorgluons will change the parton distribution functions. The regularisation of the splitting amplitudes developed above (20)–(22) will allow for solving the generalised DGLAP evolution Equation (10) for the parton distribution functions that takes into account the processes of emission of tensorgluons by gluons. The integration can now be performed using the algorithms developed in [50–53] and to find out the ratio of densities between gluons and tensorgluons.

Author Contributions: All authors have read and agreed to the published version of the manuscript.

Funding: This research received no external funding.

Conflicts of Interest: The authors declare no conflict of interest.

References

1. Bjorken, J.D.; Paschos, E.A. Inelastic Electron Proton and gamma Proton Scattering, and the Structure of the Nucleon. *Phys. Rev.* **1969**, *185*, 1975. [CrossRef]
2. Gross, D.J.; Wilczek, F. Ultraviolet behavior of non-abelian gauge theories. *Phys. Rev. Lett.* **1973**, *30*, 1343. [CrossRef]
3. Politzer, H.D. Reliable perturbative results for strong interactions? *Phys. Rev. Lett.* **1973**, *30*, 1346. [CrossRef]
4. Altarelli, G.; Parisi, G. Asymptotic freedom in parton language. *Nucl. Phys. B* **1977**, *126*, 298–318. [CrossRef]
5. Dokshitzer, Y.L. Calculation of the structure functions for deep inelastic scattering and $e^+ e^-$ annihilation by perturbation theory in quantum chromodynamics. *Zh. Eksp. Teor. Fiz* **1977**, *73*, 1216.
6. Gribov, V.N.; Lipatov, L.N. Deep inelastic e p scattering in perturbation theory. *Sov. J. Nucl. Phys.* **1972**, *15*, 438.
7. Gribov, V.N.; Lipatov, L.N. $e^+ e^-$ pair annihilation and deep inelastic e p scattering in perturbation theory. *Sov. J. Nucl. Phys.* **1972**, *15*, 675.
8. Lipatov, L.N. The parton model and perturbation theory. *Sov. J. Nucl. Phys.* **1975**, *20*, 94.
9. Savvidy, G. Proton structure and tensor gluons. *J. Phys. A* **2014**, *47*, 355401. [CrossRef]
10. Fadin, V.S.; Kuraev, E.A.; Lipatov, L.N. On the Pomeranchuk singularity in asymptotically free theories. *Phys. Lett. B* **1975**, *60*, 50–52. [CrossRef]
11. Kuraev, E.A.; Lipatov, L.N.; Fadin, V.S. The Pomeranchuk Singularity in Nonabelian Gauge Theories. *Sov. Phys. JETP* **1977**, *45*, 199.
12. Balitsky, Y.Y.; Lipatov, L.N. The Pomeranchuk Singularity in Quantum Chromodynamics. *Sov. J. Nucl. Phys.* **1978**, *28*, 822.
13. Cabibbo, N.; Petronzio, R. Two Stage Model Of Hadron Structure: Parton Distributions In addition, Their Q^2 Dependence. *Nucl. Phys. B* **1978**, *137*, 395. [CrossRef]
14. Gribov, L.V.; Levin, E.M.; Ryskin, M.G. Singlet Structure Function at Small x: Unitarization of Gluon Ladders. *Nucl. Phys. B* **1981**, *188*, 55. [CrossRef]
15. Gross, D.J.; Wilczek, F. Asymptotically Free Gauge Theories. 1. *Phys. Rev. D* **1973**, *8*, 3633. [CrossRef]
16. Gross, D.J.; Wilczek, F. Asymptotically Free Gauge Theories. 2. *Phys. Rev. D* **1974**, *9*, 980. [CrossRef]
17. Savvidy, G. Asymptotic freedom of non-Abelian tensor gauge fields. *Phys. Lett. B* **2014**, *732*, 150–155. [CrossRef]
18. Savvidy, G. Non-Abelian tensor gauge fields: Generalization of Yang–Mills theory. *Phys. Lett. B* **2005**, *625*, 341. [CrossRef]
19. Savvidy, G. Non-abelian tensor gauge fields. I. *Int. J. Mod. Phys. A* **2006**, *21*, 4931. [CrossRef]
20. Savvidy, G. Non-abelian tensor gauge fields. II. *Int. J. Mod. Phys. A* **2006**, *21*, 4959. [CrossRef]
21. Savvidy, G. Extension of the Poincaré Group and Non-Abelian Tensor Gauge Fields. *Int. J. Mod. Phys. A* **2010**, *25*, 5765. [CrossRef]

22. Georgiou, G.; Savvidy, G. Production of non-Abelian tensor gauge bosons. Tree amplitudes and BCFW recursion relation. *Int. J. Mod. Phys. A* **2011**, *26*, 2537. [CrossRef]
23. Antoniadis, I.; Savvidy, G. Conformal invariance of tensor boson tree amplitudes. *Mod. Phys. Lett. A* **2012**, *27*, 1250103. [CrossRef]
24. Kirschner, R.; Savvidy, G. Yangian and SUSY symmetry of high spin parton splitting amplitudes in generalised Yang–Mills theory. *Mod. Phys. Lett. A* **2017**, *32*, 1750121. [CrossRef]
25. Berends, F.A.; Kleiss, R.; De Causmaecker, P.; Gastmans, R.; Wu, T.T. Single Bremsstrahlung Processes In Gauge Theories. *Phys. Lett. B* **1981**, *103*, 124. [CrossRef]
26. Kleiss, R.; Stirling, W.J. Spinor Techniques For Calculating P Anti-P $\rightarrow W^{+-}/Z^0$ + Jets. *Nucl. Phys. B* **1985**, *262*, 235. [CrossRef]
27. Xu, Z.; Zhang, D.H.; Chang, L. Helicity Amplitudes for Multiple Bremsstrahlung in Massless Nonabelian Gauge Theories. *Nucl. Phys. B* **1987**, *291*, 392. [CrossRef]
28. Gunion, J.F.; Kunszt, Z. Improved Analytic Techniques For Tree Graph Calculations In addition, The G G Q Anti-Q Lepton Anti-Lepton Subprocess. *Phys. Lett. B* **1985**, *161*, 333. [CrossRef]
29. Dixon, L.J. Calculating scattering amplitudes efficiently. *arXiv* **1996**, arXiv:hep-ph/9601359.
30. Parke, S.J.; Taylor, T.R. An Amplitude for *n* Gluon Scattering. *Phys. Rev. Lett.* **1986**, *56*, 2459. [CrossRef]
31. Berends, F.A.; Giele, W.T. Recursive Calculations for Processes with n Gluons. *Nucl. Phys. B* **1988**, *306*, 759. [CrossRef]
32. Witten, E. Perturbative gauge theory as a string theory in twistor space. *Commun. Math. Phys.* **2004**, *252*, 189. [CrossRef]
33. Cachazo, F.; Svrcek, P.; Witten, E. Gauge theory amplitudes in twistor space and holomorphic anomaly. *JHEP* **2004**, *2004*, 77. [CrossRef]
34. Cachazo, F. Holomorphic anomaly of unitarity cuts and one-loop gauge theory amplitudes. *arXiv* **2004**, arXiv:hep-th/0410077.
35. Britto, R.; Cachazo, F.; Feng, B. New recursion relations for tree amplitudes of gluons. *Nucl. Phys. B* **2005**, *715*, 499. [CrossRef]
36. Britto, R.; Cachazo, F.; Feng, B.; Witten, E. Direct proof of tree-level recursion relation in Yang–Mills theory. *Phys. Rev. Lett.* **2005**, *94*, 181602. [CrossRef] [PubMed]
37. Benincasa, P.; Cachazo, F. Consistency conditions on the S-matrix of massless particles. *arXiv* **2007**, arXiv:0705.4305.
38. Cachazo, F.; Svrcek, P.; Witten, E. MHV vertices and tree amplitudes in gauge theory. *JHEP* **2004**, *2004*, 6. [CrossRef]
39. Georgiou, G.; Glover, E.W.N.; Khoze, V.V. Non-MHV Tree Amplitudes in Gauge Theory. *JHEP* **2004**, *2004*, 48. [CrossRef]
40. Arkani-Hamed, N.; Kaplan, J. On Tree Amplitudes in Gauge Theory and Gravity. **2008**, *2008*, 76.
41. Berends, F.A.; Giele, W.T. Multiple Soft Gluon Radiation in Parton Processes. *Nucl. Phys. B* **1989**, *313*, 595 [CrossRef]
42. Mangano, M.L.; Parke, S.J. Quark-Gluon Amplitudes in the Dual Expansion. *Nucl. Phys. B* **1988**, *299*, 673. [CrossRef]
43. Konitopoulos, S.; Savvidy, G. Proton structure, its spin and tensor gluons. In *EPJ Web of Conferences*; EDP Sciences: Les Ulis, France, 2016; Volume 125, p. 4016. . [CrossRef]
44. Fayet, P. Supersymmetry and Weak, Electromagnetic and Strong Interactions. *Phys. Lett. B* **1976**, *64*, 159. [CrossRef]
45. Fayet, P. Spontaneously Broken Supersymmetric Theories of Weak, Electromagnetic and Strong Interactions. *Phys. Lett. B* **1977**, *69*, 489. [CrossRef]
46. Farrar, G.R.; Fayet, P. Phenomenology of the Production, Decay, and Detection of New Hadronic States Associated with Supersymmetry. *Phys. Lett. B* **1978**, *76*, 575. [CrossRef]
47. Fayet, P. Massive Gluinos. *Phys. Lett. B* **1978**, *78*, 417. [CrossRef]
48. Farrar, G.R.; Fayet, P. Bounds on R-Hadron Production from Calorimetry Experiments. *Phys. Lett. B* **1978**, *79*, 442. [CrossRef]
49. Antoniadis, I.; Kounnas, C.; Lacaze, R. Light Gluinos in Deep Inelastic Scattering. *Nucl. Phys. B* **1983**, *211*, 216. [CrossRef]

50. Botje, M. QCDNUM: Fast QCD Evolution and Convolution. *Comput. Phys. Commun.* **2011**, *192*, 490–532. [CrossRef]
51. Bertone, V.; Botje, M. A C++ interface to QCDNUM. *arXiv* **2017**, arXiv:1712.08162.
52. Miyama, M.; Kumano, S. Numerical solution of Q^2 evolution equations in a brute force method. *Comput. Phys. Commun.* **1996**, *94*, 185–215. [CrossRef]
53. Vogt, A. Efficient evolution of unpolarized and polarized parton distributions with QCD-PEGASUS. *Comput. Phys. Commun.* **2005**, *170*, 65–92. [CrossRef]

© 2020 by the authors. Licensee MDPI, Basel, Switzerland. This article is an open access article distributed under the terms and conditions of the Creative Commons Attribution (CC BY) license (http://creativecommons.org/licenses/by/4.0/).

MDPI
St. Alban-Anlage 66
4052 Basel
Switzerland
Tel. +41 61 683 77 34
Fax +41 61 302 89 18
www.mdpi.com

Universe Editorial Office
E-mail: universe@mdpi.com
www.mdpi.com/journal/universe

www.ingramcontent.com/pod-product-compliance
Lightning Source LLC
LaVergne TN
LVHW071307170726
843515LV00006BA/1511

* 9 7 8 3 0 3 6 5 5 8 6 9 1 *